Advanced
Experimental Inorganic Chemistry

Advanced
Experimental Inorganic Chemistry

V.K. Ahluwalia

Prof. (Retd.) Department of Chemistry
University of Delhi, Delhi-110007

Visiting Professor
Dr. B.R. Ambedkar Center for Biomedical Research
University of Delhi, Delhi-110007

◆

Sunita Dhingra

Associate Prof. (Retd.) Miranda House
University of Delhi, Delhi-110007

Manakin
PRESS

CRC Press
Taylor & Francis Group
Boca Raton London New York

CRC Press is an imprint of the
Taylor & Francis Group, an **informa** business

First published 2025
by CRC Press
4 Park Square, Milton Park, Abingdon, Oxon, OX14 4RN

and by CRC Press
2385 NW Executive Center Drive, Suite 320, Boca Raton FL 33431

CRC Press is an imprint of Informa UK Limited

British Library Cataloguing-in-Publication Data
A catalogue record for this book is available from the British Library

ISBN13: 9781032789910 (hbk)
ISBN13: 9781032789927 (pbk)
ISBN13: 9781003490142 (ebk)

DOI: 10.4324/9781003490142

Typeset in Times New Roman
by Manakin Press, Delhi

Manakin
PRESS

Brief Contents

Preface

The book entitled 'Advanced Experimental Inorganic Chemistry' has been written in order to meet the requirements of B.Sc and M.Sc students of all universities and also of Research and Development wing of the Industries.

The book is divided into four parts; Part I deals with Qualitative Inorganic Analysis. Systematic procedures of anion and cation analysis alongwith their confirmatory tests and spot tests are given. Detection of cations by flame photometry and atomic absorption spectroscopy are also incorporated. Besides chromatographic separation and identification of cations are also discussed. The analysis of insolubles, alloys and some less common elements are also incorporated.

Part II deals with volumetric analysis. The instrumental methods of volumetric analysis have also been incorporated.

Part III deals with gravimetric analysis. Estimation of one or more than one constituent in a solution and instrumental methods of quantitative analysis are also incorporated.

Part IV describes a large number of inorganic preparations. These include simple salts, double salts, complex salts, amalgams, activated metals, organometallics and some other miscellaneous inorganic preparations. The uses of various compound have also been mentioned.

A set of exercises are incorporated in each chapter. This will be of great help to the students in understanding various steps of experimental inorganic chemistry.

Any suggestions from the readers will be greatefully acknowledged. These will be helpful for incorporation in subsequent editions of the book.

Authors

PART I

Qualitative Inorganic Analysis

1. Qualitative Inorganic Analysis

Contents

Qualitative Inorganic Analysis

1.1 INTRODUCTION

Qualitative analysis deals with identification of anions (acid radicals) and cations (basic radicals) present in a salt or a mixture of inorganic salts, alloys etc. A number of procedures are used to analyse a mixture. Normally, in qualitative analysis the mixture is first subjected to preliminary examination, which involve dry tests like colour, odour, action of heat, flame test, charcoal cavity test and borax bead test. The ions (anions and cations) suspected to be present are subsequently identified by wet tests and finally confirmed by their confirmatory tests.

Qualitative analysis can be carried out in macro scale, semi-micro scale or microscale. In macro analysis, the quantity of the mixture used is of the order 3-4 g. In semi-micro analysis, about 1 g of the mixture is used. However, in micro analysis the quantity of the mixture used is still smaller 0.5 g or even less.

For semi-micro and micro work special type of operations are needed alongwith different type of apparatus needed to carry out the work.

Macro-scale Analysis

In **macro-scale analysis**, the usual apparatus are used. These involve test tubes (15×2 cm with 25 mL capacity), beakers (50, 100 and 250 mL capacity), conical or Erlenmeyer flasks (50, 100 and 250 mL capacity), stirring rods (about 20 cm long and 4 mm diameter) and a wash bottle (500 mL flat-bottomed flask and stopper carrying two holes—the shorter tube is attached to a rubber bulb). The hydrogen sulphide required is generated in the usual kipps apparatus in a fume-cupboard with good ventilation (Hydrogen sulphide is highly poisonous and appropriate precaution be taken). Alternatively a saturated aqueous solution of hydrogen sulphide can be used as a reagent. In macro-scale analysis, special attention should be given to the disposal of wastes. Strong acids and alkalies should be diluted and then disposed in the sinks and the sinks flushed with excess water. Also any process involving evaporation should be carried out only in well-ventilated fume cupboard.

Semi-micro Scale Analysis

In **semi-micro scale analysis**, the apparatus used is of much smaller capacity. These involve small test tubes (usually 75×10 mm, 4 mL capacity or 100×12 mm, 8 mL capacity), centrifuge tubes (76×11 mm, 3 mL capacity), stirring rods (12 cm×2 mm), droppers, reagent bottles of special type, a centrifuge. Different types of apparatus used for semi-micro analysis and shown below in the figure.

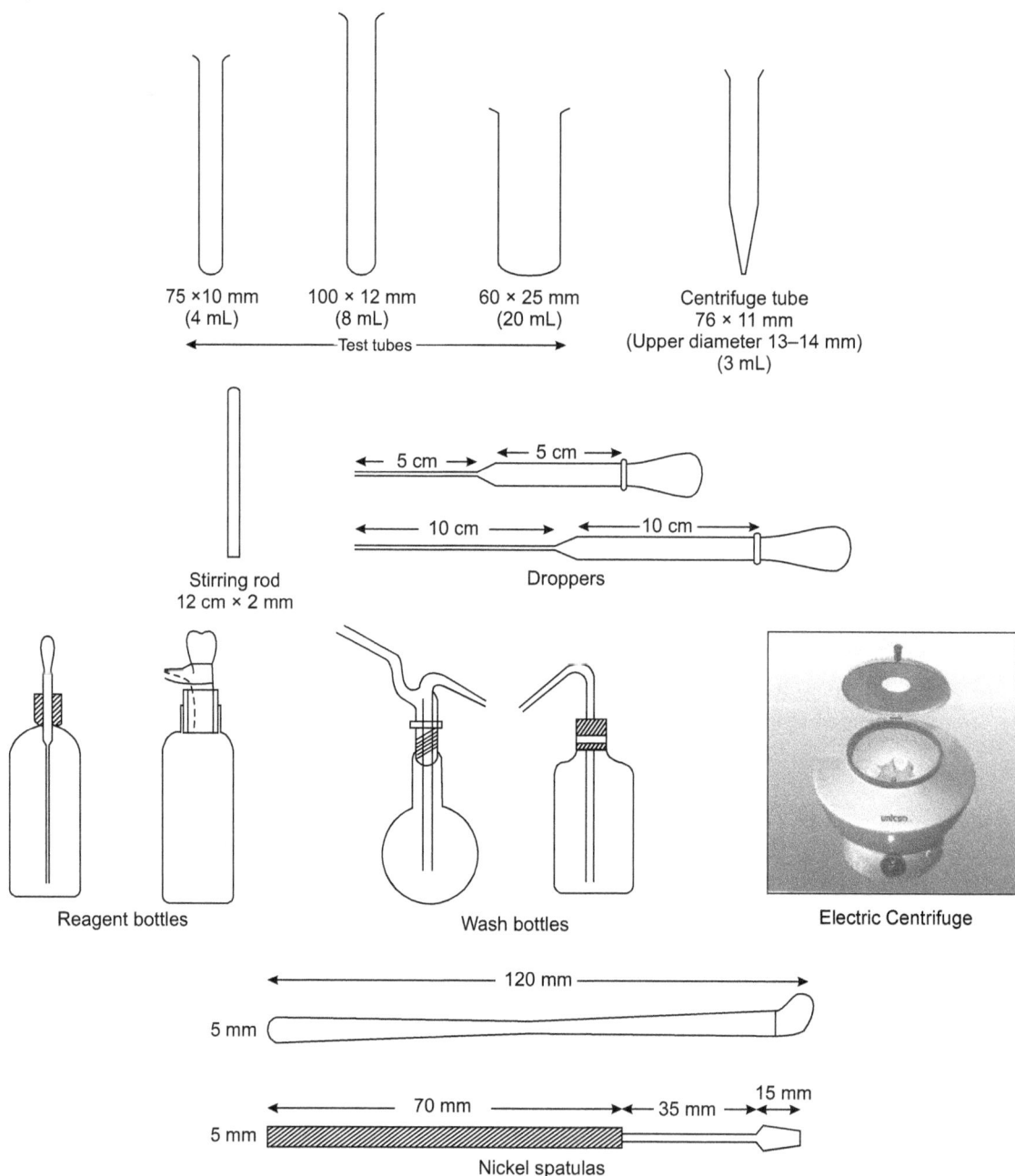

75 ×10 mm (4 mL)   100 × 12 mm (8 mL)   60 × 25 mm (20 mL)   Centrifuge tube 76 × 11 mm (Upper diameter 13–14 mm) (3 mL)

◄──────── Test tubes ────────►

← 5 cm →   ← 5 cm →

◄─── 10 cm ───►   ◄─── 10 cm ───►

Stirring rod 12 cm × 2 mm   Droppers

Reagent bottles   Wash bottles   Electric Centrifuge

◄──────── 120 mm ────────►

5 mm

◄──── 70 mm ────► ◄ 35 mm ► 15 mm

5 mm

Nickel spatulas

Choice of the Method

It is recommended that a beginner should start with macro scale procedure. Once he/she gets used to it the semi-micro scale method can be used. In fact, the semi-micro scale method is appropriate for qualitative inorganic analysis. Some of the advantages of using semi-micro method include:

- Reduction of amount of chemicals accompanied by time saving.
- Various steps involves, *viz.*, washing of the precipitate can be carried out efficiently by using a centrifuge.
- Foul smell of hydrogen sulphide, acid fumes and hazardous chemicals are reduced considerably.

1.2 SYSTEMATIC PROCEDURE FOR QUALITATIVE INORGANIC ANALYSIS

1.2.1 Dry Tests

As already stated dry tests involve colour of the mixture, odour, action of heat, flame test, charcoal cavity test and borax bead test. These tests give information which is of help in the analysis of ions present in a mixture. All these tests can be considered to be green tests since very small quantity of the mixture in required.

Colour

Some metal ions have a particular **colour**, which may be helpful to infer the presence of such ions. However, final conclusion cannot be taken on the basis of colour.

Table 1.1. *Colour of the mixture/salt.*

Colour	Inference
White	Salts of Cr, Ni, Fe, Co, Mn, Cu are absent
Dark green	Chromium salt may be present
Light pink	Manganese salt
Violet	Cobalt salt
Brown	Copper salt
Black	Sulphides of copper, silver, iron, cobalt, mercury and nickel

Note: The colours given above (Table 1.1) are for single crystalline salts. In a mixture of salts, appropriate precaution has to be taken to come to any conclusion.

Odour

The **odour** of the mixture gives some information about the nature of the salt present in it (Table 1.2).

Table 1.2. *Odour of the Mixture.*

Observation (Odour)	Inference
Smell of Vinegar	Acetate
Smell of ammonia	Ammonium salt
Smell of H_2S (rotten eggs)	Sulphides

Flame Test

For performing **flame test**, the salts must be volatilized in a non-luminous Bunsen flame and the colour imparted to the flame is noted. Since, chlorides are volatile, these are prepared *in situ* by mixing the mixture with a little concentrated hydrocloric acid and then performing flame test with the paste of the mixture in conc. HCl.

Procedure

Following steps are involved:

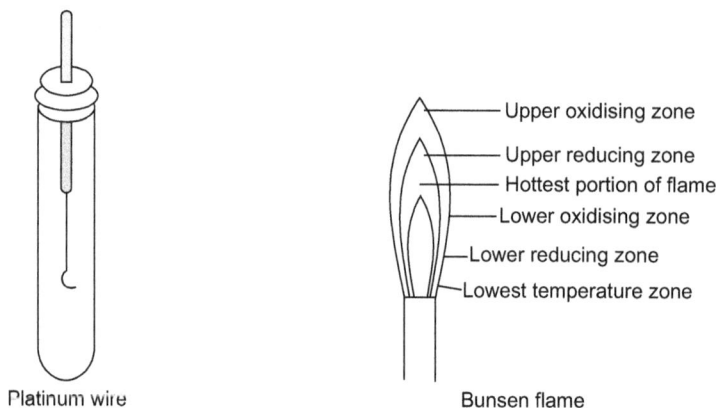

Platinum wire   Bunsen flame

(*i*) A thin platinum wire (about 5 cm long and 0.03–0.05 mm diameter), fused in a short piece of glass rod is used.

(*ii*) The platinum wire is cleaned by dipping into conc. HCl (contained in a watch glass) and heating it in a non luminous flame until it imparts no colour to the flame.

(*iii*) The cleaned platinum wire is dipped in a paste of the mixture in conc. HCl (on a watch glass) and heated again in the non luminous flame and colour imparted to the flame (if any) noted.

(*iv*) The colour imparted to the flame (by the mixture is viewed as such and then through a cobalt glass, which is helpful to differentiate between different ions. (cobalt glass absorbs radiation from sodium).

(*v*) It is recommended that the student first perform flame test with authentic samples of the chlorides of various ions and get used to recognising the various colours.

(*vi*) After the flame tests have been performed, the platinum wire should be cleaned with conc. HCl. It is advisable to keep the cleaned platinum wire suspended in conc. HCl (See Fig. given above).

Table 1.3. *Flame Tests.*

Colour of Flame as viewed from the Mixture	Colour of Flame as viewed by using a Cobalt Glass	Inference
Golden yellow	No colour	Sodium
Violet (lilac)	Crimson	Potassium
Carmine-red	—	Lithium
Brick-red	Light green	Calcium
Crimson	Purple	Strontium
Yellowish green	Bluish-green	Barium
Green	—	Borate, copper, thalium

Charcoal Cavity Test

For performing charcoal cavity test, a clean charcoal block in which a small cavity is made with a penknife is used. The test involves heating the mixture (about 0.1g) with fusion mixture (0.2–0.3g, 1:1 mixture of Na_2CO_3 and K_2CO_3) with a blow pipe in the reducing flame. By this procedure, the metallic cations are converted to carbonates, which may decompose into their oxides.

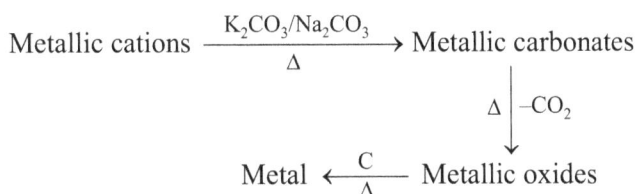

$$\text{Metallic cations} \xrightarrow[\Delta]{K_2CO_3/Na_2CO_3} \text{Metallic carbonates}$$

$$\downarrow \Delta \; -CO_2$$

$$\text{Metal} \xleftarrow[\Delta]{C} \text{Metallic oxides}$$

The final products of the above reaction are either metals or metal and their oxides, which produce specific colour in the charcoal cavity. The oxides of more electropositive metals like MgO, ZnO, Al_2O_3 etc., cannot be reduced and so do not give any characteristic colour. In such cases, a few drops of cobalt nitrate solution is added to the charcoal cavity (containing the oxides of metals) and heated in the reducing flame. This results in the formation of complex oxides with CoO which have characteristic colours

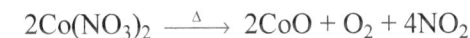

$$2Co(NO_3)_2 \xrightarrow{\Delta} 2CoO + O_2 + 4NO_2$$

$$ZnO + CoO \xrightarrow{\Delta} CoZnO_2 \text{ (green)}$$

$$MgO + CoO \xrightarrow{\Delta} CoMgO_2 \text{ (light pink)}$$

Procedure

Add the salt mixture (0.1 g) and fusion mixture (~ 0.2 g) in the cavity of the charcoal block. The mixture is heated with the help of a blow pipe in the reducing flame for about one minute. Note the colour (Table 1.4). To the cavity is added a drop of cobalt nitrate solution and heated again in the reducing flame using a blow pipe. The colour produced is noted (Table 1.5).

Table 1.4. *Charcoal Cavity Test (Mixture + Fusion mixture).*

Observation	Inference
Dark grey or brown residue	Fe, Mn, Co, Ni salts
White (infusible residue)	Mn, Fe, Ni, Cu, Co, Cd absent; Ca, Br, Sr, Zn, Al, Mg may be present
Medium brown residue	Cd
Copper red residue	Cu

Table 1.5. *Charcoal cavity test Mixture + fusion mixture* $\xrightarrow{Co(NO_3)_2}$ *colour.*

Observation	Inference
Black residue	CoO
Pink residue	Mg
Blue residue	Al
Green residue	Zn

Note: The students are advised to perform charcoal cavity test with authentic samples of salts and get used to the procedure and to colour.

Borax Bead Test

For **Borax bead test** a platinum wire, similar to that used for flame test is used. The free end of the platinum wire is coiled into a small loop. The looped end of the platinum wire is heated in the Bunsen flame until it is red hot and then dipped quickly into powdered borax ($Na_2B_4O_7.10H_2O$). The loop (having adhering borax) is heated in the hottest part of the flame. The borax sweells up loosing its water of crystallisation forming a colourless, transparent, glass-like bead which consists of a mixture of sodium borate and boric anhydride

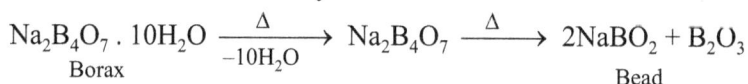

$$Na_2B_4O_7 . 10H_2O \xrightarrow[-10H_2O]{\Delta} Na_2B_4O_7 \xrightarrow{\Delta} 2NaBO_2 + B_2O_3$$
$$\text{Borax} \qquad\qquad\qquad\qquad\qquad\qquad\qquad \text{Bead}$$

The bead thus formed is dipped into the finely powdered mixture making sure that only a very small amount of the mixture adheres to the bead (in case excess of mixture gets adhered, the result will be erroneous giving a dark and opaque bead on subsequent heating). The bead alongwith the adhering mixture is heated first in the lower reducing flame, allowed to cool and colour observed. It is then heated in the lower oxidizing flame, allowed to cool and colour again observed.

The overall reactions involved in borax bead test are given below:

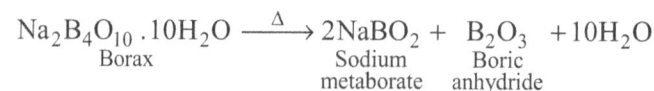

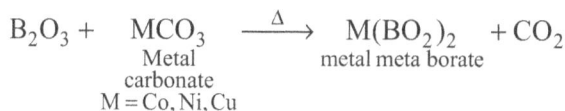

$$Na_2B_4O_{10}.10H_2O \xrightarrow{\Delta} 2NaBO_2 + B_2O_3 + 10H_2O$$
$$\text{Borax} \qquad\qquad\qquad \underset{\text{metaborate}}{\text{Sodium}} \quad \underset{\text{anhydride}}{\text{Boric}}$$

$$B_2O_3 + \underset{\substack{\text{Metal} \\ \text{carbonate} \\ M = Co, Ni, Cu}}{MCO_3} \xrightarrow{\Delta} \underset{\text{metal meta borate}}{M(BO_2)_2} + CO_2$$

The results obtained in the borax bead test are given in Table 1.6.

Table 1.6. *Colours Obtained in Borax Bead Test.*

Metal	Oxidizing Flame		Reducing Flame	
	Hot	Cold	Hot	Cold
Copper	Green	Blue	Colourless	Red-brown (opaque)
Iron	Yellowish-brown	Yellow	Green	Green
Chromium	Yellow	Green	Green	Green
Manganese	Yellow (amethyst)	Amethyst	Colourless	Colourless
Cobalt	Blue	Blue	Blue	Blue
Nickel	Violet	Reddish brown	Grey	Grey

1.3 ANION ANALYSIS

A number of schemes have been used to infer the presence of various anions in a mixture. In the following discussion, two schemes have been given for the analysis of anions. In scheme I, various anions are divided into four groups designated A, B, C and D and the group reagent is dilute H_2SO_4, dilute H_2SO_4 + $KMnO_4$, conc. H_2SO_4 for groups A, B and C respectively. The anions of group D $(SO_4^{2+}, PO_4^{3+}, BO_3^{3+})$ have to be tested individually. On the other hand in scheme II reagents like diphenyl amine, barium chloride, silver nitrate, calcium chloride are group reagents. Some anions like CO_3^{2-}, CH_3COO^-, PO_4^{3-}, BO_3^{3-} are to be tested individually.

The anions present in a mixture must be analysed first. This is because some of the anions interfere in the analysis of cations and must be removed from the solution during analysis of cations.

It is a must to proceed systematically for performing various tests since a number of anions may be present in the mixture.

Scheme I

As already stated the anions are classified into four groups (Table 1.7).

Table 1.7. *Classificnation of Anions into Four Groups (Scheme I)*

Group	Group reagent	Anions Indicated
A	Dil. H_2SO_4	CO_3^{2-}, HCO_3^-, CH_3COO^-, S^{2-}, SO_3^{2-}, $S_2O_3^{2-}$, NO_2^-
B	Dil. H_2SO_4 + $KMnO_4$	$C_2O_4^{2-}$, Cl^-, Br^-, I^-
C	Conc. H_2SO_4	NO_3^-, Cl^-, Br^-, I^-, $C_2O_4^{2-}$, $C_4H_6O_6^{2-}$, ClO_3^-
D	—	SO_4^{2-}, PO_4^{3-}, BO_3^{2-}

As seen, some of the anions are common to two groups.

1.3.1 Group A (Dilute H₂SO₄ group)

The group A includes, anions, CO_3^{2-}, HCO_3^-, NO_2^-, SO_3^{2-}, $S_2O_3^{2-}$, S^{2-} and CH_3COO^-.

To about 0.1–0.2 of the mixture is added dilute H_2SO_4 (2-3 mL) slowly. The colour and the odour of the gas evolved is noted. In case the reaction is slow, the solution is warmed (water bath). Warming is helpful for the evolution of gases which are soluble in cold. (Table 1.8).

Table 1.8 *Gases evolved by the Treatment of the Mixture with dilute H_2SO_4.*

Sl.No.	Observation	Inference
1.	Brisk effervescence with evolution of a colourless gas (possibly CO_2). The evolved gas turns lime water milky. However, on passing the gas for longer time, the turbity disappears due to the formation of bicarbonate. $$CaCO_3 + H_2SO_4 \longrightarrow CaSO_4 + CO_2\uparrow + H_2O$$ $$Ca(HCO_3)_2 + H_2SO_4 \longrightarrow CaSO_4 + 2CO_2\uparrow + 2H_2O$$ $$CO_2 + Ca(OH)_2 \longrightarrow CaCO_3\downarrow + H_2O$$ white ppt. $$CaCO_3 + CO_2 + H_2O \longrightarrow Ca(HCO_3)_2$$ soluble The apparatus used for passing CO_2 in lime water is given below. **Fig.** Apparatus for passing CO_2 into lime water. **Note:** In case the mixture contains sulphite (SO_3^-, the evolved SO_2 also turns lime water milky. So the absence of SO_3^- must be confirmed before testing CO_2. For testing CO_3^{2-} in presence of SO_3^-. (See section 1.3.7)	CO_2 from carbonate (CO_3^{2-}) or bicarbonate (HCO_3^-)
2.	Reddish brown fumes (in cold), which turn moistened acidified KI-starch paper blue or black $$2NaNO_2 + H_2SO_4 \longrightarrow Na_2SO_4 + 2HNO_2$$ $$3HNO_2 \longrightarrow HNO_3 + H_2O + 2NO$$ $$2NO + O_2 \longrightarrow 2NO_2$$ (reddish brown fumes) $$2NO_2^- + 4H^+ + 2I^- \longrightarrow 2NO + I_2 + 2H_2O$$ $$NO_2 + I_2 + starch \longrightarrow \text{iodo-starch complex}$$ blue colour This reaction should be done only in cold. Do not heat, because if the mixture contains NO_3^-, it also reacts with hot dilute H_2SO_4 to form HNO_3 which decomposes to form brown fumes of NO_2.	NO_2 from nitrite (NO_2^-)

3.	Colourless gas (in hot) having smell of burning sulphur (pungent odour), which turns acidified $K_2Cr_2O_7$ paper green $Na_2SO_3 + H_2SO_4 \longrightarrow Na_2SO_4 + SO_2\uparrow + H_2O$ $3SO_2 + K_2Cr_2O_7 + H_2SO_4 \longrightarrow K_2SO_4 + Cr_2(SO_4)_3 + H_2O$ green In case, there is white deposit of sulphur in the solution the presence of thiosulphate is indicated $Na_2S_2O_3 + H_2SO_4 \longrightarrow Na_2SO_4 + SO_2\uparrow + S + H_2O$	SO_2 from sulphite (SO_3^{2-}) SO_2 from thiosulphate $(S_2O_3^{2-})$
4.	Colourless gas (in hot) smelling like rotten eggs, which turns moist lead acetate paper black $Na_2S + H_2SO_4 \longrightarrow Na_2SO_4 + H_2S\uparrow$ $H_2S + Pb(CH_3COO)_2 \longrightarrow PbS\downarrow + 2CH_3COOH$ black	H_2S from sulphide (S^{2-})
5.	Colourless gas (in hot) having smell of vinegar $2CH_3COONa + H_2SO_4 \longrightarrow 2CH_3COOH + Na_2SO_4$ acetic acid smell of vinegar	CH_3COOH from acetate (CH_3COO^-)

For confirmatory tests of group *A* anions (dilute H_2SO_4, group) *see Section 1.3.6.*

For confirmation of carbonate in presence of bicarbonate and carbonate in presence of sulphite and thiosulphate and sulphide, sulphite, sulphate and thiosulphate in presence of each other *see Section 1.3.7.*

1.3.2 Group B (Dilute H_2SO_4 + KMnO$_4$ Solution Group)

Group *B* includes the anion $C_2O_4^{2-}$, Cl^-, Br^- and I^-.

To about 0.1–0.2 *g* of the mixture is added dilute sulphuric acid and the mixture warmed to react group *A* anions. This is followed by adding a few drops of dilute KMnO$_4$ solution. In case decolourisation takes place more KMnO$_4$ solution is added till there is no decolourisation (Table 1.9).

Table 1.9. *Gases evolved by treatment of the Mixture with Dilute H_2SO_4 (hot) + KMnO$_4$ solution.*

Sl.No.	Observation	Inference
1.	CO_2 gas is evolved which turns lime water milkey $5(COO)_2^{2-} + 2MnO_4^- + 16H^+ \longrightarrow 10CO_2 + 2Mn^{2+} + 8H_2O$	CO_2 from oxalate $(C_2O_4^{2-})$
2.	Violet vapours of iodine is evolved $10I^- + 2MnO_4^- + 16H^+ \longrightarrow 5I_2\uparrow + 2Mn^{2+} + 8H_2O$	Iodine from iodide (I^-)
3.	Brown vapour of bromine is evolved $10Br^- + 2MnO_4^- + 16H^+ \longrightarrow 5Br_2 + 2Mn^{2+} + 8H_2O$	Bromine from bromide (Br^-)
4.	Gas (chlorine) having a strong smell is evolved $10Cl^- + 2MnO_4^- + 16H^+ \longrightarrow 5Cl_2 + 2Mn^{2+} + 8H_2O$	Chlorine from chloride (Cl^-)

Notes: 1. Tests for $C_2O_4^{2-}$, Cl^-, Br^- and I^- are also given in group C. Both groups B and C are complementary to each other.

　　　　2. For testing presence of a halide in presence of other halides (*e.g.,* presence of Cl^-, Br^-, I^- in a mixture) *see in Section 1.3.7.*

1.3.3   Group C–Conc. H_2SO_4 Group

Group C includes NO_3^-, Cl^-, Br^-, I^-, F^-, $C_2O_4^{2-}$, tartarate ($C_4H_4O_6^{2-}$) and chlorate (ClO_3^-).

To about 0.2 *g* of the mixture is added conc. H_2SO_4 dropwise allowing group A aniont to react. To the solution is added more conc. H_2SO_4 (1–2 mL). The solution is first warmed and then heated. The evolved gases are tested and examined (Table 1.10).

Table 1.10. *Gases Evolved by Treatment of the Mixture with Conc. H_2SO_4*

Sl.No.	Observation	Inference
(*i*)	Colourless gas (HCl) having pungent odour is evolved. A glass rod dipped in NH_4OH solution on bringing near the mouth of the test tube gives white fumes of NH_4Cl. $NaCl + H_2SO_4 \longrightarrow NaHSO_4 + HCl\uparrow$ $HCl + NH_4OH \longrightarrow NH_4Cl + H_2O$ white fumes	HCl from chloride (Cl^-)
(*ii*)	Brown vapours of bromine is evolved. On passing the gas through CCl_4 a brownish yellow solution is obtained $NaBr + H_2SO_4 \longrightarrow NaHSO_4 + HBr$ $2HBr + H_2SO_4 \longrightarrow Br_2\uparrow + SO_2 + 2H_2O$	Br_2 from bromide (Br^-)
(*iii*)	Violet coloured iodine gas is evolved. On passing the gas through CCl_4 a violet colour is obtained $NaI + H_2SO_4 \longrightarrow NaHSO_4 + HI$ $2HI + H_2SO_4 \longrightarrow I_2\uparrow + SO_2 + 2H_2O$	I_2 from iodide (I^-)
(*iv*)	Oily appearance in the test tube in cold. On warming a pungent gas is evolved. A glass rod moistured with water on bringing in contact with the vapour gives a gelatinous white precipitate of hydrated silica which is deposited on moistured glass rod. $CaF_2 + H_2SO_4 \longrightarrow CaSO_4 + 2HF\uparrow$ $SiO_2 + 4HF \longrightarrow SiF_4 + 2H_2O$ (from test　　 oily tube)　　 appearance $3SiF_4 + 4H_2O \longrightarrow 2H_2SiF_6 + H_4SiO_4$ hydrated silica (white)	HF from fluoride (F^-)

(v)	Brown vapour of NO_2 gas evolved. The colour deepens on heating the reaction mixture with copper turnings $NO_3^- + H^+ \longrightarrow HNO_3$ $4HNO_3 \longrightarrow H_2O + 4NO_2\uparrow + O_2$ $Cu + 4HNO_3 \longrightarrow Cu(NO_3)_2 + 2H_2O +$ $2NO_2\uparrow$	NO_2 from nitrate (NO_3^-)
(vi)	A colourless gas (mixture of CO_2+CO) is evolved. On passing through lime water a turbidity is obtained indicating the presence of CO_2. And on burning the evolved gas burns with a blue flame indicating the presence of CO $(COONa)_2 + H_2SO_4 \longrightarrow Na_2SO_4 + CO_2\uparrow + CO\uparrow +$ H_2O	CO_2 and CO from oxalate ($C_2O_4^{2-}$)
(vii)	Charring (blackening) takes place (due to formation of cabron) with smell of burnt sugar $Na_2C_4H_4O_6 + 2H_2SO_4 \longrightarrow C_4H_6O_6 + Na_2SO_4$ Sod. tartarate Tartaric acid $C_4H_6O_6 \xrightarrow[\Delta]{2H_2SO_4} C + CO + 2CO_2 + 2O_2 + 5H_2O$ Tartaric Blackening acid of mixture	Tartarate ($C_4H_4O_6^{2-}$)
(viii)	A yellowish gas accompanied by cracking sound; the solution turns orange yellow $3KClO_3 + 3H_2SO_4 \longrightarrow 3KHSO_4 + 2ClO_2\uparrow + HClO_4$ $+ H_2O$	Chlorate (ClO_3^-)

For confirmatory tests of cations of group C (conc. H_2SO_4 group) see section 1.3.6.)

For confirmation Cl$^-$, Br$^-$, I$^-$ in presence of each other, nitrate in presence of nitrate see Section 1.3.7.

1.3.4 Group D–Special Group

Group D includes sulphate (SO_4^{2-}), phosphate (PO_4^{3-}) and borate (BO_3^{3-}). Each of these anions have to be tested individually since these do not react with H_2SO_4 (Table 1.11).

Table 1.11. *Individual Tests for SO_4^{2-}, PO_4^{3-} and BO_3^{3-}.*

Anion	Procedure for Testing
(i) Sulphate (SO_4^{2-})	The mixture (0.1–0.2 g) is heated with dilute HCl and filtered. To the filtrate is added $BaCl_2$ solution. A white precipitate of $BaSO_4$, insoluble in HCl is obtained. $Ba^{2+} + SO_4^{2-} \longrightarrow BaSO_4\downarrow$ white ppt.

...(Contd.)

(ii) Phosphate (PO_4^{3-})	The mixture is boiled with conc. HNO_3 and then treated with ammonium molybdate solution. A yellow ppt. of ammonium phospho molybdate is obtained. $$Na_3PO_4 + 24HNO_3 + 12(NH_4)_2MoO_4 \xrightarrow{\Delta}$$ $$\longrightarrow 3NaNO_3 + 21NH_4NO_3 + 12H_2O + (NH_4)_3PO_4.12MoO_3$$ yellow ppt. This test is also given by arsenic and so PO_4^{3-} must be tested in sodium carbonate extract.
(iii) Borate (BO_3^{3-})	To 0.2 g of the mixture (taken in a china dish) is added conc. H_2SO_4 (3 mL) and ethanol (2 mL). The mixture is heated and the vapours ignited. A green edged flame is obtained. $$BO_3^{3-} + 3H^+ \longrightarrow H_3BO_3$$ $$H_3BO_3 + 3C_2H_5OH \longrightarrow (C_2H_5)_3BO_3 + 3H_2O$$ burns with green flame

1.3.5 Scheme II

In an alternative scheme of analysis of anions (scheme II), the common anions are divided into five groups (Table 1.12). In case, the mixture is insoluble in water, the sodium carbonate extract (*see section, 1.3.6*) is used for carrying out the tests.

Table 1.12. *Scheme II*

Group (Name)	Group reagent: Procedure and Observation	Inference (Anions Indicated)
1. Diphenylamine group	Diphenylamine in conc. H_2SO_4 (Reagent). To 2 drops of Na_2CO_3 extract (in a test tube) is added diphenylamine reagent (2-3 drops). Blue colour is obtained.	NO_2^-, NO_3^-
2. Barium chloride group	$BaCl_2$ solution (reagent). To 1 mL of Na_2CO_3 extract is added dil. HCl followed by addition of $BaCl_2$ soln. (1 mL) White ppt. is obtained.	SO_4^{2-}
	The ppt is filtered and to the filtrate is added conc. HNO_3. A white ppt. is obtained.	SO_3^{2-}
3. Silver nitrate group	$AgNO_3$ solution (reagent) Na_2CO_3 extract (1 mL) is acidified with dil. HNO_3 and treated with $AgNO_3$ solution (0.5 mL). A white or yellow ppt. is obtained.	Cl^-, Br^-, I^-
4. Calcium chloride group	$CaCl_2$ solution (reagent). To Na_2CO_3 extract (0.5 mL) is added dil. acetic acid and then $CaCl_2$ solution. A white ppt. is obtained.	F^-, $C_2O_4^{2-}$
5. Special group	CO_3^{2-}, CH_3COO^-, PO_4^{3-}, BO_3^{3-} etc. have to be individually tested.	

For confirmation of SO_4^{2-} in presence of F^- and $C_2O_4^{2-}$ is presence of CO_3^{2-} see section 1.3.7.

On the basis of the results obtained by either of the schemes I or II (infact scheme I is preferred), the anions indicated to be present are confirmed by the confirmatory test of each anion.

1.3.6 Confirmatory Tests for Individual Anions

For confirmatory tests, the mixture is dissolved in water (if soluble) and the water solution is used for confirmatory tests. In case, the mixture is partially soluble or insoluble, the anions in the mixture are brought into solution by preparation of sodium carbonate extract of the mixture.

Preparation of Water Extract

Water extract is prepared by boiling the mixture (about 1 g) with distilled water (5-10 mL). The solution is filtered and the filtrate (water extract) is used for testing anions. The anions NO_3^-, NO_2^-, $S_2O_3^{2-}$ and CH_3COO^- can be tested using the water extract.

Preparation of Na_2CO_3 Extract

Sodium carbonate extract is prepared by heating the mixture (about 1 g) with Na_2CO_3 (5-6 g) and distilled water (15-20 mL) in a china dish till the volume is reduced to about 10 mL. The mixture is cooled and filtered. The filterate is the sodium carbonate extract. In this process the anions present in the mixture form the corresponding sodium salts which are soluble in water.

Confirmatory Testes for Aniom of Group A (Dilute H_2SO_4 group)

1. Carbonate (CO_3^{2-})

(*a*) To the mixture (about 0.1 g) is added dil. H_2SO_4 (1 mL). Brisk effervescence indicates carbonate. The evolved gas is passed into lime water. Formation of a white precipitate confirms the presence of CO_2.

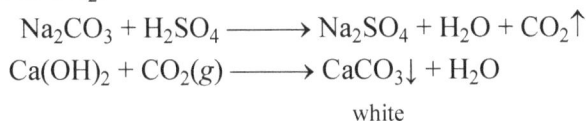

$$Na_2CO_3 + H_2SO_4 \longrightarrow Na_2SO_4 + H_2O + CO_2\uparrow$$
$$Ca(OH)_2 + CO_2(g) \longrightarrow CaCO_3\downarrow + H_2O$$
white

In case, the gas is passed through for long time the white ppt. of $CaCO_3$ dissolves due to the formation of calcium bicarbonate which is soluble in water

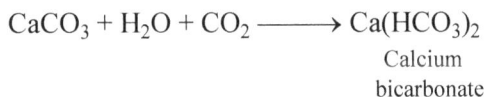

$$CaCO_3 + H_2O + CO_2 \longrightarrow Ca(HCO_3)_2$$
Calcium
bicarbonate

(*b*) In case the mixture contains sulphite and or thiosulphate, the formed SO_2 also turns lime water milky. For testing CO_3^{2-} in presence of sulphite the SO_2 (from SO_3^{2-}) must be removed before testing for CO_3^{2-}. This can be done by adding $K_2Cr_2O_7$ solution to the test tube before acidifing. The $K_2Cr_2O_7$ oxidises SO_2 without affecting CO_2.

2. Bicarbonate (HCO_3^-)

(*a*) Like carbonates, bicarbonates also gives CO_2 on treatment with dil. sulphuric acid (see test a in carbonate)

$$NaHCO_3 + H_2SO_4 \longrightarrow NaHSO_4 + CO_2\uparrow + H_2O$$

(b) Bicarbonate can be tested in presence of carbonates as follows. To the sodium carbonate extract (or aqueous extract) is added excess of calcium chloride. The carbonate gets quantitatively precipitated as $CaCO_3$.

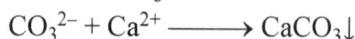

$$CO_3^{2-} + Ca^{2+} \longrightarrow CaCO_3\downarrow$$

The mixture is filtered. The filterate containing bicarbonate is treated with ammonia. Formation of a while precipitate or cloudiness confirms the presence of bicarbonate

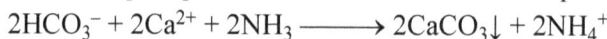

$$2HCO_3^- + 2Ca^{2+} + 2NH_3 \longrightarrow 2CaCO_3\downarrow + 2NH_4^+$$

Note: The student can perform the above test with a 1:1 mixture Na_2CO_3 and $NaHCO_3$ for practice.

3. Acetate (CH_3COO^-)

(a) To the mixture (0.1 g) is added dilute H_2SO_4 and heated. A vinegar like smell is produced

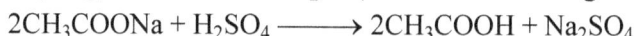

$$2CH_3COONa + H_2SO_4 \longrightarrow 2CH_3COOH + Na_2SO_4$$

In the presence of SO_3^{2-} or S^{2-} it is not possible to detect the smell of vinegar since the SO_2 or H_2S generated mars the odour of vinegar. In such cases the CH_3COO^- is conformed by the following tests.

(b) **Ferric chloride test:** To water extract (1 mL) of the mixture is added a drop of neutral ferric chloride solution. A brownish red colour appears

$$3CH_3COO^- + Fe^{3+} \longrightarrow \underset{\substack{\text{brownish} \\ \text{red colour}}}{Fe(CH_3COO)_3}$$

Note: Neutral ferric chloride solution is obtained by adding NH_4OH solution dropwise to a solution of ferric chloride in water (Shelf reagent) till a small amount of precipitate (which does not dissolve on shaking) is obtained. The solution is filtred and the filtrate is used as neutral ferric chloride solution.

(c) **Ester test:** To the mixture (about 0.1 g) is added ethyl alcohol (0.5 mL) and 2-3 drops of conc. H_2SO_4 (in a dry test tube). The mixture is warmed (on a small flame) and poured onto dilute Na_2CO_3 solution (in a beaker). A fruity odour of ethyl acetate confirms the presence of acetate.

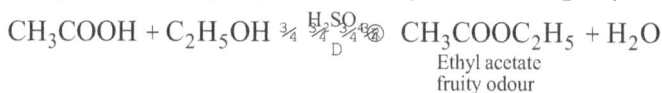

$$2CH_3COONa + H_2SO_4 \xrightarrow{\Delta} 2CH_3COOH + Na_2SO_4$$

$$CH_3COOH + C_2H_5OH \xrightarrow[D]{H_2SO_4} \underset{\substack{\text{Ethyl acetate} \\ \text{fruity odour}}}{CH_3COOC_2H_5} + H_2O$$

In the above test it is better to use amyl alcohol as the odour of resulting amyl acetate is more readily distinguished from that of alcohol. It is advisable to run a parallel test with a known acetate and compare the odour of the two products.

4. Sulphide (S^{2-})

(a) To the mixture (about 0.1 g) is added dilute H_2SO_4 (1 mL) and mixture warmed. H_2S gas having a smell of rotton eggs is evolved. On keeping a lead acetate paper near the mouth of the test tube, the paper turns black.

$$Na_2S + H_2SO_4 \longrightarrow H_2S\uparrow + Na_2SO_4$$

$$H_2S + Pb(CH_3COO)_2 \longrightarrow \underset{\text{black}}{PbS\downarrow} + 2CH_3COOH$$

(b) **Sodium nitropusside test:** To sodium carbonate extract (1 mL) is added 2-3 drops of a freshly prepared solution of sodium nitropusside. Formation of purple colour confirms the presence of S^{2-}.

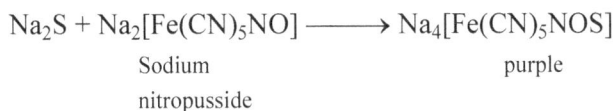

$$Na_2S + Na_2[Fe(CN)_5NO] \longrightarrow Na_4[Fe(CN)_5NOS]$$

Sodium   purple

nitropusside

The sodium nitropusside test can also be performed as spot test. It involves in adding a drop of sodium nitropusside solution to a drop of sodium carbonate extract. The test is performed on a porcelin tile. A voilet colour appears. Alternatively, filter paper moistened with ammonical solution (2 M) sodium nitropusside can be employed.

(c) **Methylene blue test (a spot test):** To a drop of the test solution placed on a spot plate is added a drop of conc. HCl and a few drops of N, N-dimethyl p-phenylene diamine (or add a drop of 1% solution of the chloride or sulphate) followed by addition of a drop of 0.5 M. $FeCl_3$ solution. A clear blue colouration is seen after 2-3 min.

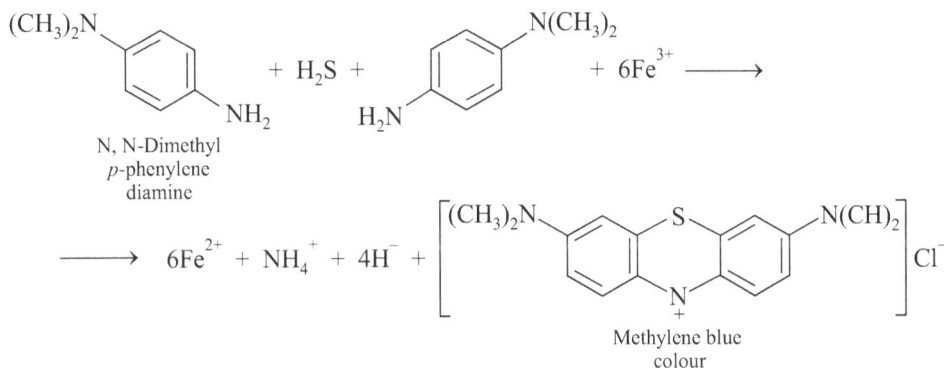

N, N-Dimethyl
p-phenylene
diamine

Methylene blue
colour

This is a very sensitive test for soluble sulphides or H_2S.

5. Sulphite (SO_3^{2-})

(a) The mixture (0.1 g) on treatment with dilute H_2SO_4 (1 mL) gives SO_2 gas which has a pungent odour. A filter paper strip moistened with acidified potassium dichromate on bringing near the mouth of the test tube is turned green.

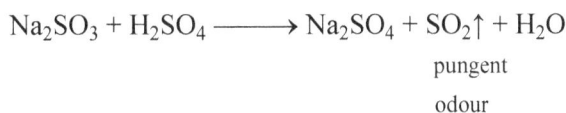

$$Na_2SO_3 + H_2SO_4 \longrightarrow Na_2SO_4 + SO_2\uparrow + H_2O$$

pungent

odour

$$3SO_2 + K_2Cr_2O_7 + H_2SO_4 \longrightarrow K_2SO_4 + Cr_2(SO_4)_3 + H_2O$$

green

(b) **Barium chloride test:** To an acidified sodium carbonate extract (1 mL) (acidified with dil. HCl) is added barium chloride solution (1 mL). Formation of a white ppt. of $BaSO_3$ indicates the presence of sulphite. The mixture is filtered (if a precipitate forms) and treated with conc. HNO_3 (1-2 drops) or chlorine or bromine water (1 mL). After 2-3 min. formation of a white ppt. confirms sulphite.

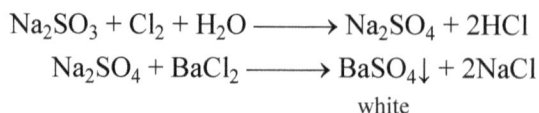

$$Na_2SO_3 + Cl_2 + H_2O \longrightarrow Na_2SO_4 + 2HCl$$
$$Na_2SO_4 + BaCl_2 \longrightarrow \underset{white}{BaSO_4\downarrow} + 2NaCl$$

6. Thiosulphate ($S_2O_3^{2-}$)

(a) The mixture (0.1 g) is heated with dilute H_2SO_4 (1 mL). Evolution of SO_2 gas (pungent odour) confirmed by acidified $K_2Cr_2O_7$ paper turning green as in case of sulphite) alongwith formation of a white or yellow precipitate of sulphur confirms the presence of $S_2O_3^{2-}$.

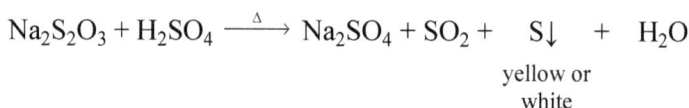

$$Na_2S_2O_3 + H_2SO_4 \xrightarrow{\Delta} Na_2SO_4 + SO_2 + \underset{\substack{yellow\ or\\white}}{S\downarrow} + H_2O$$

(b) **Silver nitrate test:** The sodium carbonate extract (0.5 mL) is acidified with dilute acetic acid and $AgNO_3$ solution (4-5 drops) added. A white precipitate which changes to yellow, orange, brown and finally black confirms the presence of thiosulphate

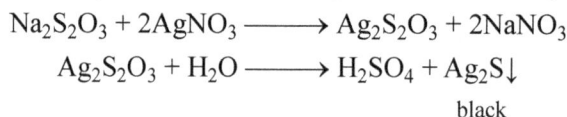

$$Na_2S_2O_3 + 2AgNO_3 \longrightarrow Ag_2S_2O_3 + 2NaNO_3$$
$$Ag_2S_2O_3 + H_2O \longrightarrow H_2SO_4 + \underset{black}{Ag_2S\downarrow}$$

(c) For testing sulphide, sulphite, sulphate and thiosulphate in presence of each other see section 1.3.7.

7. Nitrite (NO_2^-)

(a) The mixture (0.1 g) is treated with dilute H_2SO_4 (1 mL). Evolution of brown firms of NO_2 gas with a pungent odour indicates the presence of nitrite.

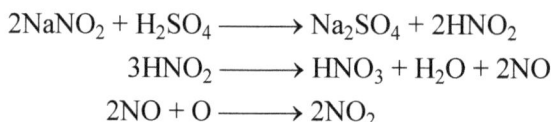

$$2NaNO_2 + H_2SO_4 \longrightarrow Na_2SO_4 + 2HNO_2$$
$$3HNO_2 \longrightarrow HNO_3 + H_2O + 2NO$$
$$2NO + O \longrightarrow 2NO_2$$

(b) **Starch-iodide test:** To the water extract of the mixture is added few crystals of KI and 2-3 drops of starch solution followed by addition of dilute H_2SO_4. A deep blue or violet colour confirms NO_2^-.

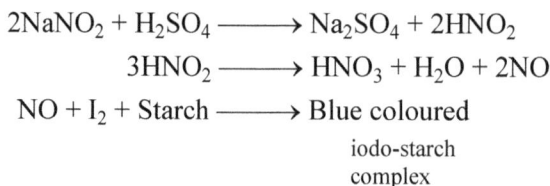

$$2NaNO_2 + H_2SO_4 \longrightarrow Na_2SO_4 + 2HNO_2$$
$$3HNO_2 \longrightarrow HNO_3 + H_2O + 2NO$$
$$NO + I_2 + Starch \longrightarrow \underset{\substack{iodo\text{-}starch\\complex}}{Blue\ coloured}$$

(c) **Ring test:** The sodium carbonate extract (0.5 mL) is acidified with dilute H_2SO_4 and solution heated to expel CO_2. To this solution is carefully added conc. solution of $FeSO_4$ acidified with dilute H_2SO_4. Formation of a brown ring at the junction of two liquids confirms NO_2^-.

$$2NaNO_2 + H_2SO_4 \longrightarrow Na_2SO_4 + 2HNO_2$$

$$3HNO_2 \longrightarrow HNO_3 + H_2O + 2NO$$

$$FeSO_4 + NO \longrightarrow \underset{\text{brown ring}}{FeSO_4 \cdot NO}$$

In case the two liquids are disturbed, brown colour is formed due to the formation of $FeSO_4 \cdot NO$.

(*d*) **Thiourea test:** The sodium carbonate extract (or water extract) is acidified with dilute acetic acid and treated with solid thiourea. This results in evolution of nitrogen and formation of thiocyanic acid which is identified by the red colour produced with dilute HCl and $FeCl_3$ solution.

$$NaNO_2 + CH_3COOH \longrightarrow HNO_2 + CH_3COONa$$

$$HNO_2 + H_2NCSNH_2 \longrightarrow N_2 + \underset{\substack{\text{Thiocyanic} \\ \text{acid}}}{HCNS} + 2H_2O$$

$$3HCNS + FeCl_3 \longrightarrow \underset{\text{red colour}}{Fe(CNS)_3} + 3HCl$$

In the above test, thiocyanate and iodide interfere. If present, these are removed first either with excess solid Ag_2SO_4 or with dilute $AgNO_3$ solution before the addition of acetic acid and thiourea.

(*e*) **m-Phenylenediamine test:** The sodium carbonate extract (1-2 mL) is acidified with dilute acetic acid and the solution heated to expel CO_2. To the remaining solution is added few drops of *m*-phenylenediamine hydrochloride and dilute HCl (few drops). Formation of a brown ppt. of bismark brown confirms nitrite.

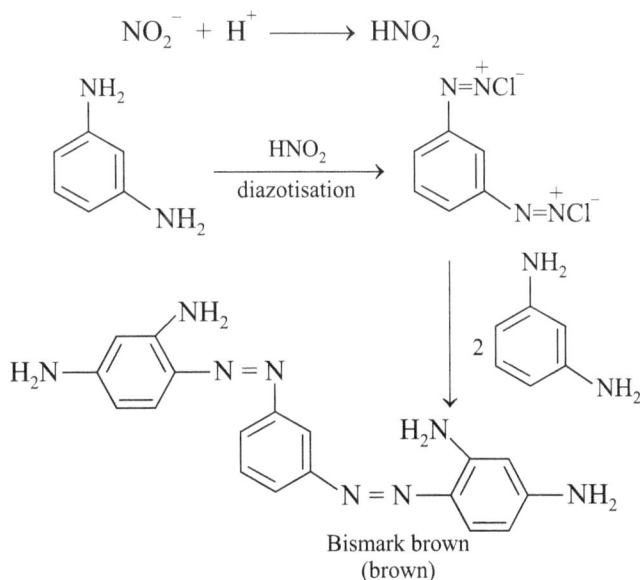

$$NO_2^- + H^+ \longrightarrow HNO_2$$

Bismark brown
(brown)

(*f*) Sulphanilic acid-1-naphthylamine reagent (Griess-Ilosvay test) (spot test).

A solution of sulphanilic acid in acetic acid on treatment with HNO_2 (generated from nitrite in the mixture and acid) gives the corresponding diazo salt, which on treatment with 1-naphthylamine forms a red azodye.

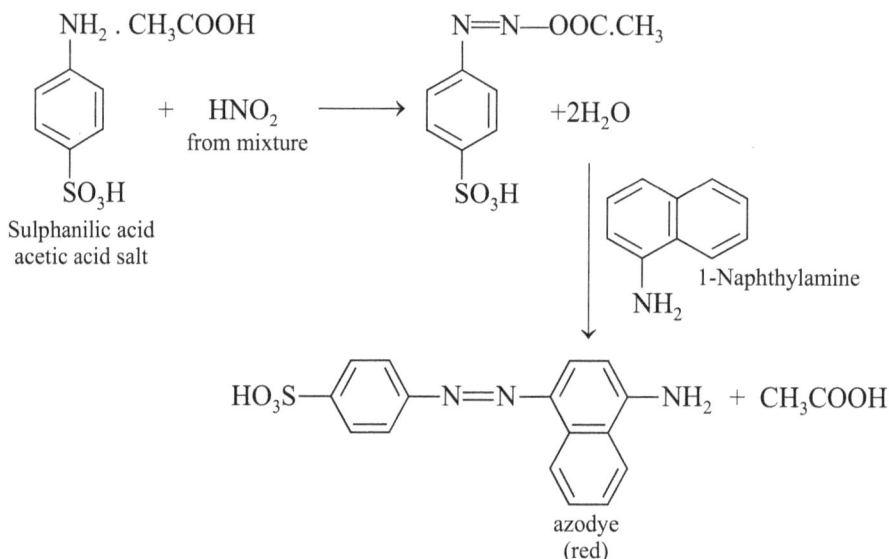

$$NH_2 \cdot CH_3COOH \quad + \quad HNO_2 \text{ (from mixture)} \quad \longrightarrow \quad N{=}N{-}OOC.CH_3 \quad + 2H_2O$$

Sulphanilic acid
acetic acid salt

(with SO_3H substituents; reacting with 1-Naphthylamine, NH_2)

$$HO_3S{-}\!\!\!\bigcirc\!\!\!{-}N{=}N{-}\!\!\!\bigcirc\!\!\!{-}NH_2 \;+\; CH_3COOH$$

azodye
(red)

The spot test is performed by placing a drop of neutral or acid test solution on a spot plate and mixing with a drop of sulphanilic acid solution. This is followed by addition of a drop of 1-naphthylamine reagent. A red colour is formed.

In the above test, if Fe(III) is present it must be masked by tartaric acid.

Confirmatory test for anions of Group B (dil. H_2SO_4.$KMnO_4$ Solution Group)

8. Oxalate ($C_2O_4^{2-}$)

(*a*) The mixture (0.1 g) is heated with conc. H_2SO_4 in a dry test tube. The mouth of the test tube is brought near the flame. The evolved CO burns with a blue flame indicating the presence of oxalate (caution: CO is poisonous)

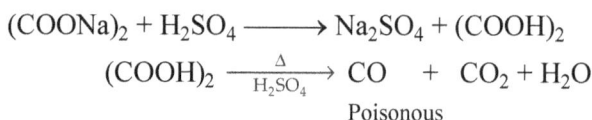

$$(COONa)_2 + H_2SO_4 \longrightarrow Na_2SO_4 + (COOH)_2$$

$$(COOH)_2 \xrightarrow[H_2SO_4]{\Delta} CO + CO_2 + H_2O$$
$$\text{Poisonous}$$

(*b*) **Calcium chloride test:** The sodium extract (1 mL) is neutralised with dilute acetic acid. It is heated to expel CO_2. This is followed by addition of dilute acetic acid (1 mL) and calcium chloride solution. The solution is shaken and allowed to stand for some time. A white precipitate of calcium oxalate is obtained.

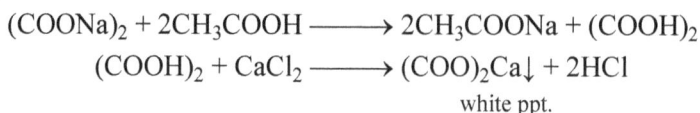

$$(COONa)_2 + 2CH_3COOH \longrightarrow 2CH_3COONa + (COOH)_2$$

$$(COOH)_2 + CaCl_2 \longrightarrow (COO)_2Ca\downarrow + 2HCl$$
$$\text{white ppt.}$$

The precipitate of calcium oxalate is filtered, washed with water and dissolved in dilute H_2SO_4. The mixture is heated and treated with few drops of $KMnO_4$ solution. The pink colour of $KMnO_4$ disappear confirming the presence of oxalate

$$(COO)_2Ca + H_2SO_4 \longrightarrow (COOH)_2 + CaSO_4$$
$$2KMnO_4 + 3H_2SO_4 + 5(COOH)_2 \longrightarrow 2MnSO_4 + K_2SO_4 + 10CO_2 + 8H_2O$$

(c) **Resorcinol test:** To sodium carbonate extract (1 mL) is added dil. H_2SO_4 followed by addition of a pinch of Mg powder. When all Mg has dissolved, solid resorcinol (0.1 g) is added, mixture shaken, cooled (tap water) and conc. H_2SO_4 added down slowly through the side of the test tube. A blue ring forms at the interface of two liquids confirming $C_2O_4^{2-}$.

Citrates do not interfere but in presence of tartarate, a blue ring is obtained in the cold or by gentle warming.

(d) **Ferrous sulphate test:** Sodium carbonate extract (1 mL) is acified with dilute acetic acid. The solution is heated to boil off CO_2. Subsequent addition of $FeSO_4$ solution gives a yellow precipitate confirming oxalate.

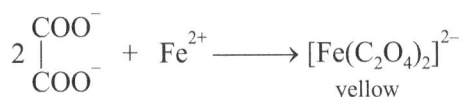

$$2 \begin{array}{c} COO^- \\ | \\ COO^- \end{array} + Fe^{2+} \longrightarrow [Fe(C_2O_4)_2]^{2-}$$
$$\text{yellow}$$

(e) **Treatment with MnO_2:** The mixture (0.1 g) is treated with dilute H_2SO_4. Any effervescence is allowed to subsidise. Subsequent addition of solid MnO_2 gives effervescence confirming oxalate. This test confirms $C_2O_4^{2-}$ in presence of CO_3^{2-}.

9. Iodide (I⁻)

(a) The mixture (0.1 g) is heated with conc. H_2SO_4. Evolution of violet vapours of I_2 due to decomposition of formed HI confirms iodide.

$$KI + H_2SO_4 \longrightarrow KHSO_4 + HI$$
$$2HI + H_2SO_4 \longrightarrow I_2\uparrow + SO_2 + 2H_2O$$

(b) **Silver nitrate test.** The sodium carbonate extract (0.5 mL) is acidified with dilute nitric acid and $AgNO_3$ solution (0.5 mL) is added. Formation of a yellow precipitate of AgI, very slightly soluble in ammonia solution confirms the presence of iodide

$$NaI + AgNO_3 \longrightarrow AgI\downarrow + NaNO_3$$
$$\text{yellow}$$

(c) **Potassium nitrite solution.** The sodium carbonate extract is acidified with acetic acid and treated with KNO_2 solution. The liberated iodine is identified by colouring starch paper blue or chloroform violet

$$3I^- + 2NO_2^- + 4H^+ \longrightarrow I_3^- + 2NO\uparrow + 2H_2O$$

(d) **Spot test:** A drop of acidified test solution is placed on a spot plate alongwith a drop of starch solution and a drop of 50% KNO_2 solution. A blue colouration is obtained.

Cyanides interface due to the formation of cyanogen iodide. If present, cyanide is removed before the test by heating with either $NaHCO_3$ solution or by acidifying and heating.

$$I_3^- + CN^- \longrightarrow 1CN\uparrow + 2I^-$$

10. Bromide (Br⁻)

(*a*) A small amount of the mixture (0.1 g) is heated with conc. H_2SO_4 (1 mL) in a dry test tube. Evolution of reddish brown fumes with pungent odour of Br_2 vapour confirms the presence of bromide. The intensity of the colour is increased if MnO_2 (0.1 g) is added to the above mixture.

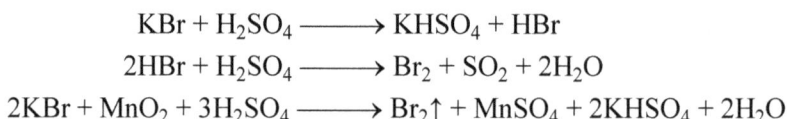

$$KBr + H_2SO_4 \longrightarrow KHSO_4 + HBr$$
$$2HBr + H_2SO_4 \longrightarrow Br_2 + SO_2 + 2H_2O$$
$$2KBr + MnO_2 + 3H_2SO_4 \longrightarrow Br_2\uparrow + MnSO_4 + 2KHSO_4 + 2H_2O$$

(*b*) **Silver nitrate test:** The sodium carbonate extract (0.5 mL) is acidified with dilute HNO_3 (0.5 mL) and then $AgNO_3$ solution (0.5 mL) added. Formation of a yellow precipitate of AgBr, which is sparingly soluble in NH_4OH solution confirms bromide.

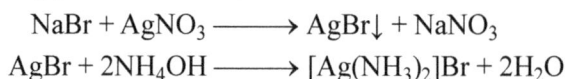

$$NaBr + AgNO_3 \longrightarrow AgBr\downarrow + NaNO_3$$
$$AgBr + 2NH_4OH \longrightarrow [Ag(NH_3)_2]Br + 2H_2O$$

The solution of AgBr in ammonia should be discarded as early as possible in order to avoid explosion.

(*c*) **Layer test:** The sodium carbonate extract (1 mL) is acidified with dil. HNO_3 and solution heated in expel CO_2. To this solution is added $CHCl_3$ or CCl_4 (0.5 mL) followed by addition of chlorine water or conc. HNO_3. The formation of orange organic layer confirms bromide.

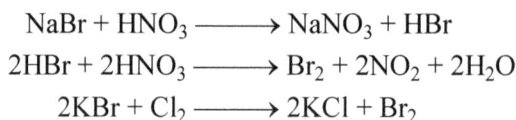

$$NaBr + HNO_3 \longrightarrow NaNO_3 + HBr$$
$$2HBr + 2HNO_3 \longrightarrow Br_2 + 2NO_2 + 2H_2O$$
$$2KBr + Cl_2 \longrightarrow 2KCl + Br_2$$

(*d*) The presence of bromide in presence of iodide and chloride can be confirmed by a special test (see Section 1.3.7).

11. Chloride (Cl⁻)

(*a*) The mixture (0.1 g) is heated with conc. H_2SO_4 (1 mL). Evolution of a colourless, pungent smelling HCl gas indicates the presence of chloride. A glass rod dipped in NH_4OH solution brought near the mouth of the test tube gives dense white fumes of NH_4Cl.

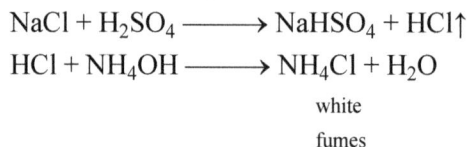

$$NaCl + H_2SO_4 \longrightarrow NaHSO_4 + HCl\uparrow$$
$$HCl + NH_4OH \longrightarrow NH_4Cl + H_2O$$
$$\text{white}$$
$$\text{fumes}$$

(*b*) **Manganese dioxide test:** The mixture (0.1 g) is heated with MnO_2 (0.1 g) and conc. H_2SO_4 (1 mL). Evolution of greenish-yellow chlorine confirms the presence of chloride.

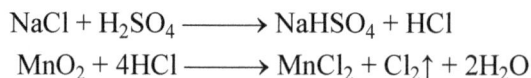

$$NaCl + H_2SO_4 \longrightarrow NaHSO_4 + HCl$$
$$MnO_2 + 4HCl \longrightarrow MnCl_2 + Cl_2\uparrow + 2H_2O$$

(*c*) **Silver nitrate test:** Sodium carbonate extract (0.5 mL) is acidified with dilute HNO_3 and $AgNO_3$ solution (0.5 mL) added. Formation of a white ppt. soluble in NH_4OH confirms the presence of chloride.

$$AgNO_3 + NaCl \longrightarrow AgCl\downarrow + NaNO_3$$
White ppt.

$$AgCl + 2NH_4OH \longrightarrow [Ag(NH_4)_2]Cl + 2H_2O$$
Soluble complex

The soluble complex obtained above must be discarded at the earliest in order to avoid explosion.

(d) **Chromyl chloride test:** The mixture (0.1 g) is heated in a dry test tube with solid $K_2Cr_2O_7$ (0.2 g) and conc. H_2SO_4 (2 mL). The evolved gas is passed into a test tube containing NaOH solution (about 2 mL). The alkaline solution is acidified with dil. CH_3COOH and lead acetate solution (1 mL) added. Formation of a yellow precipitate of $PbCrO_4$ confirms chloride.

$$K_2Cr_2O_7 + 4NaCl + 6H_2SO_4 \longrightarrow 2CrO_2Cl_2\uparrow + 2KHSO_4 + 4NaHSO_4 + 3H_2O$$
Chromyl
chloride

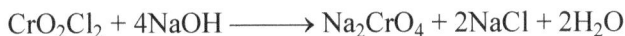

$$CrO_2Cl_2 + 4NaOH \longrightarrow Na_2CrO_4 + 2NaCl + 2H_2O$$
$$Na_2CrO_4 + Pb(CH_3COO)_2 \longrightarrow PbCrO_4 + 2CH_3COONa$$
Lead chromate
(yellow)

Chromyl chloride test should be performed only in the absence of bromide. This test is not given by chlorides of silver, lead and mercury.

(e) For testing presence of chloride in presence of bromide and iodide see Section 1.3.7.

Confirmatory tests for anions of group C (conc. H_2SO_4 group)

Confirmatory tests for some of the anions present in group C have already been discussed in group B anions. These anions include oxalate, iodide, bromide and chloride.

12. Nitrate (NO_3^-)

(a) To the mixture (0.1 g) is added conc. H_2SO_4 (1 mL) and mixture warmed. Evolution of brown fumes of NO_2 indicates the presence of nitrate. The fumes become intense if few chips of copper are added. In case the mixture also contains bromide, the yellow vapours of bromide condense on the sides of the test tube while NO_2 does not.

$$NaNO_3 + H_2SO_4 \longrightarrow NaHSO_4 + HNO_3$$
$$4HNO_3 \longrightarrow 4NO_2\uparrow + O_2 + 2H_2O$$
$$3Cu + 8HNO_3 \longrightarrow 3Cu(NO_3)_2 + 2NO + 4H_2O$$
$$2NO + O_2 \longrightarrow 2NO_2$$

(b) **Ring test:** The sodium carbonate extract (1 mL) is acidified with dilute H_2SO_4 and solid $FeSO_4$ (about 0.1 g) added and dissolved. This is followed by careful addition of conc. H_2SO_4 (1 mL) along the sides of the test tube. Formation of a brown ring at the junction of two liquids confirms the presence of nitrate.

$$NaNO_3 + H_2SO_4 \longrightarrow NaHSO_4 + HNO_3$$
$$2HNO_3 + 6FeSO_4 + 3H_2SO_4 \longrightarrow 3Fe_2(SO_4)_3 + 2NO + 4H_2O$$
$$FeSO_4 + NO \longrightarrow FeSO_4 . NO$$
(brown ring)

In this test, in place of sodium carbonate extract, aqueous solution of the mixture can also be used.

The test does not work in case the mixture contains nitrite. This is because the NO (evolved from nitrite) gives black colour when $FeSO_4$ is added.

For testing NO_3^- in presence of NO_2^- (as indicated by dilute H_2SO_4 test), the sodium carbonate extract (1 mL) is acidified with dilute H_2SO_4 and the solution heated with urea (0.2 g). By this procedure, nitrite is decomposed. Finally, the presence of nitrate is confirmed by the ring test.

The presence of bromide and iodide also interferes in the detection of nitrate. For testing nitrate in presence of bromide and iodide (see Section 1.3.7).

13. Fluoride (F^-)

(a) The mixture (0.1 g) on treatment with conc. H_2SO_4 (1 mL) gives oily droplets which are formed along the sides of the test tube

$$2NaF + H_2SO_4 \longrightarrow Na_2SO_4 + H_2F_2\uparrow$$

(b) **Formation of silic acid:** A mixture of the given mixture (0.1 g), conc. H_2SO_4 (1 mL) and sand (0.2 g) is heated and a moistened glass rod is brought over the mouth of the test tube. Deposition of a gelatinous while precipitate of silicic acid on the glass rod confirms fluoride.

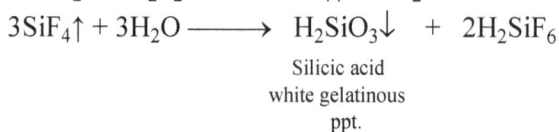

$$2NaF + H_2SO_4 \longrightarrow Na_2SO_4 + H_2F_2\uparrow$$
$$SiO_2 + 2H_2F_2 \longrightarrow SiF_4\uparrow + 2H_2O$$
$$3SiF_4\uparrow + 3H_2O \longrightarrow \underset{\substack{\text{Silicic acid} \\ \text{white gelatinous} \\ \text{ppt.}}}{H_2SiO_3\downarrow} + 2H_2SiF_6$$

(c) **Calcium chloride test:** To sodium carbonate extract (1 mL) is added dilute acetic acid (till acidic) and the solution boiled to expel CO_2. To the remaining solution is added $CaCl_2$ solution (1 mL). Formation of a which precipitate of CaF_2 which is sparingly soluble in acetic acid but slightly more soluble in dil. HCl confirms the presence of fluoride.

$$Ca^{2+} + 2F^- \longrightarrow CaF_2\downarrow$$

(d) **Spot test:** Alizarin red S (sodium alizarin solution (0.1% solution, 2 drops) and zirconyl chloride solution (0.1%, 2 drops) are mixed on a spot plate. To the mixture is added fluoride solution (1-2 drops). Decolourisation of the zirconium lake (a yellow solution in obtained) confirms the presence of fluoride.

14. Chlorate (ClO_3^-)

(a) The mixture (0.1 g) is heated (carefully) with conc. H_2SO_4 (0.5 mL). A greenish yellow gas (ClO_2) accompanied by cracking sound is obtained

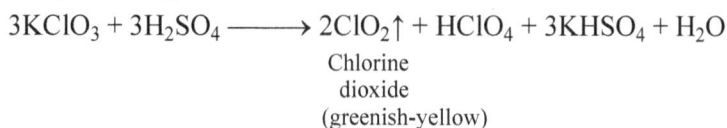

$$3KClO_3 + 3H_2SO_4 \longrightarrow \underset{\substack{\text{Chlorine} \\ \text{dioxide} \\ \text{(greenish-yellow)}}}{2ClO_2\uparrow} + HClO_4 + 3KHSO_4 + H_2O$$

The formed chlorine dioxide dissolves in H_2SO_4 to give an orange-yellow solution.

While carring out this test, the mouth of the test tube should be directed away from the experimenter.

(b) **Silver nitrate test:** The mixture (0.1 g) and a pinch of sodium nitrite is dissolved in dilute HNO_3 (by warming) and few drops of $AgNO_3$ solution added. Formation of a curdy white precipitate of AgCl confirms the presence of chlorate. The function of $NaNO_2$ is to reduce chlorate to chloride.

$$KClO_3 + 3NaNO_2 \longrightarrow 3NaNO_3 + KCl$$
$$KCl + AgNO_3 \longrightarrow AgCl\downarrow + KNO_3$$

(c) **Indigo test:** The solution containing chlorate is treated with a dilute solution of indigo in conc. H_2SO_4 until a pale-blue colour is obtained. Subsequent slow addition of dilute. sulphurous acid or sodium sulphite solution the blue colour disappears. In this test, sulphurous acid reduces chlorate to chlorine which bleaches the indigo.

15. Tartarate ($C_4H_4O_6^{2-}$)

(a) The solid mixture (0.1 g) is heated with conc. H_2SO_4. Charring occurs immediately due to the separation of carbon and is accompanied by evolution of CO_2, CO and SO_2 (SO_2 is evolved by the reduction of H_2SO_4 by carbon). An odour of burnt sugar is noticed.

$$\underset{\text{Sod. tartarate}}{Na_2C_4H_4O_6} + H_2SO_4 \longrightarrow \underset{\text{Tartaric acid}}{C_4H_6O_6} + Na_2SO_4$$

$$2H_2SO_4 + \underset{\text{Tartaric acid}}{C_4H_6O_6} \xrightarrow{\Delta} \underset{\substack{\text{blackening}\\\text{of mixture}\\\text{(odour of}\\\text{burnt sugar)}}}{C} + CO\uparrow + 2CO_2\uparrow + 2SO_2\uparrow + 5H_2O$$

(b) **Fenton's test:** To neutral or acidic sodium carbonate extract (0.5 mL) is added a drop of saturated ferrous sulphate solution followed by addition of 2-3 drops of H_2O_2 solution (3 percent). A deep-violet colour develops on addition of excess NaOH solution. The colour becomes intense by addition of a drop of $FeCl_3$ solution. The violet colour is due to the formation of a salt of dihydroxymaleic acid, HOOC . C(OH) = C(OH) COOH. This test is not given by citrates, malates or succinates.

(c) **Resorcinol test:** The sodium carbonate extract (0.5 mL) is acidified with dilute H_2SO_4 followed by addition of very small amount of magnesium powder. After the magnesium has dissolved, resorcinol (0.1 g) is added, mixture shaken (to dissolve resorcinol). Conc. H_2SO_4 (3-4 mL) is added carefully through the sides of the test tube. The test tube is carefully gently warmed at the bottom. A red layer or ring is formed at the junction of the two liquids. On heating the colour spreads and whole of H_2SO_4 layer becomes red.

Citrates do not interfere in this test. However, if oxalate is present, a blue ring is formed in the cold. On gentle warming, the blue colour spreads downwards and a red ring is formed at the interface.

The colour is due to the formation of a condensation product of resorcinol, glycolic aldehyde (which is obtained by the action of H_2SO_4 and tartaric acid). The condensation product is $[CH_2OH . CH[C_6H_3(OH)_2]_2$.

(d) **1, 1-Bis-2-naphthol test:** An acidified sodium carbonate extract (or the solid mixture) on heating with a solution of 1, 1-Bis-2-naphthol in conc. H_2SO_4, a green colouration results. Oxalic, citric, succinic and cinnamic acids do not interfere.

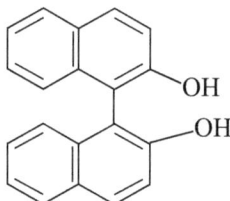

Confirmatory Tests of Anions of D Group (Special Group)

16. Sulphate (SO_4^{2-})

(a) To sodium carbonate extract (0.5 mL) is added dilute HCl (1 mL) followed by addition of barium chloride solution. Formation of a white ppt. of $BaSO_4$ insoluble in conc. HCl confirms sulphate

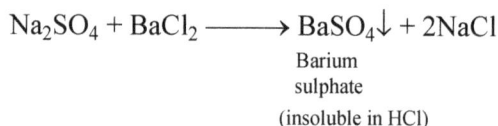

$$Na_2SO_4 + BaCl_2 \longrightarrow BaSO_4\downarrow + 2NaCl$$

Barium
sulphate
(insoluble in HCl)

(b) **Lead acetate test:** To sodium carbonate extract (0.5 mL) is added dil. acetic acid (1 mL) followed by addition of lead acetate solution (1 mL). A white ppt. of $PbSO_4$ is formed. The mixture is centrifugaled and to the residue is added ammonium acetate solution and sodium hydroxide solution. A clear solution confirms sulphate.

(c) **Sodium rhodizonate test:** Spot test. To a drop of $BaCl_2$ solution placed on a spot plate is added sodium rhodizonate solution (0.5% aqueous solution, one drop). A red colour is produced. To this is added a drop to Na_2CO_3 extract acidified with dil. HCl. The disappearance of red colour confirms sulphate.

17. Phosphate (PO_4^{3-})

(a) The sodium carbonate extract (1 mL) is acidified with dilute HNO_3 (1 mL) and treated with ammonium molybdate solution (1 mL) and conc. HNO_3 (1 mL). A yellow ppt. is formed on heating.

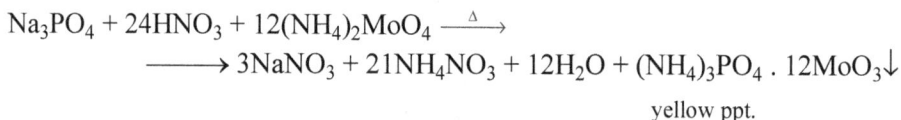

$$Na_3PO_4 + 24HNO_3 + 12(NH_4)_2MoO_4 \xrightarrow{\Delta}$$
$$\longrightarrow 3NaNO_3 + 21NH_4NO_3 + 12H_2O + (NH_4)_3PO_4 . 12MoO_3\downarrow$$

yellow ppt.

This test is also given by arsenate and or arsenite. To test phosphate in presence of arsenate/arsenite, the above test is performed in presence of a tartarate.

(b) **Silver nitrate test:** To sodium carbonate extract (0.5 mL) is added dilute acetic acid (1 mL) followed by addition of $AgNO_3$ solution (1 mL). A yellow precipitate of Ag_3PO_4 is obtained, which is soluble in dilute ammonia solution and in dilute HNO_3.

18. Borate (BO_3^{3-})

(*a*) To the mixture (0.1 g) taken in a dry test tube is added conc. H_2SO_4 (1 mL) and ethyl alcohol (1-2 mL). The mixture is heated cautiously and the vapours coming out of the test tube are brought near a flame. A green edged flame confirms borate.

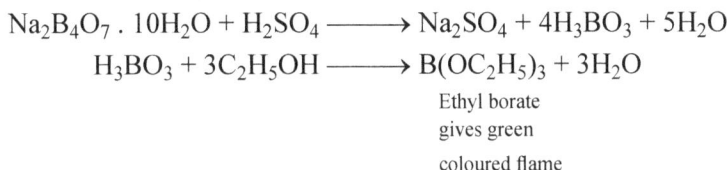

$$Na_2B_4O_7 . 10H_2O + H_2SO_4 \longrightarrow Na_2SO_4 + 4H_3BO_3 + 5H_2O$$

$$H_3BO_3 + 3C_2H_5OH \longrightarrow B(OC_2H_5)_3 + 3H_2O$$

Ethyl borate
gives green
coloured flame

Copper and barium salts interfere in the above test. It is advisable to perform the above test is a test tube (not in a china dish) so that only the vapour of ethyl borate reach the mouth of the test tube and Cu^{2+} and Ba^{2+} remain behind in the solution.

(**Caution:** Vapours of ethyl borate and poisonous).

(*b*) The mixture (0.1 g) is mixed with CaF_2 (0.1 g) and conc. H_2SO_4 (few drops) (in a watch glass). A small amount of the paste on the edge of a glass rod (or on the loop of a platinum wire) is brought near the flame (do not touch the flame). The formed volatile BF_3 turns the flame green, confirming the presence of borate in the mixture.

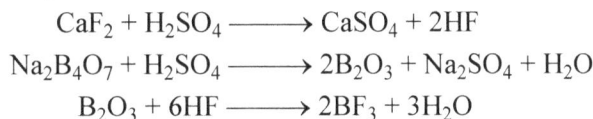

$$CaF_2 + H_2SO_4 \longrightarrow CaSO_4 + 2HF$$

$$Na_2B_4O_7 + H_2SO_4 \longrightarrow 2B_2O_3 + Na_2SO_4 + H_2O$$

$$B_2O_3 + 6HF \longrightarrow 2BF_3 + 3H_2O$$

Confirmatory Test of some other Anions

19. Cyanide ($\overline{CN}$)

All cyanides are extremely posionous and so all tests must be carried out in well ventilated fume cupboard.

(*a*) The mixture on treating in cold with dilute HCl gives HCN having odour reminiscent of bitter almonds. (Do not smell the gas)

(*b*) **Conc. sulphuric acid:** The mixture on heating with conc. H_2SO_4 gives carbon monoxide gas which is poisonous and on igniting burns with a blue flame. All cyanides (except complex cyanides) are decomposed by heating with conc. H_2SO_4

$$2KCN + 2H_2SO_4 + 2H_2O \longrightarrow 2CO\uparrow + K_2SO_4 + (NH_4)_2SO_4$$

(*c*) **Prussian blue test:** To sodium carbonate extract (0.5 mL) rendered alkaline by addition dilute NaOH solution is added a freshly prepared solution of $FeSO_4$ (1 mL). The mixture is heated and then acidified (dil HCl) and treated with a little of ferric chloride solution. A green solution is obtained which on standing deposits prussian blue.

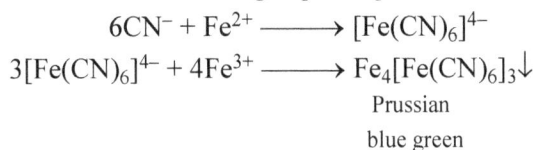

$$6CN^- + Fe^{2+} \longrightarrow [Fe(CN)_6]^{4-}$$

$$3[Fe(CN)_6]^{4-} + 4Fe^{3+} \longrightarrow Fe_4[Fe(CN)_6]_3\downarrow$$

Prussian
blue green

20. Cyanate (OCN^-)

(a) The solid mixture (0.1 g) on treatment with dilute H_2SO_4 gives vigorous effervescence due to evolution of CO_2. The free cyanic acid, initially produced decomposes into carbon dioxide and ammonia, which combines with H_2SO_4 to give ammonium sulphate. The resulting solution on heating with NaOH liberates ammonia

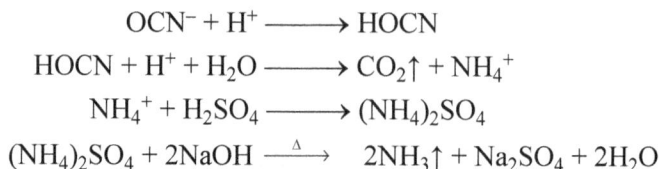

$$OCN^- + H^+ \longrightarrow HOCN$$
$$HOCN + H^+ + H_2O \longrightarrow CO_2\uparrow + NH_4^+$$
$$NH_4^+ + H_2SO_4 \longrightarrow (NH_4)_2SO_4$$
$$(NH_4)_2SO_4 + 2NaOH \xrightarrow{\Delta} 2NH_3\uparrow + Na_2SO_4 + 2H_2O$$

(b) Copper sulphate-pyridine test. The cyanate solution (neutral) on adding to a dilute solution of copper sulphate containing a little pyridine, a lilac blue precipitate of composition $[Cu(C_5H_5N)_2](OCN)_2$ is formed, which is soluble in $CHCl_3$ producing a sapphire-blue solution.

$$2OCN^- + Cu^{2+} + 2C_5H_5N \longrightarrow [Cu(C_5H_5N)_2](OCN)_2$$

21. Thiocyanate (SCN^-)

(a) The mixture (0.1 g) on treatment in cold with conc. H_2SO_4 gives a yellow colouration, which on warming (caution) gives carbonyl sulphide, which burns with a blue flame

$$SCN^- + H_2SO_4 + H_2O \longrightarrow COS\uparrow + NH_4^+ + SO_4^{2-}$$

(b) **Ferric chloride test:** Addition of ferric chloride solution (2-3 drops) to a solution of thiocyanate gives blood-red colouration

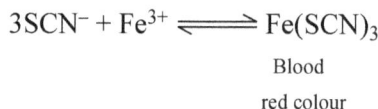

$$3SCN^- + Fe^{3+} \rightleftharpoons Fe(SCN)_3$$
$$\text{Blood}$$
$$\text{red colour}$$

21. Hypochlorite (OCl^-)

(a) The mixture containing hypochlorite on treatment with dilute HCl gives yellow solution followed by effervescence liberating chlorine (irritating odour) which is identified by potassium iodide-starch paper (giving black colour)

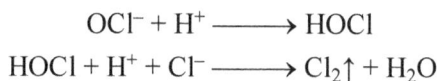

$$OCl^- + H^+ \longrightarrow HOCl$$
$$HOCl + H^+ + Cl^- \longrightarrow Cl_2\uparrow + H_2O$$

(b) **Potassium iodide-starch test:** A bluish-black colour is formed is weakly alkaline or neutral solution containing hypochloride due to separation of iodine

$$OCl^- + 3I^- + H_2O \longrightarrow I_3^- + 2OH^- + Cl^-$$

The colour disappear if the solution is too alkaline

$$I_3^- + 2OH^- \longrightarrow OI^- + 2I^- + H_2O$$

22. Bromate (BrO_3^-)

(a) The mixture (0.1 g) on treatment with conc. H_2SO_4 in cold gives bromine and oxygen

$$4KBrO_3 + 2H_2SO_4 \longrightarrow 2Br_2\uparrow + 5O_2 + 2K_2SO_4 + 2H_2O$$

(b) **Silver nitrate solution:** Treatment of bromate with $AgNO_3$ solution gives a white crystalline ppt. of $AgBrO_3$, which is soluble in hot water and readily soluble in NH_4OH solution forming a complex salt

$$BrO_3^- + Ag^+ \longrightarrow AgBrO_3\downarrow$$

23. Iodate (IO_3^-)

(a) On treatment with conc. H_2SO_4 in presence of ferrous sulphate, a brown precipitate is formed

$$IO_3^- + 6Fe^{2+} + 6H^+ \longrightarrow I^- + 6Fe^{3+} + 3H_2O$$

(b) **Potassium iodide:** Mixing potassium iodide and potassium iodate (1:1) with dilute HCl liberates iodine

$$IO_3^- + 5I^- + 6H^+ \longrightarrow 3I_2 + 3H_2O$$

24. Silicate (SiO_3^{2-})

(a) The mixture on heating with dilute HCl gives a gelatinous ppt. of meta silicic acid

$$SiO_3^{2-} + 2H^+ \longrightarrow H_2SiO_3\downarrow$$

(b) **Silver nitrate test.** Yellow ppt. of silver silicate soluble in dil acids and NH_4OH is obtained

$$SiO_3^{2-} + 2Ag^+ \longrightarrow Ag_2SiO_3\downarrow$$

25. Chromate (CrO_4^{2-})

(a) The sodium carbonate extract on acidification with dilute acetic acid followed by addition of $BaCl_2$ solution gives a pale yellow precipitate of barium chromate, insoluble in acetic acid but soluble in dilute mineral acid.

$$CrO_4^{2-} + Ba^{2+} \longrightarrow BaCrO_4\downarrow$$

(b) **Silver nitrate solution:** Brownish red ppt. of silver chromate soluble in dil. HNO_3 and NH_4OH but insoluble in acetic acid is produced

$$CrO_4^{2-} + 2Ag^+ \longrightarrow Ag_2CrO_4\downarrow$$

26. Formate ($HCOO^-$)

(a) The mixture (0.1 g) on warming with dilute H_2SO_4 gives formic acid having pungent odour

$$HCOO^- + H^+ \longrightarrow HCOOH$$

(b) On heating the mixture with conc. H_2SO_4 gives carbon monoxide identified by igniting

$$HCOONa + H_2SO_4 \longrightarrow CO\uparrow + NaHSO_4 + H_2O$$

(c) **Ester test:** On heating with alcohol in presence of conc. H_2SO_4 pleasant odour of ethyl formate is obtained

$$HCOONa + H_2SO_4 \longrightarrow HCOOH + Na^+ + HSO_4^-$$

$$HCOOH + C_2H_5OH \longrightarrow \underset{\substack{\text{pleasant}\\\text{odour}}}{HCOOC_2H_5\uparrow} + H_2O$$

27. Citrate ($C_6H_5O_7^{3-}$)

The mixture (0.2 g) on heating with conc. H_2SO_4 darkens due to separation of carbon.

$$\underset{\substack{\text{CH}_2\text{COOH} \\ | \\ \text{HO}-\text{C}-\text{COOH} \\ | \\ \text{CH}_2\text{COOH} \\ \text{Citric acid}}}{} + \xrightarrow[\text{Conc. H}_2\text{SO}_4]{\Delta} CO\uparrow + H_2O\uparrow + \underset{\substack{\text{CH}_2 \\ | \\ \text{CO} \\ | \\ \text{CH}_2 \\ | \\ \text{COOH} \\ \text{Acetone} \\ \text{dicarboxylic} \\ \text{acid}}}{\overset{\substack{\text{COOH} \\ | \\ \text{CH}_2 \\ |}}{}}$$

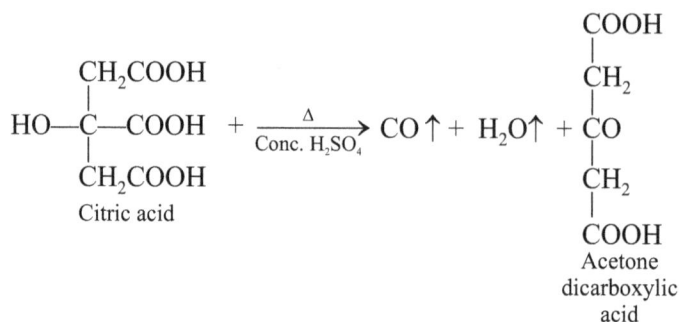

In case of citric acid, the formed acetone dicarboxylic acid is made use of for confirmation of citrate. Thus, the mixture (0.2 *g*) containing citrate is treated with conc. H_2SO_4 for 1 minute, mixture cooled, diluted with water (caution) and rendered alkaline (with NaOH solution) and few millilitres of freshly prepared solution of sodium nitropusside added. An intense red colour is obtained.

1.3.7   Tests for Anions which Interfere in Presence of each other

A number of anions which are present alongwith certain other anions are difficult to detect. For such combination of anions special procedures are to be adopted. Following are given the combination of anions with other anions which interfere in their detection.

1. Carbonate in presence of bicarbonate.
2. Carbonate in presence of sulphite and thiosulphate.
3. Nitrate in presence of nitrite.
4. Sulphide, sulphite, sulphate and thiosulphate in presence of each other.
5. Chloride, bromide and iodide in presence of each other.
6. Nitrate in presence of bromide and iodide.
7. Sulphate in presence of fluoride.
8. Oxalate in presence of carbonate.

The above combination of anions are detected by special procedures as discussed below:

Carbonate in Presence of Bicarbonate

Both carbonate and bicarbonate give CO_2 on treatment with dil. H_2SO_4, bicarbonate can be tested in presence of carbonate by adding excess of calcium chloride to aqueous extract of the mixture when CO_3^{2-} gets quantitatively precipitated as $CaCO_3$

$$CO_3^{2-} + Ca^{2+} \longrightarrow CaCO_3\downarrow$$

The mixture is filtered and to the filtrate is added ammonia. Formation of a white ppt. or cloudiness confirms the presence of bicarboante.

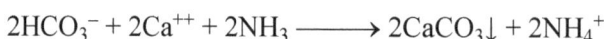

$$2HCO_3^- + 2Ca^{++} + 2NH_3 \longrightarrow 2CaCO_3\downarrow + 2NH_4^+$$

Carbonate in Presence of Sulphite and Thiosulphate

Carbonate and sulphite (or thiosulphate) on treatment with dil. H_2SO_2 give CO_2 and SO_2 respectively. Both CO_2 and SO_2 give milky white ppt. with lime water. Though the dichromate paper test for SO_2 can be carried out even in the presence of CO_2, but SO_2 (from SO_3^{2-} and $S_2O_3^{2-}$) make the detection of CO_2 (from CO_3^{2-}) difficult. For testing CO_2 in presence of SO_2, the mixture is treated with solid $K_2Cr_2O_7$ and then dilute H_2SO_4 added. By this procedure SO_2 reacts with $K_2Cr_2O_7$ and the gas evolved contains only CO_2 which turns lime water milky.

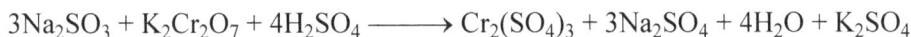

$$3Na_2SO_3 + K_2Cr_2O_7 + 4H_2SO_4 \longrightarrow Cr_2(SO_4)_3 + 3Na_2SO_4 + 4H_2O + K_2SO_4$$
$$CaCO_3 + H_2SO_4 \longrightarrow CaSO_4 + CO_2\uparrow + H_2O$$
$$CO_2 + Ca(OH)_2 \longrightarrow \underset{\text{milky}}{CaCO_3\downarrow} + H_2O$$

On passing excess CO_2 in lime water, the milkyness disappears due to formation of calcium bicarbonate which is water soluble

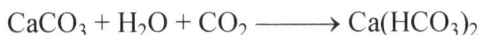

$$CaCO_3 + H_2O + CO_2 \longrightarrow Ca(HCO_3)_2$$

Nitrate in Presence of Nitrite

It is difficult to detect nitrate in presence of nitrite. This is because nitrite on treatment with H_2SO_4 gives nitric oxide, which interferes in the ring test for nitrate. For testing NO_3^- in presence of NO_2^-, the sodium carbonate extract is treated with dilute acetic acid or dilute H_2SO_4 and the solution heated gently. Addition of freshly prepared starch and KI solution (0.1 mL each) gives a blue colour confirming the presence of nitrite.

$$2NaNO_2 + H_2SO_4 \longrightarrow Na_2SO_4 + 2HNO_2$$
$$H_2SO_4 + 2HNO_2 + 2KI \longrightarrow 2NO + I_2 + K_2SO_4 + 2H_2O$$
$$I_2 + Starch \longrightarrow blue \ colour$$

For testing the presence of NO_3^- in presence of NO_2^-, the NO_2^- is first decomposed completely by warming with dilute H_2SO_4. Subsequently, NO_3^- is converted into NO_2^- which is confirmed as follows: Na_2CO_3 extract (2 mL) is acidified with acetic acid and solution boiled with solid urea (to decompose nitrite to N_2). The mixture is heated till all NO_2^- is removed. This is ascertained by taking a few drops of the solution and treating with KI-starch solution–no blue colour is produced. To the nitrite free solution is added granulated zinc and dilute H_2SO_4 is order to reduce NO_3^- to NO_2^-. To the resulting solution is added starch-KI solution. Appearance of blue colour confirms the presence of NO_3^- in the pre-reduced solution.

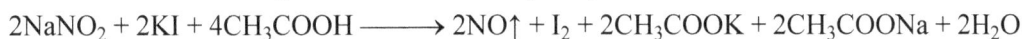

$$Zn + 2CH_3COOH \longrightarrow (CH_3COO)_2Zn + 2H$$
$$NaNO_3 + 2H \longrightarrow NaNO_2 + H_2O$$
$$2NaNO_2 + 2KI + 4CH_3COOH \longrightarrow 2NO\uparrow + I_2 + 2CH_3COOK + 2CH_3COONa + 2H_2O$$
$$I_2 + Starch\text{-}KI \longrightarrow Blue \ colour$$

Alternatively, NO_3^- can be tested in presence of NO_2^- by first eleminating NO_2^- as in the above case and then confirming NO_3^- by the ring test.

Sulphide, Sulphite, Sulphate and Thiosulphate in Presence of each other

The presence of sulphide, sulphite, sulphate and thiosulphate in presence each other is tested as follows:

To sodium carbonate extract is added dilute acetic acid (till acidic) followed by addition of sodium nitropusside solution (0.5 mL). A purple colour indicates the presence of sulphide. In case S^{2-} is present, take another portion of sodium cabronate extract (0.5 mL) and add excess of solid cadmium carbonate or copper carbonate (0.3 g). The sulphide gets precipitated as CdS or CuS

$$Na_2S + CdCO_3 \longrightarrow CdS\downarrow + Na_2CO_3$$
$$Na_2S + 3CuCO_3 \longrightarrow CuS\downarrow + Na_2CO_3$$

The precipitate of CdS or CuS is filtered and to the filtrate is added conc. HNO_3 till effervescense ceases and the solution is rendered alkaline by addition of NH_4OH solution. Excess ammonia is boiled off and $BaCl_2$ solution (2-3 mL) added. A white precipitate indicates the presence of SO_3^{2-} and or SO_4^{2-}.

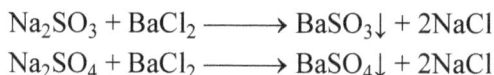

$$Na_2SO_3 + BaCl_2 \longrightarrow BaSO_3\downarrow + 2NaCl$$
$$Na_2SO_4 + BaCl_2 \longrightarrow BaSO_4\downarrow + 2NaCl$$

The precipitate obtained is filtred (the filterate, A is used for testing thiosulphate), washed (on the filter paper) with water and tested for SO_4^{2-} and SO_3^{2-} as follows. To the precipitate (on the filter paper) is added cold dilute HCl. In case some of the precipitate remains undissolved, the presence of SO_4^{2-} is confirmed. To the filtrate is added conc. HNO_3 or Br_2 water (3-4 drops) and solution warmed. In this way SO_3^{2-} gets converted into SO_4^{2-}. Appearance of white precipitate confirms the presence of sulphite.

Thiosulphate is tested in the filtrate (A) (obtained above) by adding $AgNO_3$ solution (0.5 mL). A white ppt. turning yellow, orange and finally black confirms $S_2O_3^{2-}$.

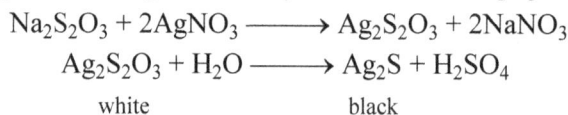

$$Na_2S_2O_3 + 2AgNO_3 \longrightarrow Ag_2S_2O_3 + 2NaNO_3$$
$$Ag_2S_2O_3 + H_2O \longrightarrow Ag_2S + H_2SO_4$$
$$\text{white} \qquad\qquad\qquad \text{black}$$

In case the mixture also contains Cl^-, Br^- or I^-, only a white or yellow ppt. will be obtained. However, the $S_2O_3^{2-}$ gives black ppt. of Ag_2S.

Chloride, Bromide and Iodide in Presence of each other

Chloride, bromide and iodide in presence of each other are tested as follows:

To sodium carbonate extract (1 mL) is added dilute H_2SO_4 (till acidic) followed by addition of solid $NaNO_2$ (0.1 g) and gentle heating. Evolution of violet vapours of iodine (confirmed by CCl_4 test or starch paper) confirms the presence of iodide in the mixture. In case no violet vapours and only brown fumes of NO_2 gas are obtained then I^- is absent.

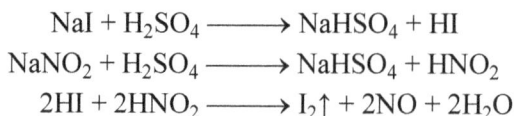

$$NaI + H_2SO_4 \longrightarrow NaHSO_4 + HI$$
$$NaNO_2 + H_2SO_4 \longrightarrow NaHSO_4 + HNO_2$$
$$2HI + 2HNO_2 \longrightarrow I_2\uparrow + 2NO + 2H_2O$$

In case iodide is absent, the remaining solution is used as such for further testing. In case iodide is present (as seen above), the above solution is boiled to expel all the iodide. The solution is boiled and more of $NaNO_2$ (0.2 g) added. The heating is continued till the solution is clear. To the remaining solution is added dilute H_2SO_4 (0.5 mL) and conc. HNO_3 (0.5 mL) and the solution boiled. A brown coloured solution and brown vapours of bromine (confirmed by CCl_4 test) confirms the presence of bromide in the mixture

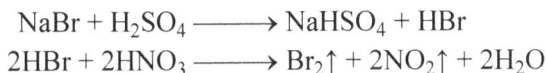

$$NaBr + H_2SO_4 \longrightarrow NaHSO_4 + HBr$$
$$2HBr + 2HNO_3 \longrightarrow Br_2\uparrow + 2NO_2\uparrow + 2H_2O$$

Finally, chloride is tested by heating the above solution (after removal of iodide and bromide) with conc. HNO_3 (1 mL) till all the bromide is removed. To the remaining solution is added $AgNO_3$ solution (0.5 mL). Formation of white ppt. of AgCl soluble in NH_4OH confirms the presence of Cl^-

$$NaCl + AgNO_3 \longrightarrow NaNO_3 + AgCl \downarrow$$
$$AgCl + 2NH_4OH \longrightarrow [Ag(NH_3)_2]Cl + 2H_2O$$

Nitrate in Presence of Bromide and Iodide

A mixture containing NO_3^-, Br^-, and I^- on treatment with conc. HNO_3 gives a mixture of NO_2, Br_2 and I_2 gases which are difficult to differentiate. Also the ring test for NO_3^- cannot be performed in presence of Br^- and I^-. So the presence of NO_3^- in presence of Br^- and I^- can be tested as follows.

The sodium cabronate extract (2 drops) are taken in a groove tile and sulphanilic acid solution (1 drop), acetic acid (1 drop), α-naphthylamine solution (1 drop) and a pinch of zinc dust added. Appearance of a rose red colour confirms the presence of nitrate.

In this test NO_3^- is reduced by zinc dust to nitrite, which gives a red azo dye

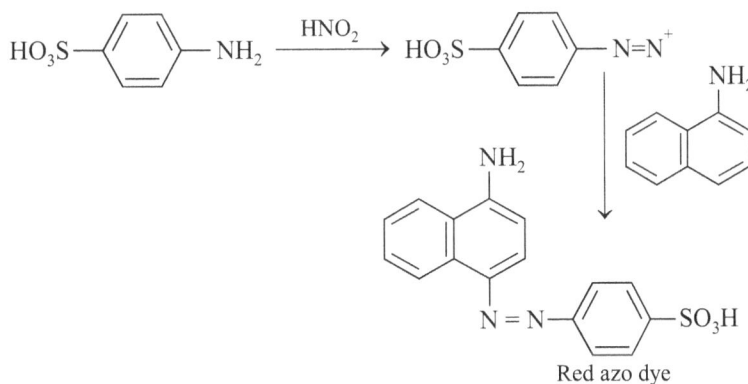

$$NaNO_2 + CH_3COOH \longrightarrow HNO_2 + CH_3COONa$$

Red azo dye

Sulphate in Presence of Fluoride

Both SO_4^{2-} and F^- give white ppt. of $BaSO_4$ and BaF_2 respectively on treatment with $BaCl_2$ solution. The white ppt. of $BaSO_4$ and BaF_2 are both insoluble in HCl. So SO_4^{2-} is tested in presence of F^- as follows.

The sodium carbonate extract (0.5 mL) is acidified with dil. acetic acid and the solution heated to expell CO_2. This is followed by addition of lead acetate solutions. Appearance of a white ppt. confirms the presence of SO_4^{2-} (PbF_2 is soluble in acetic acid)

$$Na_2SO_4 + Pb(CH_3COO)_2 \longrightarrow PbSO_4\downarrow + 2CH_3COONa$$
$$\text{white}$$

Oxalate in Presence of Carbonate

A mixture containing both oxalate and carbonate on heating with conc. H_2SO_4 gives CO_2. It is not possible to know whether CO_2 is coming from $C_2O_4^{2-}$ or CO_3^-. Of course, in case of oxalate CO is also evolved which can be confirmed. To procedure for testing oxalate in presence of carbonate involves treatment of the mixture (0.2 g) with dilute H_2SO_4. If there is effervescence due to evolution of CO_2 (can be confirmed by the lime water test), the mixture is warmed till there is no effervescence. This is followed by addition of a pinch of MnO_2 and warming the mixture. In case there is further effervescence due to the formation of CO_2 gas, then the presence of oxalate is confirmed.

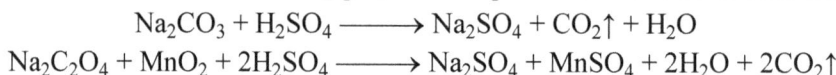

$$Na_2CO_3 + H_2SO_4 \longrightarrow Na_2SO_4 + CO_2\uparrow + H_2O$$
$$Na_2C_2O_4 + MnO_2 + 2H_2SO_4 \longrightarrow Na_2SO_4 + MnSO_4 + 2H_2O + 2CO_2\uparrow$$

1.4   CATION ANALYSIS

A number of procedure are available for the analyse of cations present in a mixture. These include
1. Systematic qualitative analysis of cations using hydrogen sulphide.
2. Analysis of cations using potassium thiocarbonate reagent.
3. Greener alternative to qualitative analysis of cations.
4. Spot test for cations.
5. Detection of metals (cations) by flame photometry.
6. Detection of metals (cations) by Atomic Absorption Spectroscopy.
7. Chromatography of cations.
8. Analysis of insoluble substances.
9. Analysis of a metal or alloy.
10. Analysis of rare or trace elements.

The choice of the method for the analysis of cations depends on whether the cations are to be determined in a mixture or soluble salts, insoluble salts, metal or alloy or cations of rare elements. The method selected should not as far as possible have any environmental effects. Though there is no absolutely green procedure but the analysis should be carried out on a semi-micro scale so that the effect on the environment is bare minimum and also the use and evolution of hazardous products minimised. During analysis, centrifugation (for separation of ppt.) be used in place of usual filteration.

The Dry tests (Section 1.2.1) give some idea about the presence of certain cations.

1.4.1 Systematic Qualitative Analysis of Cations using Hydrogen Sulphide

The systematic qualitative analysis of cations was developed by Fresenius (H.Z. Fresenius, Analyst. Chem. 1897, 30, 10). The hydrogen sulphide required is generated in the Kipps apparatus from iron sulphide and sulphuric acid.

$$FeS(s) + H_2SO_4(aq) \longrightarrow FeSO_4(aq) + H_2S(g)$$

Hydrogen sulphide in concentration of as low as 0.05 percent, by volume in air may be fatal. Appropriate precautions must be taken for handling H_2S. The best that can be done is to use H_2S in well ventilated fume cupboard.

In this scheme of analysis, the common cations which may be present in a mixture are divided into seven groups (zero to six) on the basis of the reagent used for the precipitation and separation of cations from a solution of the mixture (Table 1.13).

Table 1.13. *Division of cations into 7 groups*

Group	Cations	Group reagent(s)
Zero	NH_4^+, K^+	Tested individually in the mixture
I	Ag^+, Hg_2^{2+}, Pb^{2+}	dilute HCl
II A II B	Hg^{2+}, Pb^{2+}, Bi^{3+}, Cu^{2+}, Cd^{2+} Sb^{3+}, As^{3+}, Sn^{2+}	H_2S in presence of dilute HCl
III	Al^{3+}, Cr^{3+}, Fe^{3+}	$NH_4Cl + NH_4OH$
IV	Zn^{2+}, Mn^{2+}, Co^{2+}, Ni^{2+}	H_2S is presence of NH_4Cl and NH_4OH
V	Ba^{2+}, Ca^{2+}, Sr^{2+}	$NH_4Cl + NH_4OH + (NH_4)_2CO_3$
VI	Mg^{2+}, K^+	

The separation of cations into various groups using appropriate reagents is schematically given in scheme 1.1.

The solution of the mixture (called original solution, O.S. is first prepared (*see page 42*) and then used for separation into various groups.

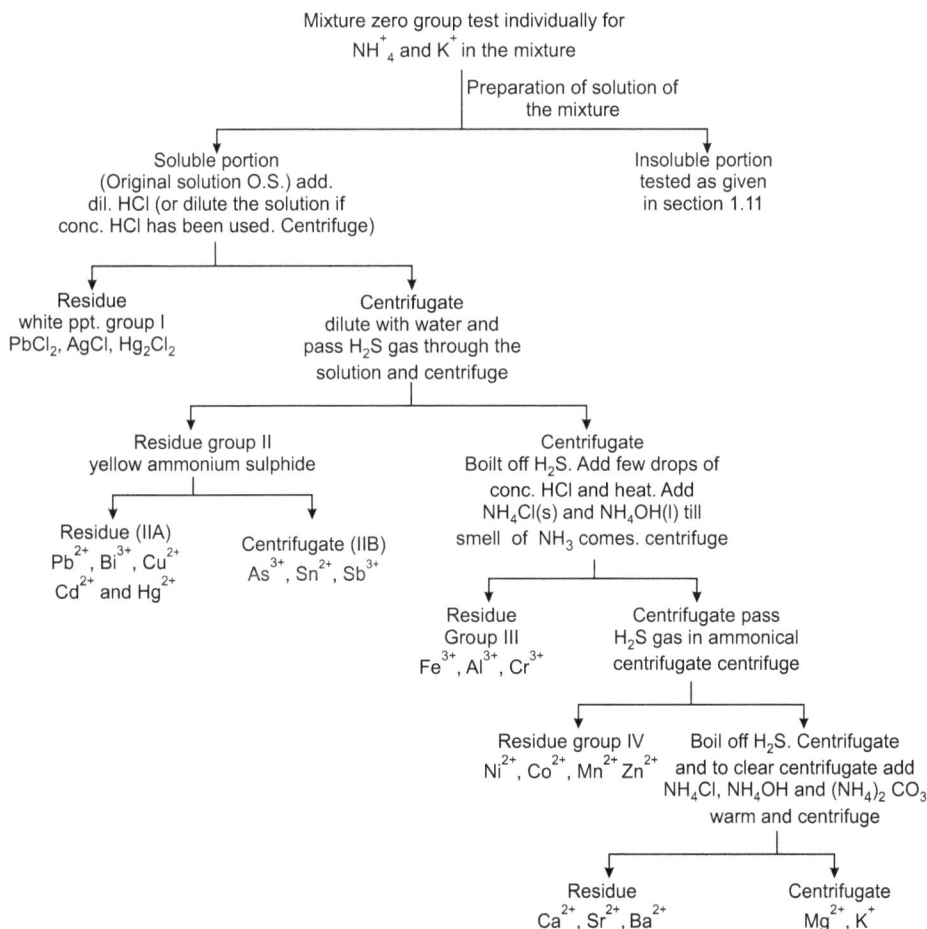

Scheme 1.1. Separation of Mixture into Various Groups of cations.

Following steps are involved in the systematic analysis of cations:

1. Preparation of solution
2. Analysis of different groups.

Preparation of Solution of the Mixture

The solution of the mixture is prepared using one of the appropriate solvent in the order given below:

(*a*) **Water.** De-ionised or distilled water should be used since ordinarily water contains Na^+, Ca^{2+} and Mg^{2+} etc. The process involves treating the mixture (0.1 g) with water (5 mL) in a test tube. Test the solubility first in cold and then on heating. If the mixture is soluble in water, the aqueous solution [called the original solution (O.S.)] is used. However, if the mixture is insoluble in water, proceed as given below.

(*b*) **Dilute HCl.** To about 0.1 g of the mixture is added 2-3 mL dilute HCl. Test the solubility first in cold and then on heating. If the mixture is soluble in cold dil. HCl the solution so obtained is the O.S. and cations of first groups are absent. However, if the mixture is soluble in hot dil. HCl but on cooling gives white crystalline ppt. then presence of lead is indicated. The white ppt. may also be due to the presence of Bi^{3+}, Sn^{2+} or Sb^{3+}, which form crystalline oxychlorides on hydrolysis. In such a situation, the mixture is treated with 3-4 drops of conc. HCl. If the ppt. dissolves group I is absent and Bi^{3+}, Sn^{2+} or Sb^{3+} may be present.

(*c*) **Conc. HCl.** To about 0.1 g of the mixture in a test tube is added conc. HCl (2-3 mL). Check the solubility of the mixture first in cold and then on heating. In case the mixture is soluble in conc. HCl, dilute with water (4-5 mL) and the solution is used as original solution. However, if the mixture is insoluble, proceed as given below.

(*d*) **HNO₃.** To about 0.1 g of the mixture taken in a test tube is added dil. HNO_3 (2-3 mL) and check the solubility first in cold and then on heating. In case the mixture is soluble in dil. HNO_3 this forms the original solution. In case the mixture in insoluble in dil. HNO_3 try the solubility in conc. HNO_3.

In case the mixture is soluble in HNO_3 (dilute or concentrated), the solution is evaporated to almost dryness (in a china dish) and water (10 mL added). The solution is again evaporated to almost dryness and finally dissolved in water. This process is necessary to remove HNO_3 which interferes in group II (on passing H_2S a ppt. of sulphur is obtained).

Once a suitable solvent has been found, the mixture (2-3 g) is dissolved in that solvent and then used as original solution.

1.4.1.1 Analysis of Group Zero Cations

As already stated (Scheme 1.1), the zero group cations (NH_4^+ and K^+) and are tested individually in the mixture itself.

Ammonium (NH_4^+)

(i) To the mixture (0.1 g) is added NaOH solution (2 mL, 10 percent). The mixture is heated. Smell of ammonia indicates the presence of ammonium (NH_4^+).

$$NH_4Cl + NaOH \xrightarrow{\Delta} NH_3\uparrow + H_2O + NaCl$$

A glass rod dipped in conc. HCl on bringing to the mouth of the test tube gives dense white fumes of NH_4Cl confirming the presence of NH_4^+ in the mixture.

(ii) The gas ($NH_3\uparrow$) evolved in the above test is passed through Nessler's reagent. Formation of a brown ppt. confirms NH_4^+.

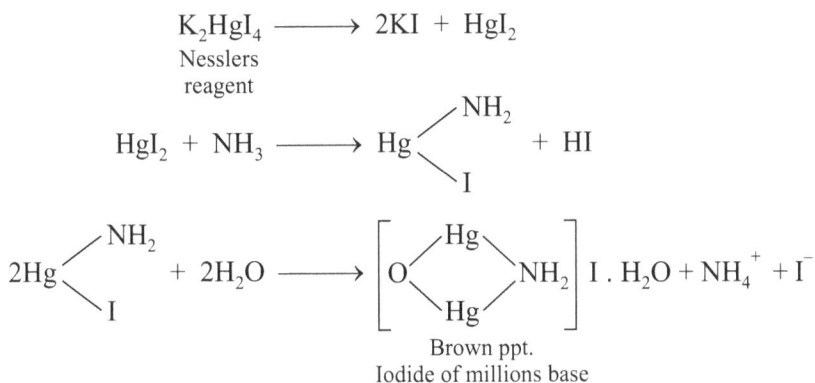

$$K_2HgI_4 \longrightarrow 2KI + HgI_2$$

Nesslers reagent

$$HgI_2 + NH_3 \longrightarrow Hg\begin{smallmatrix} NH_2 \\ \\ I \end{smallmatrix} + HI$$

$$2Hg\begin{smallmatrix} NH_2 \\ \\ I \end{smallmatrix} + 2H_2O \longrightarrow \left[O\begin{smallmatrix} Hg \\ \\ Hg \end{smallmatrix} NH_2 \right] I . H_2O + NH_4^+ + I^-$$

Brown ppt.
Iodide of millions base

(iii) The gas ($NH_3\uparrow$) evolved (in test (i) above) turns a filter paper moistened with a solution of $MnCl_2$ and H_2O_2 brown due to oxidation of manganese by alkaline solution thus formed

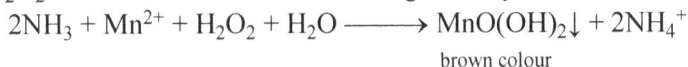

$$2NH_3 + Mn^{2+} + H_2O_2 + H_2O \longrightarrow MnO(OH)_2\downarrow + 2NH_4^+$$

brown colour

(iv) **Spot test:** To a drop of the aqueous extract of the mixture is added a drop of conc. NaOH solution on a watch glass. A drop of this resulting solution is transferred to a filter paper strip dipped in Nessler's reagent. A yellow or orange red spot is produced.

Potassium K^+

(i) Flame test directly with the mixture (for details see Table 1.3).

(ii) To aqueous extract (1 mL) of the mixture is added dilute acetic acid (1 mL) and 3-4 drops of freshly prepared solution of sodium cobalti nitrite. A yellow ppt confirms K^+.

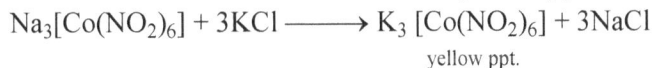

$$Na_3[Co(NO_2)_6] + 3KCl \longrightarrow K_3[Co(NO_2)_6] + 3NaCl$$

yellow ppt.

1.4.1.2 Analysis of Group I Cations (Pb^{2+}, Ag^+, Hg_2^{2+})

If the mixture is soluble in cold dil. HCl, group I cations are absent and proceed for group II. If the solution of the mixture is prepared in conc. HCl, then dilute the solution with twice its volume of water (if the concentration of HCl is too much $BaCl_2$ also gets precipitated due to common ion effect, in such a case the solution is diluted so that $BaCl_2$ dissolves). The solution on cooling gives a white ppt. of group I cations ($PbCl_2$, $AgCl$, Hg_2Cl_2). In case, the solution of the mixture has been prepared in hot dil. HCl, formation to a white ppt. on cooling also indicates the presence of group I cations ($PbCl_2$, $AgCl$, Hg_2Cl_2). Centrifuge the mixture. The centrifuge (residue) is analysed for group I cations and the centrifugate (filtrate) is used for analysis of subsequent groups (Scheme 1.2).

OS + dil HCl

Centrifuge or filter

Residue
Chloride of group I cations
($PbCl_2$, AgCl, Hg_2Cl_2)
Wash the ppt. with water (2mL)
Boil with 2-3 mL water
Filter or centrifuge in hot
(Filtering in hot is necessary
otherwise $PbCl_2$ crystals remain
in the residue

**Filtrate or
centrifugate.**
For subsequent
groups (II to VI)

Filtrate or centrifugate
may contain $PbCl_2$
divide in two parts
(i) To one part add dilute
acetic acid (3-4 drops) and
K_2CrO_4 solution (2-3 drops).
A yellow ppt. of $PbCrO_4$
confirms Pb^{2+} (the ppt. is
soluble in NaOH solution)
$PbCl_2 + K_2CrO_4 \rightarrow$
$\qquad PbCrO_4 \downarrow + 2KCl$
$\qquad$ yellow
$PbCrO_4 + 4NaOH \rightarrow$
$\qquad Na_2[Pb(OH)_4] + Na_2CrO_4$

(ii) To the second part add
KI solution. A yellow ppt.
which dissolves on heating
and reappears on cooling
confirms Pb^{2+}
$PbCl_2 + 2KI \rightarrow$
$\qquad PbI_2 \downarrow + 2KCl$
$\qquad$ yellow

Residue
Chlorides of Ag and Hg
(AgCl, Hg_2Cl_2)
Wash the residue with hot
water, warm it with 1-2 ml
NH_4OH. Filter or centrifuge

Filtrate or centrifugate
may contain $[Ag(NH_4)_2]Cl$
$AgCl + 2NH_4OH \rightarrow$
$\qquad [Ag(NH_4)_2]Cl + 2H_2O$
Divide in two parts
(i) To one part add
dil. HNO_3. White
ppt. in obtained
$[Ag(NH_4)_2]Cl + 2HNO_3$
$\rightarrow AgCl\downarrow + 2NH_4NO_3$
$\qquad$ White
(ii) To the other part
add K_2CrO_4 solution
Red ppt. is obtained
$2[Ag(NH_4)_2]Cl + K_2CrO_4$
$\rightarrow Ag_2CrO_4\downarrow + 2KCl + 4\overset{+}{N}H_4$
$\qquad$ red

Residue
(test for Hg)
Dissolve in
aqua regia
$Hg_2Cl_2 + 2NH_4OH \rightarrow$
$\qquad Hg(NH_2)Cl + Hg +$
$\qquad NH_4Cl + 2H_2O$
$HNO_3 + 3HCl \rightarrow$
$\qquad NOCl + Cl_2 + 2H_2O$

$2Hg(NH_2)Cl + 3Cl_2 \rightarrow$
$\qquad 2HgCl_2 + 4HCl + N_2$
$Hg + Cl_2 \rightarrow HgCl_2$
Dilute the solution with
water and divide in two
parts
(i) Soln. + $SnCl_2$ $\rightarrow$ white ppt
$\qquad\qquad\qquad$ turning black
$2HgCl_2 + SnCl_2 \rightarrow Hg_2Cl_2 + SnCl_4$
$\qquad\qquad\qquad\qquad$ white
$Hg_2Cl_2 + SnCl_2 \rightarrow 2Hg + SnCl_4$
$\qquad\qquad\qquad\qquad$ black
(ii) Soln. + KI $\rightarrow$ Scarlet
$\qquad\qquad\qquad$ ppt.
$HgCl_2 + 2KI \rightarrow HgI_2 + 2KCl$
$HgI_2 + 2KI \rightarrow K_2HgI_4\downarrow$
$\qquad\qquad\qquad$ scarlet

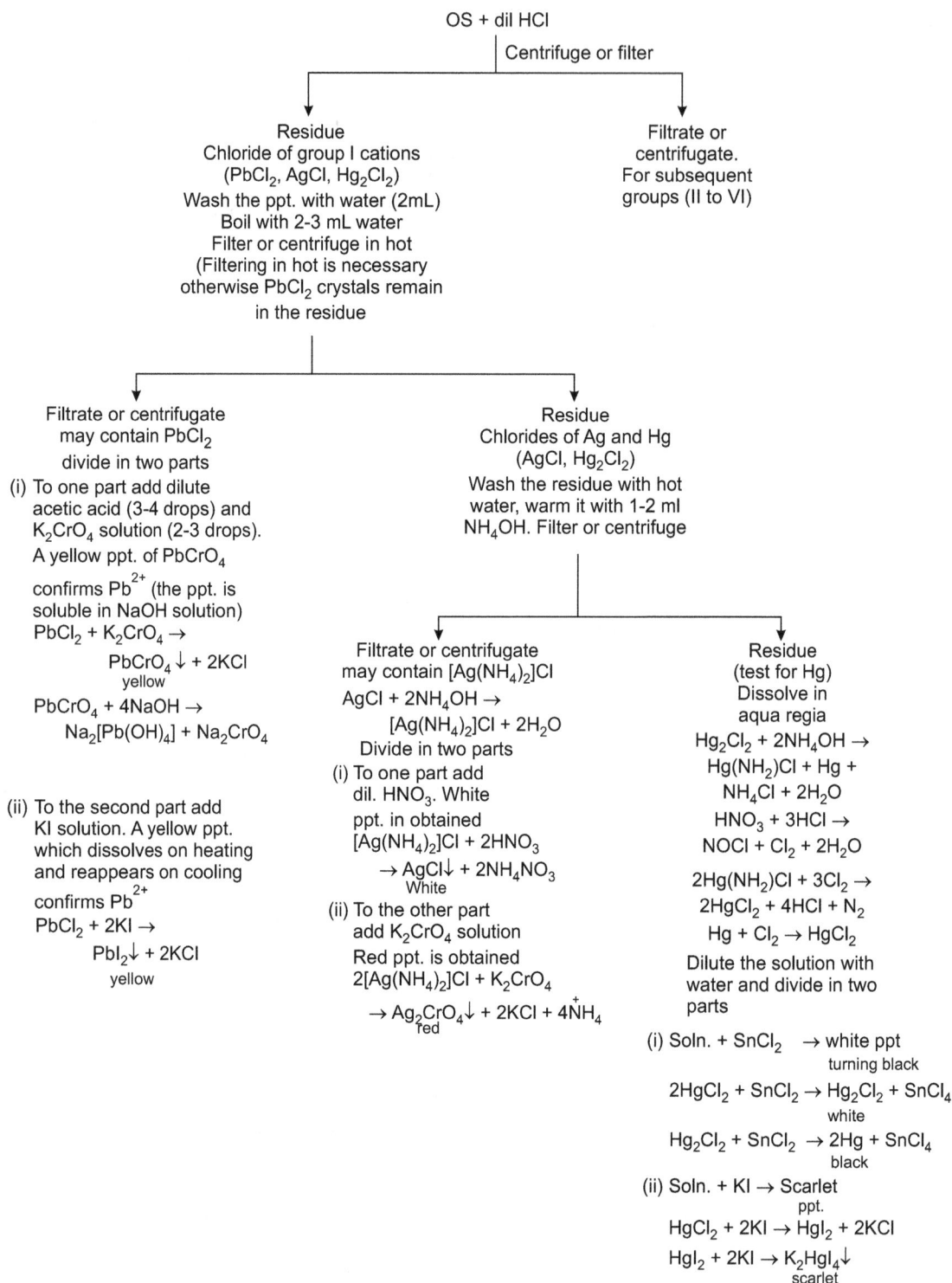

Scheme 1.2: Analysis of group I cations.

Confirmatory tests for cations of Group I

Confirmatory tests for cations of group I are performed as follows:

Lead: The filtrate or centrifugate containing $PbCl_2$ (Scheme 1.2) is treated with dilute HNO_3 and the solution of lead nitrate thus obtained is used for confirming lead.

(*i*) **Ammonia solution:** Addition of NH_4OH to the solution (containing $PbNO_3$) gives a white ppt. of $Pb(OH)_2$ which is insoluble in excess NH_4OH.

$$Pb^{2+} + 2NH_3 + 2H_2O \longrightarrow Pb(OH)_2\downarrow + 2NH_4^+$$

(*ii*) **Sodium hydroxide solution:** White ppt. of $Pb(OH)_2$ is formed, which is soluble is excess of the reagent. The solution on treatment with H_2O_2 gives black ppt.

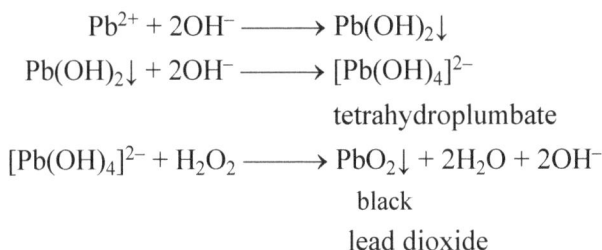

$$Pb^{2+} + 2OH^- \longrightarrow Pb(OH)_2\downarrow$$
$$Pb(OH)_2\downarrow + 2OH^- \longrightarrow [Pb(OH)_4]^{2-}$$
tetrahydroplumbate
$$[Pb(OH)_4]^{2-} + H_2O_2 \longrightarrow PbO_2\downarrow + 2H_2O + 2OH^-$$
black
lead dioxide

(*iii*) **Sodium carbonate.** White ppt. of a mixture of lead carbonate and lead hydroxide is formed

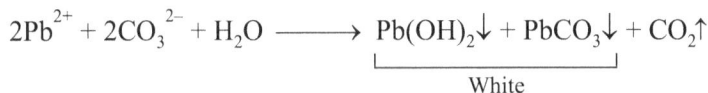

$$2Pb^{2+} + 2CO_3^{2-} + H_2O \longrightarrow \underbrace{Pb(OH)_2\downarrow + PbCO_3\downarrow}_{White} + CO_2\uparrow$$

(*iv*) **Di-(4-dimethylaminodiphenyl) methane:** To 1 mL of the test solution is added 1 mL, 2M KOH solution and 1 mL 3% H_2O_2 solution. The mixture is allowed to stand for 5 min and centrifuged. The solid residue is washed with cold water and 2 mL of the reagent is added. A blue colour is formed (CH_2 of the reagent gets oxidised to $CH(OH)$)

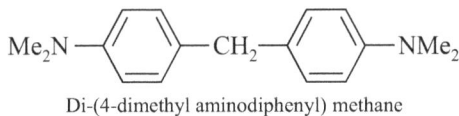

Di-(4-dimethyl aminodiphenyl) methane

(*v*) **Diphenyl thiocarbazone (or Dithiazone)**

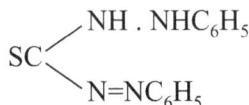

To the test solution (neutral or faintly alkaline) is added a small crystal of KCN (extremely poisonous) and then the reagent (2 drops). After about 30 sec. the green colour of the reagent changes to red.

Mercury

The residue of Hg obtained in scheme 1.2 is dissolved in aqua regia, clear solution concentrated to almost dryness and dissolved in dilute HNO_3. The mercury nitrate solution thus obtained is used for testing mercury.

(*i*) **NH₄OH solution.** A black ppt. (mixture of mercury and basic mercury (II) amido nitrate (white ppt.) is obtained

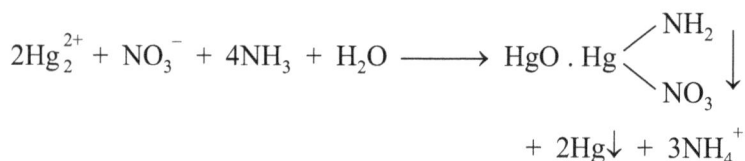

$$2Hg_2^{2+} + NO_3^- + 4NH_3 + H_2O \longrightarrow HgO \cdot Hg \begin{cases} NH_2 \\ \\ NO_3 \end{cases} \downarrow$$

$$+ 2Hg\downarrow + 3NH_4^+$$

(*ii*) **Sodium hydroxide.** Black ppt. of Hg_2O is obtained

$$Hg_2^{2+} + 2OH^- \longrightarrow Hg_2O\downarrow + H_2O$$

(*iii*) **Potassium chromate.** Red ppt. of Hg_2CrO_4 is obtained. On heating, it gives a brown amorphous ppt.

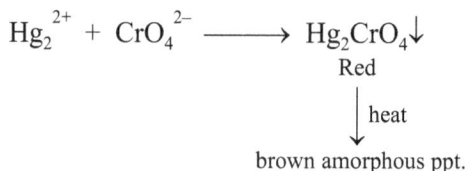

$$Hg_2^{2+} + CrO_4^{2-} \longrightarrow Hg_2CrO_4\downarrow$$
$$\text{Red}$$
$$\downarrow \text{heat}$$
$$\text{brown amorphous ppt.}$$

Caution: Mercury is extremely poisonous especially as mercury vapour.

Silver

The filtrate or centrifugate containing $[Ag(NH_3)_2]Cl$ (Scheme 1.2) is treated with dilute nitric acid and the solution of $AgNO_3$ thus obtained is used for confirming Ag.

(*i*) **Ammonia solution.** A brown ppt. of Ag_2O is formed

$$2Ag^+ + 2NH_3 + H_2O \longrightarrow Ag_2O\downarrow + 2NH_4^+$$

The formed Ag_2O is soluble in excess of the reagent

$$Ag_2O\downarrow + 4NH_3 + H_2O \longrightarrow 2[Ag(NH_3)_2]^+ + 2OH^-$$

(*ii*) **Sodium hydroxide.** A brown ppt. of Ag_2O is obtained

$$2Ag^+ + 2OH^- \longrightarrow Ag_2O\downarrow + H_2O$$

(*iii*) **Potassium chromate.** A red ppt. of Ag_2CrO_4 is obtained

$$2Ag^+ + CrO_4^{2-} \longrightarrow Ag_2CrO_4\downarrow$$

Group I cations (Pb^+, Ag^+ and Hg^+) can also be separated and identified by paper chromatography (see Section 1.10.1).

1.4.1.3 Analyses of Group II Cations

Separation of group II cations into group II A and group II B cation. To the original solution (if group I is absent) or the filtrate or centrifugate obtained after removal of group I cations (Scheme 1.2) is added few drops of HCl (if the solution does not contain HCl), the solution heated and H_2S passed. The solution is again diluted and H_2S again passed (this is necessary to precipitate CdS which gets precipitated in very dilute HCl solution). If a coloured (black, yellow or orange) ppt. is obtained, group II is present. The mixture is centrifuged or filtered [(Scheme 1.2(a)].

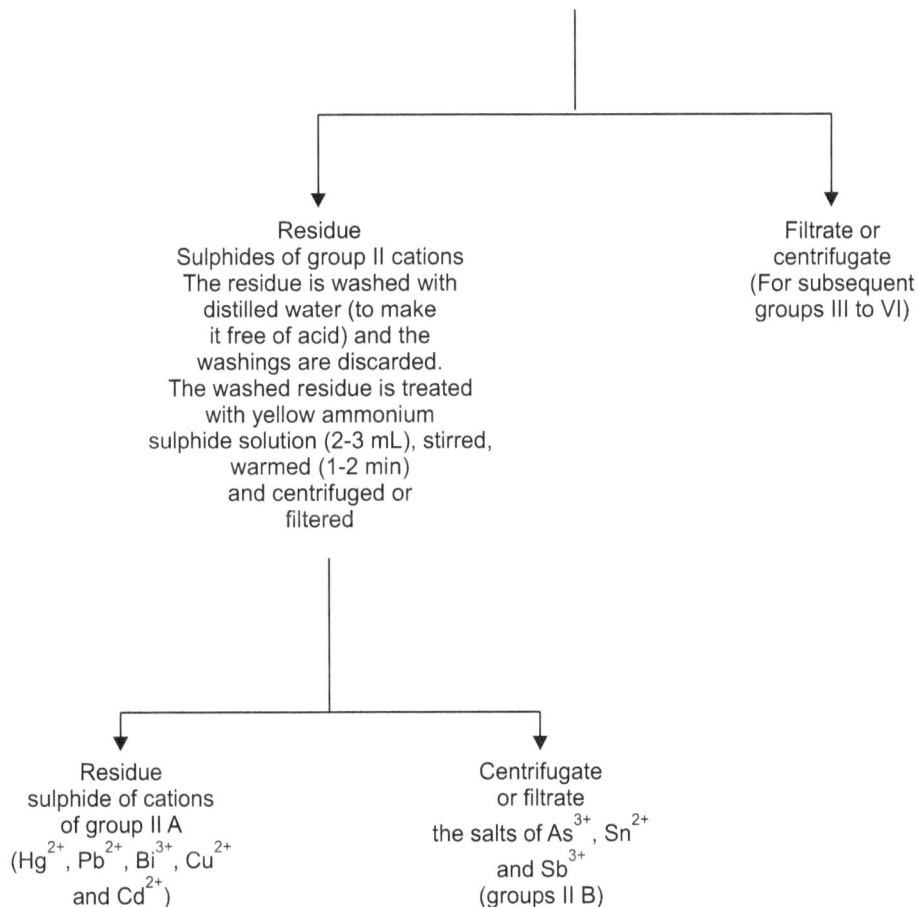

Residue
Sulphides of group II cations
The residue is washed with
distilled water (to make
it free of acid) and the
washings are discarded.
The washed residue is treated
with yellow ammonium
sulphide solution (2-3 mL), stirred,
warmed (1-2 min)
and centrifuged or
filtered

Filtrate or
centrifugate
(For subsequent
groups III to VI)

Residue
sulphide of cations
of group II A
(Hg^{2+}, Pb^{2+}, Bi^{3+}, Cu^{2+}
and Cd^{2+})

Centrifugate
or filtrate
the salts of As^{3+}, Sn^{2+}
and Sb^{3+}
(groups II B)

Scheme 1.2(a): Separation of Group II cations into Group IIA and IIB cations.

Analysis of Group II A cations

The residue obtained after washing the precipitate of sulphides of group II cation with yellow ammonium sulphide contains sulphides of group II A cations (*viz.*, HgS, PbS, Bi_2S_3, CuS and CdS). The residue is treated with 33% HNO_3 (1 part conc. HNO_3 + 2 parts distilled water), heated (2-3 min) and then centrifuged or filtered (Scheme 1.3).

Residue
Black HgS
The residue is dissolved
in aqua regia, solution
diluted with water and
divided into 3 parts

(i) One part + $SnCl_2$ soln.
→ white ppt. turning
grey is obtained
$2HgCl_2 + SnCl_2 \rightarrow$
$\quad Hg_2Cl_2 + SnCl_4$
$Hg_2Cl_2 + SnCl_2 \rightarrow$
$\quad SnCl_4 + 2Hg$

(ii) Second part + copper
chip → silvery white
deposit on copper chip
$HgCl_2 + Cu \rightarrow Hg$
$\quad\quad\quad + CuCl_2$

(iii) Third part + KI →
Red coloured ppt. of
$HgCl_2$ soluble in
excess KI
$HgCl_2 + 2KI \rightarrow$
$\quad \underset{red}{HgI_2\downarrow} + 2KCl$
$HgI_2 + 2KI \rightarrow$
$\quad\quad K_2[HgI_4]$

Centrifugate
or filtrate contains
nitrates of Cu^{2+}, Cd^{2+}, Pb^{2+} and Bi^{3+}
The filtrate is rendered distinctly
alkaline with excess NH_4OH and
then centrifuged or filtered

Centrifugate or filtrate
contains soluble
complexes of copper
$[Cu(NH_3)_4](NO_3)_2$ or
cadmium $[Cd(NH_3)_4](NO_3)_2$
If the solution is blue Cu is
present (otherwise not)

(i) If Cu is present acidifiy
a portion with CH_3COOH
add $K_4Fe(CN)_6$ soln.
→ chocolate colour
ppt. confirms Cu
$[Cu(NH_3)_4]^{2+} + 2CH_3COO^-$
$\rightarrow Cu(CH_3COO)_2 + 4NH_3$
$2Cu(CH_3COO)_2 +$
$\quad\quad K_4Fe(CN)_6$
$\rightarrow Cu_2[Fe(CN)_6] +$
$\quad\quad 4CH_3COOK.$

(ii) If Cu is absent (as
seen by absence of
blue colour of solution)
the solution is acidified
(dil. H_2SO_4), and H_2S
passed Yellow ppt.
confirms Cd
$CdCl_2 + H_2S \rightarrow$
$\quad\quad CdS\downarrow + 2HCl$
$\quad\quad\quad$ yellow

(iii) If Cu^{2+} and Cd^{2+} both are present,
the solution is neutralised (dil. HCl)
and treated with Zn dust or Fe
powder to remove Cu ions. The
solution is heated and H_2S passed.
Yellow ppt. confirms the presence of Cd^{2+}

Residue
Test for Bi^{3+} and Pb^{2+}
warm with NaOH soln.
(5 mL) and filter

Residue
(white) $Bi(OH)_3$
Dissolve in
dil HCl and
divide in 2 parts

(i) One part +
excess H_2O
→ turbidity
$BaCl_2 + H_2O \rightarrow$
$BiOCl + 2HCl$

(ii) 2nd part +
excess sod.
stannite →
Black ppt.
$Bi(OH)_3 + HCl \rightarrow$
$BiCl_3 + 3H_2O$
$2BiCl_3 + 3Na_2SnO_2$
$+ 3H_2O \rightarrow 3Na_2SnO_3$
$+ 6HCl + 2Bi\downarrow$
$\quad\quad\quad$ black

Filtrate
contains Pb as
sod. plumbate
$Pb(NO_3)_2 + 2NaOH$
$\rightarrow Pb(OH)_2 + 2NaNO_3$
$Pb(OH)_2 + 2NaOH$
$\rightarrow Na_2(PbO_2) + 2H_2O$
Acidity the solution
(CH_3COOH) and add
K_2CrO_4 solution
yellow ppt. soluble
in NaOH solution
confirms Pb^{2+}
$Pb^{2+} + K_2CrO_4 \rightarrow$
$PbCrO_4\downarrow + 2K^+$
$\quad$ yellow
(If Pb^{2+} has been
detected in group I,
this test need
not be done)

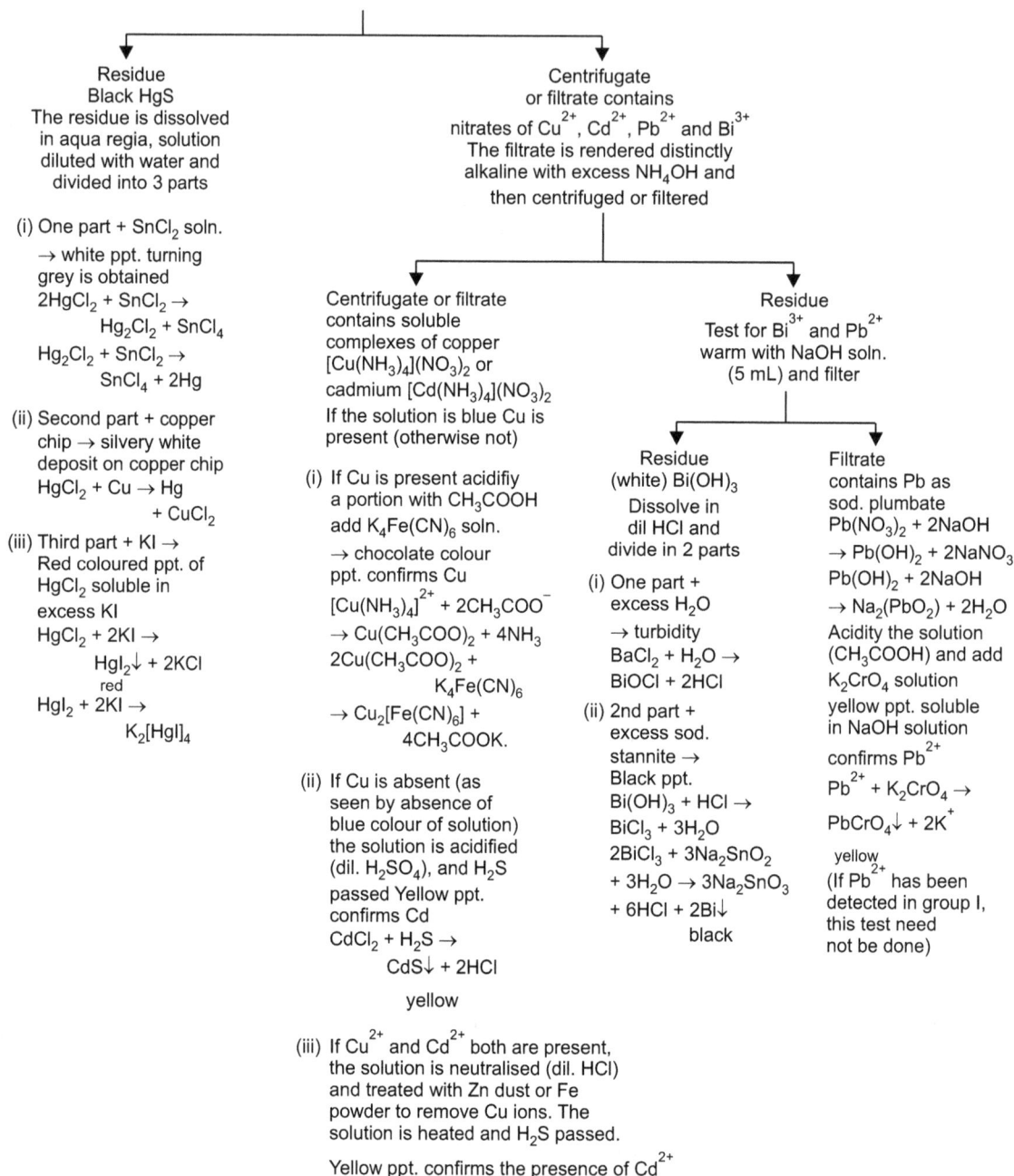

Scheme 1.3. Analysis of group II A cations.

Confirmatory Tests for Group II A cations

Confirmatory tests for group II A cations are performed as follows:

Note: All compounds of mercury are extremety toxic.

Mercury

The residue of Hg obtained in Scheme 1.3 is dissolved in aqua regia, clear solution concentrated to almost dryness and dissolved in dil. HNO_3. The mercury nitrate solution thus obtained is used for testing mercury.

(*i*) **Ammonium solution.** White ppt. is obtained

$$2Hg^{2+} + NO_3^- + 4NH_3 + H_2O \longrightarrow HgO . Hg(NH_2)NO_3\downarrow + 3NH_4^+$$
$$\text{white}$$

(*ii*) **NaOH solution.** Addition of small amount of NaOH solution gives yellow ppt. of HgO. The ppt. is insoluble in excess NaOH solution

$$Hg^{2+} + 2OH^- \longrightarrow HgO\downarrow + H_2O$$
$$\text{yellow}$$

This reaction is characteristic for Hg^{2+} ions and is useful to differentiate Hg^{2+} from Hg^+.

(*iii*) **Cobalt thiocyanate.** Addition of equal volume of cobalt thiocyanate reagent (10%, freshly prepared) gives a deep blue ppt. of cobalt tetrathio cyanatomerurate

$$Hg^{2+} + Co^{2+} + 4SCN^- \longrightarrow Co[Hg(SCN)_4]\downarrow$$
$$\text{deep blue}$$

Copper

The filtrate or centrifugate containing Cu and Cd (Scheme 1.3) is treated with dilute H_2SO_4 and the copper sulphate solution thus obtained is used for testing copper.

(*i*) **NH$_4$OH solution.** Blue ppt. of basic salt of copper sulphate is obtained. The ppt. is soluble in excess NH_4OH due to the formation of tetra-aminocuprate (II) complex ion.

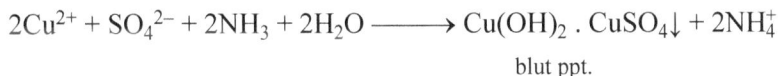

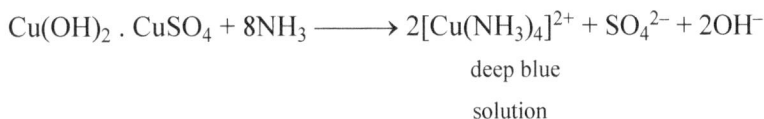

$$2Cu^{2+} + SO_4^{2-} + 2NH_3 + 2H_2O \longrightarrow Cu(OH)_2 . CuSO_4\downarrow + 2NH_4^+$$
$$\text{blut ppt.}$$

$$Cu(OH)_2 . CuSO_4 + 8NH_3 \longrightarrow 2[Cu(NH_3)_4]^{2+} + SO_4^{2-} + 2OH^-$$
$$\text{deep blue}$$
$$\text{solution}$$

(*ii*) **NaOH solution.** A blue ppt. of $Cu(OH)_2$ is obtained

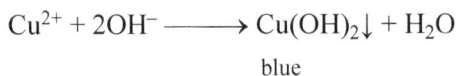

$$Cu^{2+} + 2OH^- \longrightarrow Cu(OH)_2\downarrow + H_2O$$
$$\text{blue}$$

In presence of tartaric acid or critic acid, $Cu(OH)_2$ is not precipitated in the above test. Only a blue coloured solution is obtained.

(*iii*) **Potassium thiocyanate.** A black ppt. of $Cu(SCN)_2$ is obtained

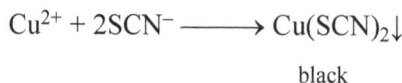

$$Cu^{2+} + 2SCN^- \longrightarrow Cu(SCN)_2\downarrow$$
$$\text{black}$$

(*iv*) Copper can be confirmed in presence of cadmium by paper chromatography (see Section 1.10.4).

Cadmium

The filtrate or centrifugate containing Cd and Cu (Scheme 1.3) is treated with dilute H_2SO_4 and the solution thus obtained is used for testing cadmium.

(*i*) **NH₄OH solution.** Addition of NH_4OH solution (dropwise) gives a white ppt. of $Cd(OH)_2$, which is soluble in excess NH_4OH solution

$$Cd^{2+} + 2NH_3 + 2H_2O \rightleftharpoons Cd(OH)_2\downarrow + 2NH_4^+$$
$$\text{white}$$

$$Cd(OH)_2\downarrow + 4NH_3 \longrightarrow [Cd(NH_3)_4]^{2+} + 2OH^-$$
$$\text{colourless}$$

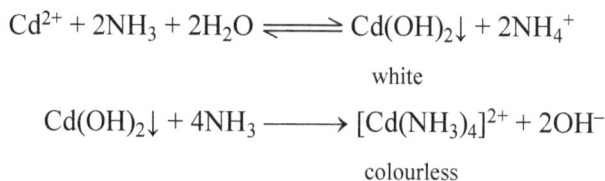

(*ii*) **NaOH solution.** A white ppt. of $Cd(OH)_2$ is formed, which is insoluble in excess NaOH

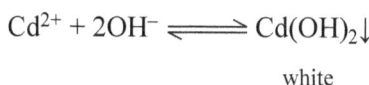

$$Cd^{2+} + 2OH^- \rightleftharpoons Cd(OH)_2\downarrow$$
$$\text{white}$$

(*iii*) **KI solution.** No precipitation (distinction from copper)

(*iv*) Cd can be tested in presence of copper by paper chromatography (Section 1.10.4).

Bismuth

The residue containing Bi^{3+} and Pb^{2+} (Scheme 1.3) is dissolved in dilute HNO_3 and the solution is used for testing bismuth.

(*i*) **NH₄OH solution.** A white ppt. of basic salt, insoluble in excess NH_4OH is obtained

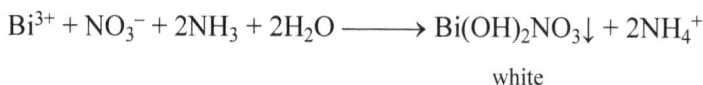

$$Bi^{3+} + NO_3^- + 2NH_3 + 2H_2O \longrightarrow Bi(OH)_2NO_3\downarrow + 2NH_4^+$$
$$\text{white}$$

(*ii*) **NaOH solution.** White ppt. of $Bi(OH)_3$ is obtained

$$Bi^{3+} + 3OH^- \longrightarrow Bi(OH)_3\downarrow$$
$$\text{white}$$

On boiling, the precipitate of $Bi(OH)_3$ looses water and gives yellowish-white ppt. of BiO.OH

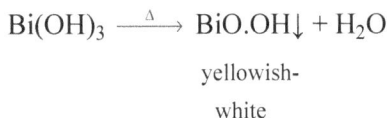

$$Bi(OH)_3 \xrightarrow{\Delta} BiO.OH\downarrow + H_2O$$
$$\text{yellowish-}$$
$$\text{white}$$

(*iii*) **KI solution.** Dropwise addition of KI solution gives a black ppt. of BiI_3. The precipitate dissolves in excess reagent to give orange coloured tetraiodobismuthate ions.

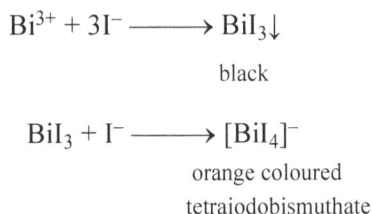

$$Bi^{3+} + 3I^- \longrightarrow BiI_3\downarrow$$
$$\text{black}$$

$$BiI_3 + I^- \longrightarrow [BiI_4]^-$$
$$\text{orange coloured}$$
$$\text{tetraiodobismuthate}$$

(*iv*) **Pyrogallol solution.** Addition of 10 percent freshly prepared pyrogallol solution gives a yellow ppt. of bismuth pyrogallate

$$Bi^{3+} + C_6H_3(OH)_3 \longrightarrow Bi(C_6H_3O_3)\downarrow + 3H^+$$
$$\text{yellow}$$

In the above test, the test solution is first neutralised by ammonium hydroxide and then 2-3 drops of dilute HNO_3 added.

(*v*) **8-Hydroxyquinoline and potassium iodide.** Addition of 8-hydroxyquinoline (5%) and KI (6 M) solution to the test solution gives a red precipitate of 8-hydroxyquinoline-tetraiodobismuthate

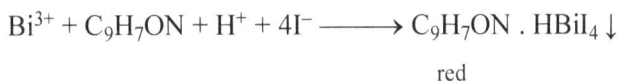

$$Bi^{3+} + C_9H_7ON + H^+ + 4I^- \longrightarrow C_9H_7ON . HBiI_4 \downarrow$$
$$\text{red}$$

Lead see page 45.

Group II A cations (Hg^{2+}, Cu^{2+}, Cd^{2+} and Bi^{3+}) can also be separated and identified by paper chromatography (see Sections 1.10.2 and 1.10.3).

Analysis of Group II-B cations

The group II-B cations are tested in the filtrate obtained by treating group II ppt. with yellow ammonium sulphide solution (Scheme 1.2). This filtrate contains thiosalts of As^{3+}, Sn^{2+} and Sb^{3+}. The above filtrate is treated with dilute HCl (till distinctly acidic to litmus paper) to give coloured ppt. of group II-B cations. The mixture is filtered and the filtrate discarded (Scheme 1.4).

Filtrate
(Yellow ammonium sulphide
soluble portion) (Scheme 1.2)

dil HCl (till
distinctly acedic

Residue
Heat with conc. HCl (3-4 mL)
and filter or centrifuge

**Filtrate or
centrifugate discard**

Residue
Contains As_2S_5 or AsS_3 (yellow)
Dissolve in conc. HNO_3.
Heat and filter or centrifuge.
To the filtrate or centrifugate
is added ammonium molybdate
and heated. Yellow ppt. confirms
the presence of As^{3+}
$3As_2S_5 + 10HNO_3 + 4H_2O \rightarrow$
$6H_3AsO_4 + 15S + 10NO$
$H_3AsO_4 + 12(NH_4)_2MoO_4$
$+ 21HNO_3 \rightarrow$
$(NH_4)_3AsMo_{12}O_{40}\downarrow + 21NH_4NO_3$
$\quad$ yellow ppt. $\qquad + 12H_2O$

Filtrate or centrifugate
contains chlorides of tin
and antimony.
Divide in two parts.
(i) To one part add NH_4OH
    solution (till ammonical),
    add 1 g. solid oxalic acid,
    heat to boiling and pass H_2S.
    Formation of orange ppt.
    Of Sb_2S_3 confirms Sb^{3+}
    $SnCl_4 + 2H_2C_2O_4 \rightarrow$
    $\quad Sn(C_2O_4)_2 + 4HCl$
    $\quad\quad$ Stable
    $\quad$ not decomposed
    $\quad\quad$ by H_2S
    $2SbCl_3 + 3H_2C_2O_4 \rightarrow$
    $\quad Sb_2(C_2O_4)_3 + 6HCl$
    $\quad\quad$ Unstable
    $\quad$ decomposed by H_2S

$\Big\downarrow H_2S$

$Sb_2S_3 \downarrow$
orange ppt.

(ii) To the second part add
     granulated Zn metal
     (2-3 pieces), allow the
     reaction to complete
     $SnCl_4 + Zn \rightarrow ZnCl_2 + SnCl_2$
     Filter and to the filtrate
     add $HgCl_2$ solution. White or
     grey ppt. confirms tin
     $SnCl_2 + 2HgCl_2 \rightarrow SnCl_4 + Hg_2Cl_2$
     $Hg_2Cl_2 + SnCl_2 \rightarrow SnCl_4 + 2Hg$
     $\quad\quad\quad\quad\quad\quad\quad\quad$ grey

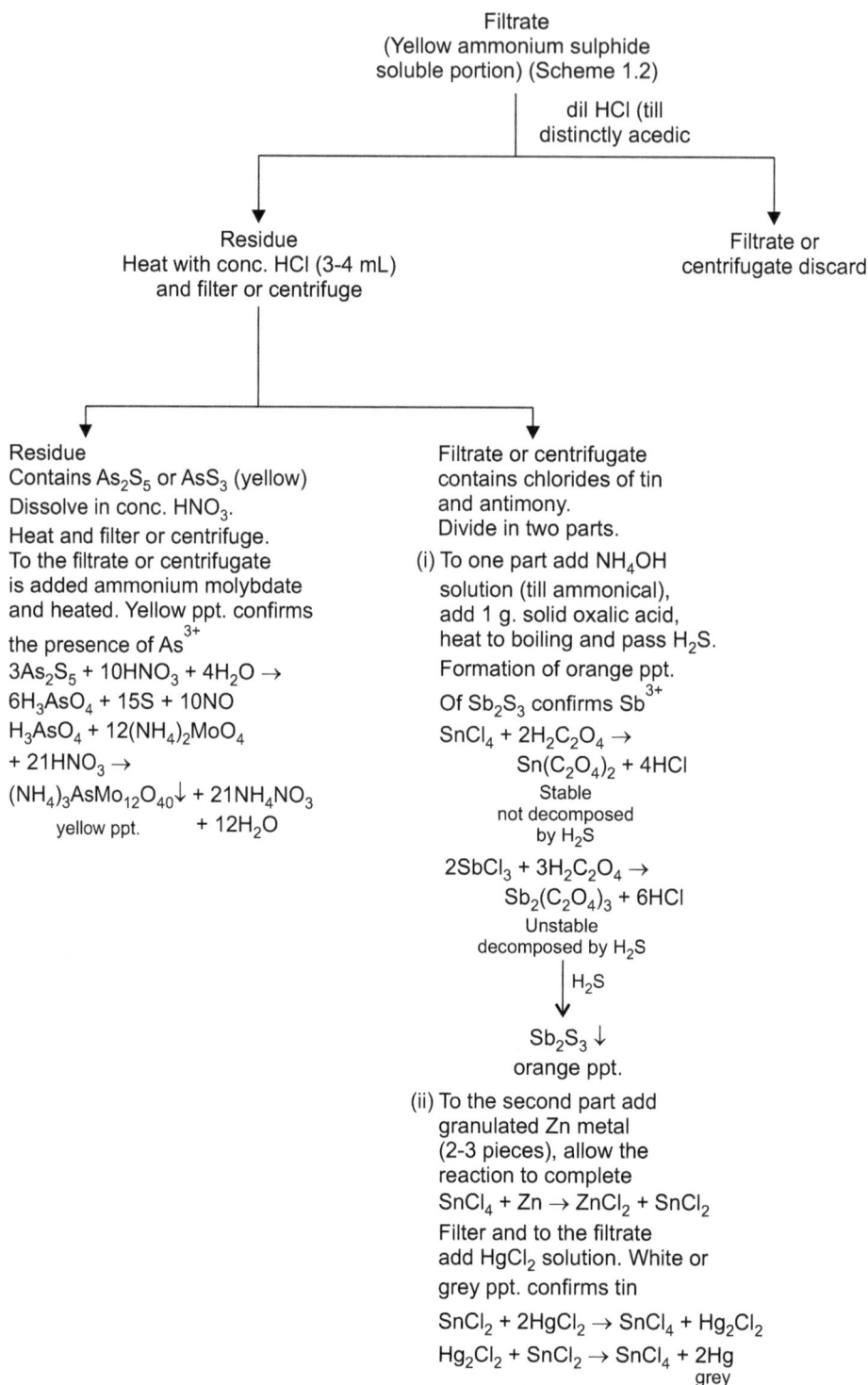

Scheme 1.4. Analysis of group II B cations.

Confirmation of Group II B cations

Confirmatory tests for group II B cations are performed as follows:

Arsenic

(*i*) **Silver nitrate solution.** Addition of $AgNO_3$ solution to neutral solution containing As gives a yellow ppt. of silver arsenite.

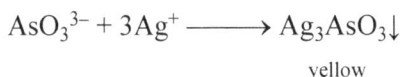

$$AsO_3^{3-} + 3Ag^+ \longrightarrow Ag_3AsO_3\downarrow$$
$$\text{yellow}$$

The ppt. is soluble in NH_4OH and also in HNO_3.

(*ii*) **Copper sulphate solution.** Addition of $CuSO_4$ solution to neutral solution containing As gives a green ppt. of copper arsenite [$CuHAsO_3$ and $Cu_3(AsO_3)_2 . xH_2O$]. Which is soluble in acids and NH_4OH giving a blue solution.

(*iii*) **Potassium triodate.** Potassium tri-iodate (a solution of iodine in potassium iodide) is decoloured due to oxidation of arsenite ions

$$AsO_3^{3-} + I_3^- + H_2O \rightleftharpoons AsO_4^{3-} + 3I^- + 2H^+$$

(*iv*) **Bettendorffs test.** Addition of few drops of arsenite solution to 2 mL conc. HCl and 0.5 mL sat'd $SnCl_2$ solution and the solution on warming gives dark brown colour turning to black due to separation of elemental Arsenic

$$2Ag^{3+} + 3Sn^{2+} \longrightarrow 2As\downarrow + 3Sn^{4+}$$

(*v*) **Flectmann's test.** It consist in treatment of As (III) compound with nacent hydrogen generated in alkaline solution (*e.g.*, from aluminium or zinc and NaOH solution). Under these conditions arsenic (III) compound is converted to arsene ($AsH_3 \uparrow$) which gives a black stain on a filter paper strip impregnated with 20% $AgNO_3$ solution. The apparatus set-up is as shown in the figure.

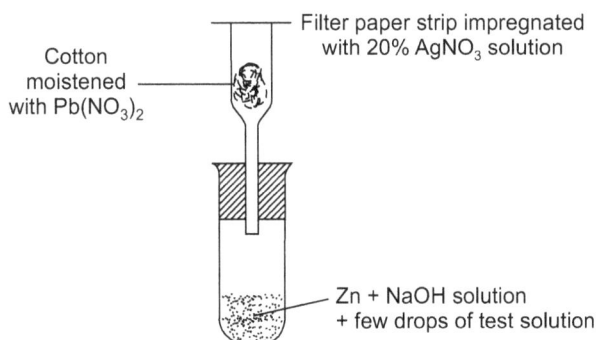

Antimony

(*i*) To antimony trichloride solution (as obtained in Scheme 1.4) is added water. A white ppt. of antimonyl chloride (SbO.Cl) is formed, which is soluble in HCl and in tartaric acid solution (different from bismuth). However, with excess of water, the hydrated oxide $SbO_3 . xH_2O$ is obtained.

(*ii*) **Sodium hydroxide or ammonium hydroxide.** A white ppt. of hydrated antimony (III) oxides $Sb_2O_3 . xH_2O$, dissolves in conc. solution (5 M) of caustic alkali giving antimonites.

$$2Sb^{3+} + 6OH^- \longrightarrow Sb_2O_3\downarrow + 3H_2O$$

$$Sb_2O_3 + 2OH^- \longrightarrow 2SbO_2^- + H_2O$$

(*iii*) **KI solution.** Yellow colouration due to the formation of a complex salt.

$$2Sb^{5+} + 9I^- \longrightarrow [SbI_6]^{3-} + I_3^-$$

yellow

colour

Tin

(*i*) **NaOH solution.** To tin chloride solution (as obtained in scheme 1.4) is added NaOH solution. A white ppt. of tin (II) hydroxide (which is soluble in excess alkali) is obtained

$$Sn^{2+} + 2OH^- \rightleftharpoons Sn(OH)_2\downarrow$$

white

$$Sn(OH)_2\downarrow + 2OH^- \rightleftharpoons [Sn(OH)_4]^{2-}$$

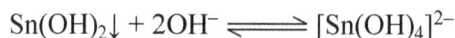

(*ii*) **NH₄OH solution.** With NH_4OH, white ppt. of tin (II) hydroxide (which is insoluble in excess NH_4OH is obtained.)

1.4.1.4 Analysis of Group III Cations

Before preceeding for the analysis of group III cations, any anion which interfers in the analysis of group III cations must be removed. The interfering anions include oxalate, tartarate, borate, fluoride and phosphate.

It is of interest to understand why these interferring cations do not interfere during the analysis of group I and group II cations and interfere only during the analysis of group III cations.

During the analysis of group I and group II cations, the medium is acidic. The acids corresponding to the interferring anions maintain the following equalibria.

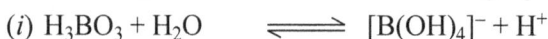

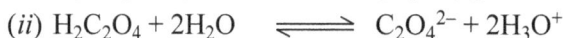

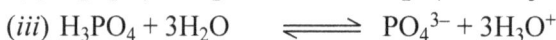

(*i*) $H_3BO_3 + H_2O \rightleftharpoons [B(OH)_4]^- + H^+$

(*ii*) $H_2C_2O_4 + 2H_2O \rightleftharpoons C_2O_4^{2-} + 2H_3O^+$

(*iii*) $H_3PO_4 + 3H_2O \rightleftharpoons PO_4^{3-} + 3H_3O^+$

(*iv*) $HF + H_2O \rightleftharpoons F^- + H_3O^+$

(*v*) $H_2C_4H_4O_6 + 2H_2O \rightleftharpoons C_4H_4O_6^{2-} + 2H_3O^+$

The ionisation of the acids corresponding to these interfering anions is suppressed due to common ion effect due to high concentration of H^+ ions from the ionisation of HCl. In this way, the concentration of these ions in the solution is reducd and the solubility products of the corresponding salts with the interferring ions is not exceeded. Thus, no precipitation of insoluble salts occur. However, if the medium becomes alkaline (on addition of NH_4OH for analysis of group III cations), the availability of OH^- ions removes the H^+ ions from the solution and due to this there is increased ionisation of the acids of all the interferring ions. Thus, the interferring ions are available in abundance. As an example,

$$HF + OH^- \longrightarrow H_2O + F^-$$

Thus, due to the presence of OH^- ions, the concentration of the interferring anions and the solubility product of insoluble fluoride is exceeded resulting in their precipitation.

Removal of interferring anions

Following is given the method for the removal of interferrng anions.

Removal of oxalate and tartarate

For the removal of oxalate and tartarates the filtrate from group II (Scheme 1.2) is heated with conc. HNO_3 in a china dish to near dryness. By this procedure oxalate is oxidised to CO_2 which escapes in the air. A similar reaction occurs between tartaric acid and HNO_3.

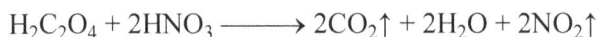

$$H_2C_2O_4 + 2HNO_3 \longrightarrow 2CO_2\uparrow + 2H_2O + 2NO_2\uparrow$$

Finally, the residue is dissolved in conc. HCl and then diluted with water and the solution so obtained is used for the analysis of group III cations.

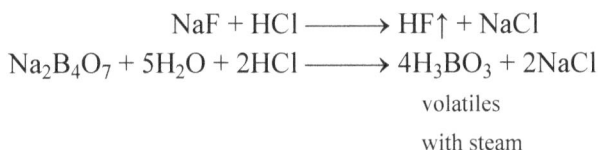

Removal of Borate and Fluoride

For removal of borate and fluoride, the filtrate from group II (Scheme 1.2) is evaporated in a china dish to dryness. The process is repeated 2-3 times with conc. HCl. Finally, the residue is dissolved in conc. HCl, solution diluted with water and the solution so obtained is used for the analysis of group III cations.

$$NaF + HCl \longrightarrow HF\uparrow + NaCl$$
$$Na_2B_4O_7 + 5H_2O + 2HCl \longrightarrow 4H_3BO_3 + 2NaCl$$
volatiles
with steam

In case only borate is present (and Fluoride is absent), the filtrate from group II is evaporated to dryness. The residue is treated with ethyl alcohol (4-5 mL) and conc. HCl (2-3 mL) and the solution again evaporated to almost dryness. The residue is dissolved in conc. HCl, diluted with water and used for group III analysis.

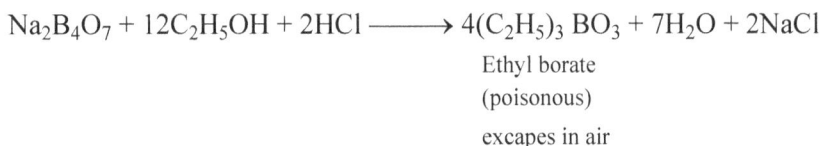

$$Na_2B_4O_7 + 12C_2H_5OH + 2HCl \longrightarrow 4(C_2H_5)_3 BO_3 + 7H_2O + 2NaCl$$
Ethyl borate
(poisonous)
excapes in air

Removal of phosphate

The phosphate is removed by boiling the filtrate from group II in order to remove H_2S. Conc. HNO_3 (2-3 drops) are added, solution boiled (to convert ferrous into ferric) and zirconium nitrate solution (10 mL) added. The solution is again boiled, cooled and centrifuged. Few drops of zirconium nitrate solution are again added (excess avoided) to ensure complete precipitation. The solution is centrifuged. The residue of zirconium phosphate is discarded and the filtrate or the centrifugate is used for group III analysis.

To the solution (obtained after removal of interferring anions) is added solid NH_4Cl and NH_4OH. In case interferring anions are absent, the filtrate from group II can be used directly. The cations of group III are precipitated as hydroxides. These are filtered, the filtrate used for subsequent groups (Scheme 1.5).

Filtrate of group II
(after removal of interferring anions)

Solid NH_4Cl + NH_4OH
Filter or centrifuge

Residue
group III cations
Hydroxides of Fe, Al, Cr
along with some (MnO_2 . xH_2O)
wash the ppt. with NH_4Cl solution (3-4 mL)
Boil the residue with NaOH soln. (5 mL)
and H_2O_2 solution (3-4 mL).
Heat the solution and centrifuge or filter

Centrifugate
or filtrate
used for group IV

Residue
May contain ferric
hydroxide and some
MnO_2 . xH_2O
Divide in two parts

(i) To one part, add dil. HNO_3
(1-2 mL) and H_2O_2 (4-5 drops).
Boil the soln. to decompose
H_2O_2.Cool and add a pinch
of sodium bismuthate or lead
peroxide. Formation of violet
colour confirms Mn^{2+}
(see Scheme 1.7 for reaction)

(ii) To the second part of the residue
add dil. HCl (2 mL) warm and
divide in two parts
$Fe(OH)_3 + 3HCl \rightarrow FeCl_3 + 3H_2O$

(a) To one part add a few drops
of KCNS solution. Formation
of blood red colour confirms Fe^{3+}.
$FeCl_3 + 3KCNS \rightarrow Fe(CNS)_3 + 3KCl$
        blood red
        colour

(b) To the other part, add few drops
of $K_4Fe(CN)_6$ solution. Formation
of blue ppt. or blue colour
confirms Fe^{3+}.
$4FeCl_3 + 3K_4Fe(CN)_6 \rightarrow$
        $Fe_4[Fe(CN)_6]_3 + 12KCl$
        Blue colour

Centrifugate or filtrate
May contain Al^{3+}, Cr^{3+}
$Al(OH)_3 + NaOH \rightarrow NaAlO_2 + 2H_2O$
$2Cr(OH)_3 + 4NaOH + 3H_2O_2$
                $\rightarrow 2Na_2CrO_4 + 8H_2O$

If filtrate is colourless, it may
contain Al^{3+} as $NaAlO_2$ (but not Cr^{3+}).
However, it is yellow it may also
contain Cr^{3+}. Divide the filtrate
is 4 parts.

(i) One part + dil. HCl + NH_4OH (till
alkaline). Boil, formation of gelatenous
white ppt. shows the presence of Al^{3+}.
$NaAlO_2 + H_2O + HCl \rightarrow NaCl + H_3AlO_3$
$H_3AlO_3 + NH_4OH \rightarrow Al(OH)_3 + NH_3 + H_2O$
                        gelatinous
                        white

(ii) Second part + HCl + ammonium
acetate soln. (1-2 mL) + aluminion
reagent (0.5 mL). Formation of
red ppd. confirms Al^{3+}.

(iii) Third part + excess acetic acid
till acidic + lead acetate. Formation of
yellow ppt. confirms Cr^{3+}.
$Na_2CrO_4 + (CH_3COO)_2Pb$
        $\rightarrow PbCrO_4\downarrow + 2CH_3COONa.$
        yellow

(iv) Fourth part + 1-2 mL dil. H_2SO_4.
Cool and add alcohol (1 mL) and
H_2O_2 (4-5 drops). Allow the two
layers to separate. Blue colour
in organic layer confirms Cr^{3+}.
$2Na_2CrO_4 + H_2SO_4 \rightarrow$
        $Na_2Cr_2O_7 + Na_2SO_4 + H_3O$
$Na_2Cr_2O_7 + H_2SO_4 + 7H_2O_2$
        $\rightarrow 2H_3CrO_8 + 5H_2O + Na_2SO_4$
                (Blue)

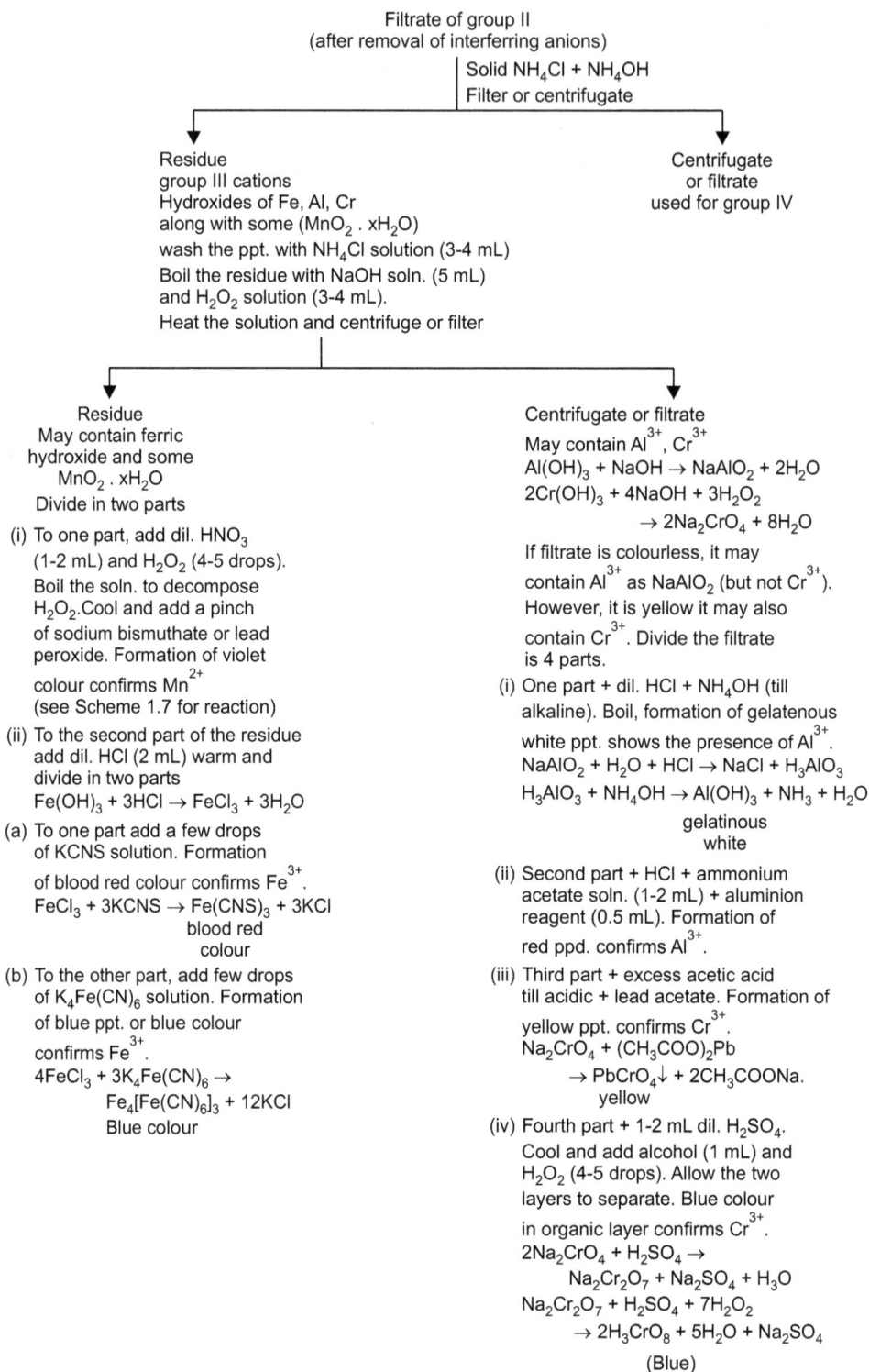

Scheme 1.5. Analysis of group III cations.

Confirmatory tests of Group III cations

The confirmatory tests of group III cation are given below:

Iron

(a) **NaOH solution.** The residue containing iron (Scheme 1.5) is treated with dil. HCl. To the clear solution is added NaOH solution. A reddish-brown ppt. of $Fe(OH)_3$ is formed, which is insoluble in excess NaOH but soluble in dil. acids.

$$FeCl_3 + 3NaOH \longrightarrow Fe(OH)_3 + 3NaCl$$

(b) **NH$_4$OH solution.** To the solution of the residue containing iron is added dil. HCl followed by addition of NH_4OH solution. A reddish-brown gelatinous ppt. of $Fe(OH)_3$ is formed, which is insoluble in excess NH_4OH but soluble is acids

$$Fe^{3+} + 3NH_3 + 3H_2O \longrightarrow Fe(OH)_3\downarrow + 3NH_4^+$$

On heating, the hydroxide is converted into oxide, which dissolves on boiling with conc. HCl.

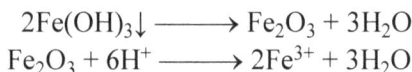

$$2Fe(OH)_3\downarrow \longrightarrow Fe_2O_3 + 3H_2O$$
$$Fe_2O_3 + 6H^+ \longrightarrow 2Fe^{3+} + 3H_2O$$

Aluminium

(a) To the solution containing aluminium (Scheme 1.5) is added sodium hydroxide solution. White ppt. of $Al(OH)_3$ is obtained which is soluble in excess NaOH solution

$$Al^{3+} + 3OH^- \longrightarrow Al(OH)_3\downarrow$$
$$Al(OH)_3\downarrow + OH^- \longrightarrow [Al(OH)_4]^-$$

However, in presence of tartaric acid there is no precipitation of $Al(OH)_3$.

(b) **Sodium acetate solution.** Addition of sodium acetate solution to neutral solution containing Al^{3+} in cold there is no reaction. However, on boiling a voluminous ppt. of basic aluminium acetate is formed.

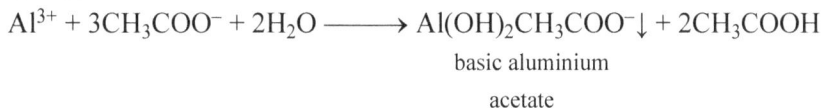

$$Al^{3+} + 3CH_3COO^- + 2H_2O \longrightarrow Al(OH)_2CH_3COO^-\downarrow + 2CH_3COOH$$

basic aluminium
acetate

(c) **Aluminon reagent**

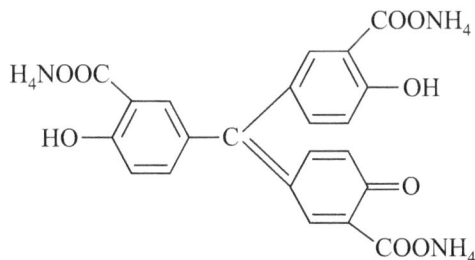

To a solution of $Al(OH)_3$ in HCl (2 M) is added ammonium acetate solution (6 M) and aluminon reagent (2 mL, 0.1% aqueous solution). The mixture on shaking and allowing to stand (5 min) and excess of ammonical ammonium carbonate solution is added. A bright red ppt. or colouration is obtained.

Chromium

Sodium hydroxide solution. Addition of sodium hydroxide solution to a solution containing chromium salt gives a green ppt. of $Cr(OH)_3$ which is soluble in excess NaOH solution

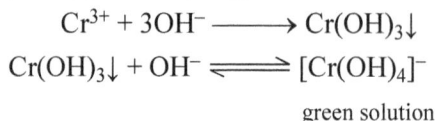

$$Cr^{3+} + 3OH^- \longrightarrow Cr(OH)_3\downarrow$$
$$Cr(OH)_3\downarrow + OH^- \rightleftharpoons [Cr(OH)_4]^-$$
$$\text{green solution}$$

Addition of H_2O_2 to the alkaline green solution gives a yellow solution.

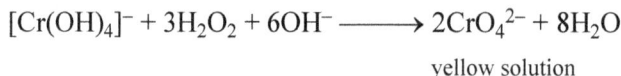

$$[Cr(OH)_4]^- + 3H_2O_2 + 6OH^- \longrightarrow 2CrO_4^{2-} + 8H_2O$$
$$\text{yellow solution}$$

The above yellow solution on acidification with acetic acid and addition of $BaCl_2$ solution gives a yellow ppt. of $BaCrO_4$

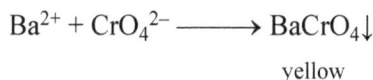

$$Ba^{2+} + CrO_4^{2-} \longrightarrow BaCrO_4\downarrow$$
$$\text{yellow}$$

Group III cations can also be separated and identified by paper chromatography (see Section 1.10.5).

1.4.1.5 Analysis of Group IV Cations

The filtrate or centrifugate obtained after separation of group III cations (Scheme 1.5) contains the cations of group IV–VI. Hydrogen sulphide gas is passed into the solution. The precipitate formed consists of sulphides of Co^{2+}, Ni^{2+}, Mn^{2+} and Zn^{2+}. The ppt. is centrifuged and the filtrate or centrifugate is used for analysis of group V and VI (Scheme 1.6).

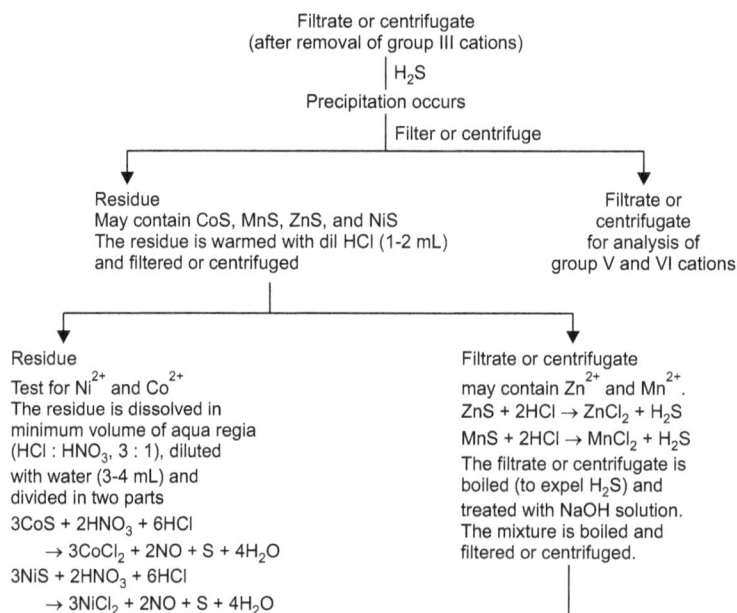

Filtrate or centrifugate
(after removal of group III cations)
| H_2S
Precipitation occurs
| Filter or centrifuge

Residue
May contain CoS, MnS, ZnS, and NiS
The residue is warmed with dil HCl (1-2 mL)
and filtered or centrifuged

Filtrate or
centrifugate
for analysis of
group V and VI cations

Residue
Test for Ni^{2+} and Co^{2+}
The residue is dissolved in minimum volume of aqua regia ($HCl : HNO_3$, 3 : 1), diluted with water (3-4 mL) and divided in two parts
$3CoS + 2HNO_3 + 6HCl$
$\rightarrow 3CoCl_2 + 2NO + S + 4H_2O$
$3NiS + 2HNO_3 + 6HCl$
$\rightarrow 3NiCl_2 + 2NO + S + 4H_2O$

Filtrate or centrifugate
may contain Zn^{2+} and Mn^{2+}.
$ZnS + 2HCl \rightarrow ZnCl_2 + H_2S$
$MnS + 2HCl \rightarrow MnCl_2 + H_2S$
The filtrate or centrifugate is boiled (to expel H_2S) and treated with NaOH solution. The mixture is boiled and filtered or centrifuged.

contd....

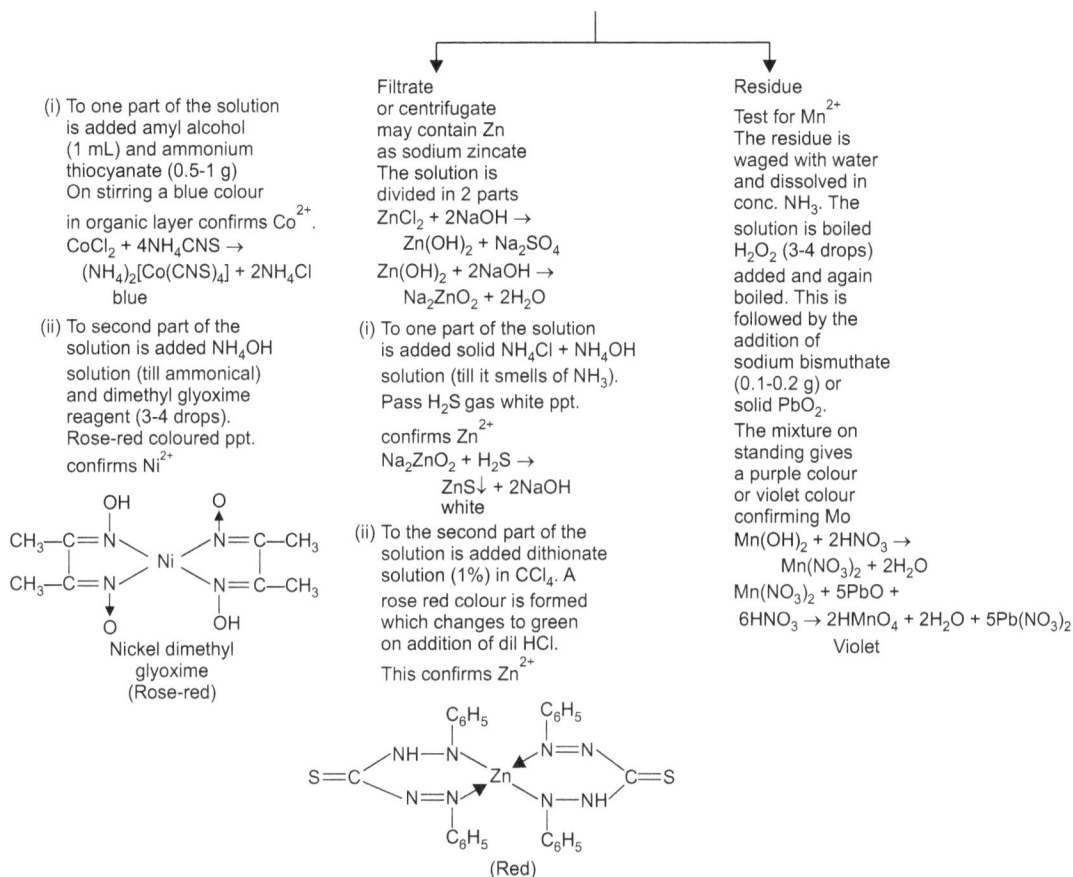

(i) To one part of the solution is added amyl alcohol (1 mL) and ammonium thiocyanate (0.5-1 g) On stirring a blue colour in organic layer confirms Co^{2+}.

$CoCl_2 + 4NH_4CNS \rightarrow$
$\quad (NH_4)_2[Co(CNS)_4] + 2NH_4Cl$
$\quad$ blue

(ii) To second part of the solution is added NH_4OH solution (till ammonical) and dimethyl glyoxime reagent (3-4 drops). Rose-red coloured ppt. confirms Ni^{2+}

Nickel dimethyl glyoxime
(Rose-red)

Filtrate or centrifugate may contain Zn as sodium zincate The solution is divided in 2 parts

$ZnCl_2 + 2NaOH \rightarrow$
$\quad Zn(OH)_2 + Na_2SO_4$
$Zn(OH)_2 + 2NaOH \rightarrow$
$\quad Na_2ZnO_2 + 2H_2O$

(i) To one part of the solution is added solid $NH_4Cl + NH_4OH$ solution (till it smells of NH_3). Pass H_2S gas white ppt. confirms Zn^{2+}

$Na_2ZnO_2 + H_2S \rightarrow$
$\quad ZnS\downarrow + 2NaOH$
$\quad$ white

(ii) To the second part of the solution is added dithionate solution (1%) in CCl_4. A rose red colour is formed which changes to green on addition of dil HCl. This confirms Zn^{2+}

(Red)

Residue
Test for Mn^{2+}
The residue is waged with water and dissolved in conc. NH_3. The solution is boiled H_2O_2 (3-4 drops) added and again boiled. This is followed by the addition of sodium bismuthate (0.1-0.2 g) or solid PbO_2. The mixture on standing gives a purple colour or violet colour confirming Mo

$Mn(OH)_2 + 2HNO_3 \rightarrow$
$\quad Mn(NO_3)_2 + 2H_2O$
$Mn(NO_3)_2 + 5PbO +$
$6HNO_3 \rightarrow 2HMnO_4 + 2H_2O + 5Pb(NO_3)_2$
$\quad$ Violet

Scheme 1.6. Analysis of group IV cations.

Confirmatory tests of group IV cations

Confirmatory tests for group IV cations are gives below.

Cobalt

The solution obtained by dissolving the residue containing Co and Ni in aqua regia (Scheme 1.6) and subsequent dilution is used for confirming cobalt.

(i) **Sodium hydroxide solution.** A blue basic salt is precipitated in cold

$$Co^{2+} + OH^- + NO_3^- \longrightarrow Co(OH)NO_3\downarrow$$
$$\text{basic salt}$$
$$\text{(blue)}$$

On addition of excess NaOH solution and on warming the blue basic salt is converted to a pink ppt. of $Co(OH)_2$.

$$Co(OH)NO_3\downarrow + OH^- \longrightarrow Co(OH)_2\downarrow \ + \ NO_3^-$$
$$\text{pink cobalt}$$
$$\text{hydroxide}$$

On exposure to air the pink $Co(OH)_2$ gives brownish-black $Co(OH)_3$.

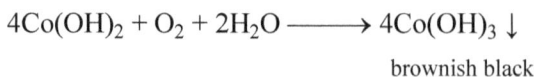

$$4Co(OH)_2 + O_2 + 2H_2O \longrightarrow 4Co(OH)_3 \downarrow$$

brownish black

(*ii*) **Ammonium hydroxide solution.** A basic salt is formed (as in case of reaction with NaOH), which is soluble in excess of the reagent

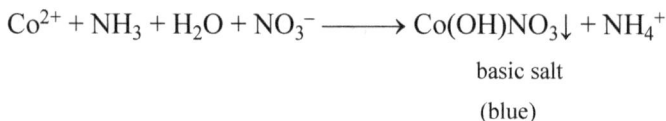

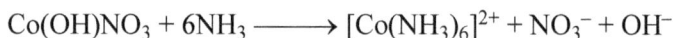

$$Co^{2+} + NH_3 + H_2O + NO_3^- \longrightarrow Co(OH)NO_3\downarrow + NH_4^+$$

basic salt
(blue)

$$Co(OH)NO_3 + 6NH_3 \longrightarrow [Co(NH_3)_6]^{2+} + NO_3^- + OH^-$$

(*iii*) **Potassium nitrite solution.** To the neutral solution containing Co^{2+} is added dilute acetic acid (0.5 mL) and then a freshly prepared saturated solution of KNO_2 gives yellow precipitate

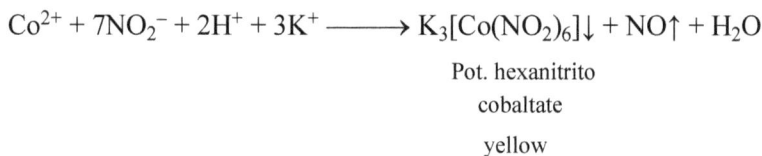

$$Co^{2+} + 7NO_2^- + 2H^+ + 3K^+ \longrightarrow K_3[Co(NO_2)_6]\downarrow + NO\uparrow + H_2O$$

Pot. hexanitrito
cobaltate
yellow

(*iv*) **Vogel reaction.** Addition of a few crystals of ammonium thiocyanate to a neutral or acidic solution of Co^{2+} gives a blue colour due to formation of tetrathiocyanatocobaltate (II) ions

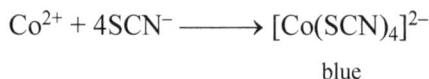

$$Co^{2+} + 4SCN^- \longrightarrow [Co(SCN)_4]^{2-}$$

blue

Nickel

(*i*) **Sodium hydroxide solution.** To a solution of nickel chloride is added sodium hydroxide solution. A green ppt. of $Ni(OH)_2$ is obtained

$$Ni^{2+} + 2OH^- \longrightarrow Ni(OH)_2\downarrow$$

green

The precipitate is insoluble in NaOH solution. No precipitation occurs in presence of citrate or tartarate.

(*ii*) **Ammonium hydroxide solution.** Green precipitate of $Ni(OH)_2$ is obtained, which is soluble in excess reagent.

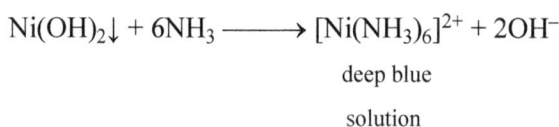

$$Ni^{2+} + NH_3 + H_2O \longrightarrow Ni(OH)_2\downarrow + 2NH_4^+$$

green

$$Ni(OH)_2\downarrow + 6NH_3 \longrightarrow [Ni(NH_3)_6]^{2+} + 2OH^-$$

deep blue
solution

(*iii*) **Potassium nitrite.** No precipitation (different from cobalt).

Manganese

(*i*) **Sodium hydroxide solution.** A white ppt. of $Mn(OH)_2$ obtained, which gets oxidised in air to form brown hydrated manganese dioxide.

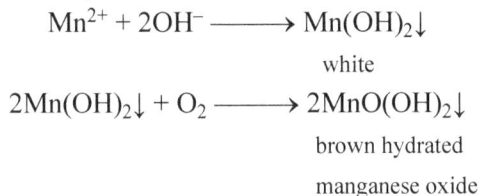

$$Mn^{2+} + 2OH^- \longrightarrow Mn(OH)_2\downarrow$$
white

$$2Mn(OH)_2\downarrow + O_2 \longrightarrow 2MnO(OH)_2\downarrow$$
brown hydrated
manganese oxide

(*ii*) **Potassium persulphate.** Addition of solid $K_2S_2O_8$ to a acidified solution (H_2SO_4) and few drops of $AgNO_3$ solution (as catalyst) gives on heating a reddish violet solution

$$2Mn^{2+} + 5S_2O_8^{2-} + 8H_2O \longrightarrow 2MnO_4^- + 10SO_4^{2-} + 16H^+$$

Zinc

(*i*) **Sodium hydroxide solution.** White gelatinous ppt. of $Zn(OH)_2$, which is soluble in excess reagent and also in acids is obtained

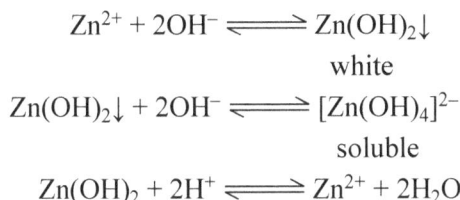

$$Zn^{2+} + 2OH^- \rightleftharpoons Zn(OH)_2\downarrow$$
white

$$Zn(OH)_2\downarrow + 2OH^- \rightleftharpoons [Zn(OH)_4]^{2-}$$
soluble

$$Zn(OH)_2 + 2H^+ \rightleftharpoons Zn^{2+} + 2H_2O$$

(*ii*) **Disodium hydrogen phosphate:** White ppt. of zinc phosphate is obtained

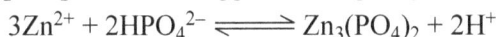

$$3Zn^{2+} + 2HPO_4^{2-} \rightleftharpoons Zn_3(PO_4)_2 + 2H^+$$

(*iii*) **Potassium hexacyanoferrate (II) solution:** White ppt. is obtained

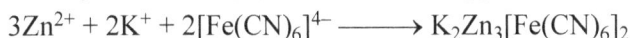

$$3Zn^{2+} + 2K^+ + 2[Fe(CN)_6]^{4-} \longrightarrow K_2Zn_3[Fe(CN)_6]_2$$

the ppt. is soluble in NaOH solution

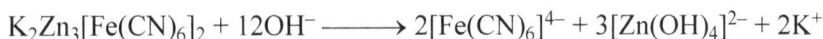

$$K_2Zn_3[Fe(CN)_6]_2 + 12OH^- \longrightarrow 2[Fe(CN)_6]^{4-} + 3[Zn(OH)_4]^{2-} + 2K^+$$

This reaction is useful to distinguish Zn from Al.

(*iv*) **Charcoal cavity test.** Table 1.5.

Group IV cations (Ni^{2+}, Co^{2+}, Mn^{2+} and Zn^{2+}) can also be separated and identified by paper chromatography (see Section 1.10.7 and 1.10.8).

1.4.1.6 Analysis of Group V Cations

The filtrate after separation of group IV cations (Scheme 1.6) is concentrated in a china dish to about 10 mL and then treated with solid NH_4Cl and 2-3 mL of NH_4OH solution and $(NH_4)_2CO_3$ (common ion effect); in the absence of NH_4Cl the magnesium (which usually is a group VI cation) also gets precipitated alongwith group V cations. To the resulting solution is added a saturated solution of ammonium carbonate. A white ppt. of carbonates of group V cations is obtained. The filtrate or centrifugate is used for group VI cations. (Scheme 1.7)

Filtrate after separation of group IV
cations (Scheme 1.4) It is concentrated

Solid NH_4Cl
+
NH_4OH soln. + $(NH_4)_2CO_3$ solution
Filter or centrifugal

Residue group V cations carbonate of
Ba^{2+}, Sr^{2+}, Ca^{2+}. The ppt. is washed
with water in cold and dissolved
in dilute acetic acid (3-4 mL). Warm
it necessary.
To a test solution (0.5 mL) obtained
above is added few drops it $K_2Cr_2O_4$
solution. Formation of yellow ppt.
indicates that Ba^{2+} in present.
In case Ba^{2+} is present, add $K_2Cr_2O_4$
solution to the whole of the solution
containing group V cations)
$BaCO_3 + 2CH_3COOH \rightarrow (CH_3COO)_2Ba + CO_2\uparrow + H_2O$
$(CH_3COO)_2Ba + K_2Cr_2O_4 \rightarrow BaCrO_4\downarrow + 2CH_3COOK$
                                          yellow
Filter or centrifuge (If no yellow ppt. is formed in
the test solution, Ba^{2+} is absent. In this case
discard the test solution and test for Ca^{2+} and Sr^{2+})

Filterate or centrifugate
use for testing group VI cations

Filter or centrifugal

Yellow residue contains $BaCrO_4$
Dissolve the precipitate in
minimum amount of conc. HCl
and perform flame test green
flame confirms Ba^{2+} (Table 1.3)

Filtrate or centrifugate
(Test for Sr^{2+} and Ca^{2+})
Add excess ammonium
sulphate solution warm
and filter or centrifuge

Residue white ($SrSO_4$) Dissolve
in conc. HCl and perform flame
test (Table 1.3) deep red colour
in the flame confirms Sr^{2+}.

Filtrate or centrifugate
$CaSO_4 + (NH_4)_2SO_7 \rightarrow (NH_4)_2[Ca(SO_4)_2]$
may contain Ca^{2+}
Add excess ammonium oxalate solution.
Formation of white ppt. shows presence it Ca^{2+}
$(NH_4)_2[Ca(SO_4)_2] + (COONH_4)_2 \rightarrow$
                                          $(COO)_2Ca\downarrow + 2(NH_4)_2SO_4$
Filter and dissolve the ppt. is conc. HCl
perform in flame test (Table 1.3)
Brick red coloured flame confirms Ca^{2+}.

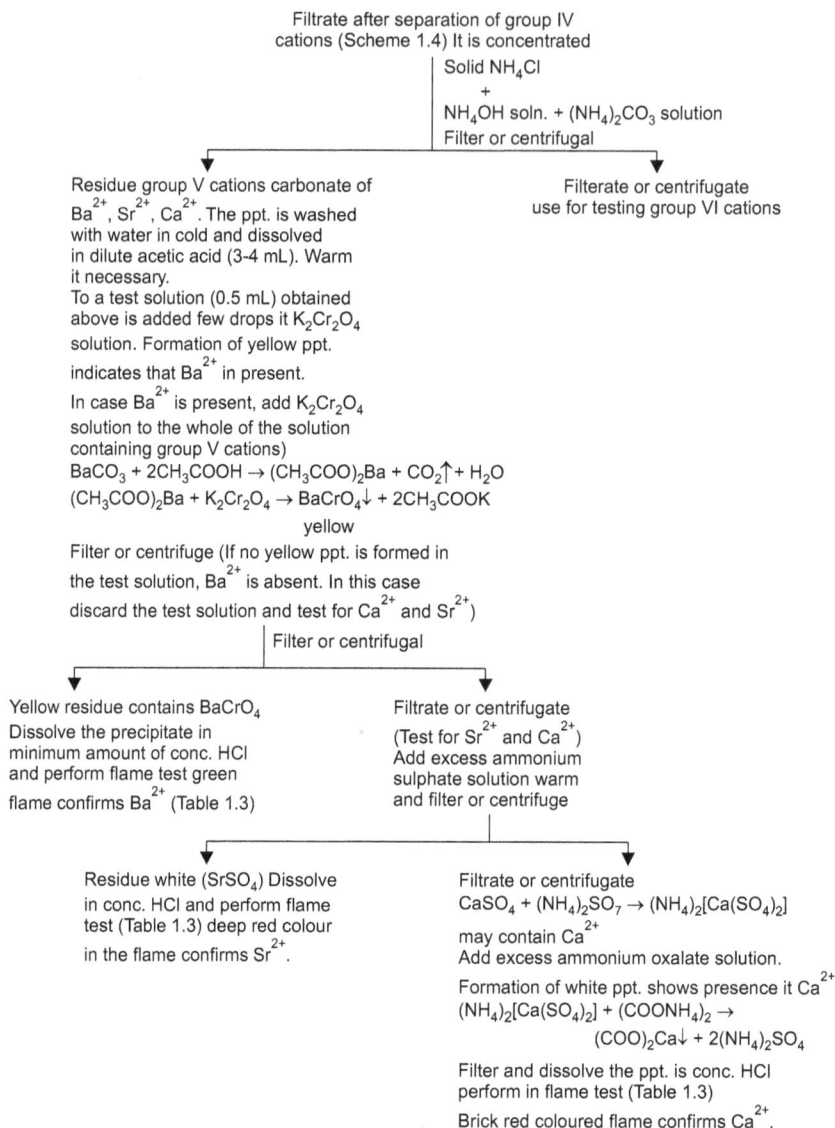

Scheme 1.7. Analysis of group V cations.

Confirmatory tests of Group V cations

Confirmatory tests for group V cations are given below:

Barium

(*a*) **Ammonium hydroxide.** No precipitation occurs due to high solubility of $Ba(OH)_2$ in the reagent.

(*b*) **Ammonium carbonate.** White ppt. of $BaCO_3$ soluble in acetic acid and dilute mineral acids is obtained

$$Ba^{2+} + CO_3^{2-} \longrightarrow BaCO_3\downarrow$$
white

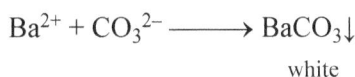

(c) **Ammonium oxalate solution.** White ppt. of $Ba(COO)_2$ is obtained.
$$Ba^{2+} + (COO)_2^{2-} \longrightarrow Ba(COO)_2\downarrow$$
white

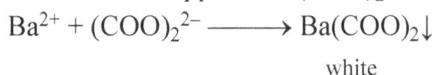

(d) **Potassium chromate solution.** A yellow ppt. of $BaCrO_4$ obtained.
$$Ba^{2+} + CrO_4^{2-} \longrightarrow BaCrO_4\downarrow$$
yellow

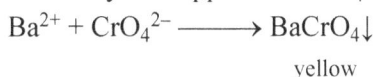

(e) **Ether-ethanol.** A 1:1 mixture of anhydrous ether-ethanol does not dissolve anhydrous $BaCl_2$ or $Ba(NO_3)_2$ (distinction from Sr^{2+} or Ca^{2+}).

Strontium

(a) **Ammonium carbonate solution.** A white ppt. of $SrCO_3$ is obtained.
$$Sr^{2+} + CO_3^{2-} \longrightarrow SrCO_3\downarrow$$
white

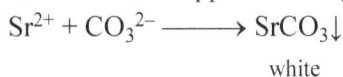

(b) **Sat'd calcium sulphate solution.** A white ppt. of $SrSO_4$ is formed slowly in cold but readily in hot (distinction from Ba^{2+}).

(c) **Ammonium oxalate solution.** A white ppt. of strontium oxalate is formed.
$$Sr^{2+} + (COO)_2^{2-} \longrightarrow Sr(COO)_2\downarrow$$
white

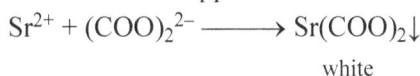

(d) **Potassium chromate solution.** A yellow ppt. of $SrCrO_4$ is obtained.
$$Sr^{2+} + CrO_4^{2-} \longrightarrow SrCrO_4\downarrow$$
yellow

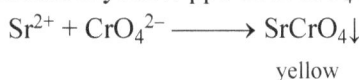

(e) **Ethanol ether mixture.** A 1:1 mixture of anhydrous ethanol and ether does not dissolve anhyd. $Sr(NO_3)_2$ but dissolves anhyd. $SrCl_2$. This test is useful for the separation of Ca^{2+}, Sr^{2+} and Ba^{2+}.

Calcium

(a) **Ammonium hydroxide.** No precipitation since $Ca(OH)_2$ is fairly soluble in the reagent.

(b) **Ammonium carbonate solution.** A white ppt. of $CaCO_3$ is obtained, which dissolves on boiling
$$Ca^{2+} + CO_3^{2-} \longrightarrow CaCO_3\downarrow$$

(c) **Saturated calcium sulphate solution.** No precipitation (different from Sr^{2+} and Ba^{2+}).

(d) **Ammonium oxalate solution.** White ppt. of $Ca(COO)_2$ is obtained
$$Ca^{2+} + (COO)_2^{2-} \longrightarrow Ca(COO)_2\downarrow$$
white

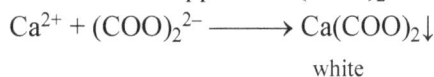

(e) **Potassium chromate solution.** No precipitation in presence of acetic acid.

(f) **Potassium hexacyanoferrate (II) solution.** White ppt. obtained
$$Ca^{2+} + 2K^+ + [Fe(CN)_6]^{4-} \longrightarrow K_2Ca[Fe(CN)_6]\downarrow$$
white

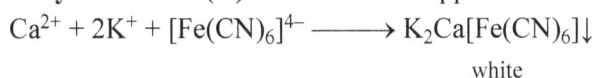

The test is more sensitive in presence of NH_4Cl. In this case K is replaced by NH_4. This test is used to distinguish Ca^{2+} from Sr^{2+}.

1.4.1.7 Analysis of Group VI Cations (Mg^{2+} and K^+)

Magnesium. The filtrate after separation of group V cations (Scheme 1.7) is heated to dryness and then heated strongly in a fume cupboard to decompose ammonium salts. The cooled residue is dissolved in dil. HCl (1-2 mL), heated and filtered (if necessary). To the clear filtrate is added few drops of Na_2HPO_4 solution and few drops of NH_4OH till the solution smells of ammonia. Formation of a white ppt. (on scratching the inner sides of the test tube with a glass rod) confirms Mg^{2+}.

$$MgCl_2 + Na_2HPO_4 + NH_4OH \longrightarrow \underset{\text{white}}{Mg(NH_4)PO_4\downarrow} + 2NaCl + H_2O$$

Potassium

(*a*) Potassium can be confirmed by the flame test either in the mixture directly or with the residue obtained by heating to dryness and subsequent decomposition of ammonium salt (by strongly heating in a fume cupboard) as already mentioned in case of magnesium.

(*b*) The final residue is dissolved in acetic acid and the solution treated with sodium cobalti nitrite solution. A yellow ppt. confirms potassium

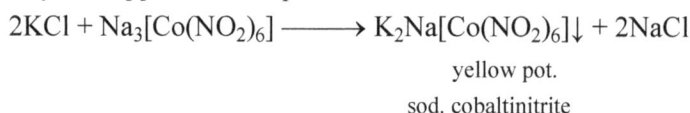

$$2KCl + \underset{\text{sod. cobaltinitrite}}{Na_3[Co(NO_2)_6]} \longrightarrow \underset{\text{yellow pot.}}{K_2Na[Co(NO_2)_6]\downarrow} + 2NaCl$$

Confirmatory tests for Group VI cations

Following are given confirmatory tests for group VI cations.

(*c*) **Tartaric acid.** A white ppt. of potassium hydrogen tartarate is obtained

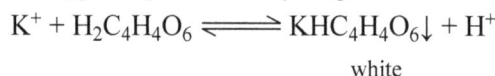

$$K^+ + H_2C_4H_4O_6 \rightleftharpoons \underset{\text{white}}{KHC_4H_4O_6\downarrow} + H^+$$

and

$$K^+ + H.C_4H_4O_6^- \rightleftharpoons \underset{\text{white}}{KHC_4H_4O_6\downarrow}$$

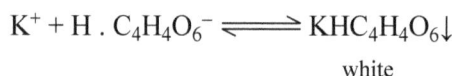

(*d*) **Perchloric acid solution.** White ppt. of pot. perchlorate is obtained

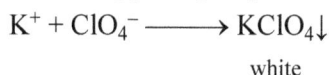

$$K^+ + ClO_4^- \longrightarrow \underset{\text{white}}{KClO_4\downarrow}$$

Caution: The solution should not be heated to avoid explosion.

(*e*) **Hexachloroplatinic (IV) acid reagent.** A yellow ppt. of pot. hexachloroplatinate is obtained

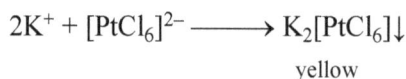

$$2K^+ + [PtCl_6]^{2-} \longrightarrow \underset{\text{yellow}}{K_2[PtCl_6]\downarrow}$$

Magnesium

(*a*) **Ammonium hydroxide.** Partial precipitation of white gelatinous $Mg(OH)_2$ takes place. The ppt. is sparingly soluble in H_2O but readily soluble in NH_4OH.

$$Mg^{2+} + 2NH_3 + 2H_2O \longrightarrow Mg(OH)_2\downarrow + 2NH_4^+$$
white

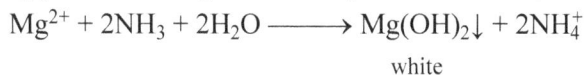

(*b*) **Sodium hydroxide solution.** A white ppt. of $Mg(OH)_2$ is obtained

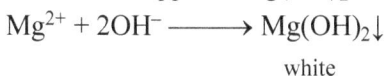

$$Mg^{2+} + 2OH^- \longrightarrow Mg(OH)_2\downarrow$$
white

To the white ppt. is added few drops of diphenyl carbazide reagent ($C_6H_5NH.NH$. $CO.NH.NH.C_6H_5$) and solution filtered. The ppt. is washed with hot water, violet-red colour is seen in the ppt. due to formation of a complex salt.

(*c*) **Oxine reagent.** Treatment of a solution of magnesium salt with a little NH_4Cl and 1-2 mL of the oxine reagent (8-hydroxyquinoline, 2% in acetic acid) and solution rendered ammonical by addition of NH_4OH (3-4 mL). On heating the solution, a yellow ppt. of the complex salt $Mg(C_9H_6ON)_2$. $4H_2O$ is obtained.

Various cations found to be present in a mixture on the basis of above tests (Section 1.4) can also be confirmed on the basis of their spot tests (see section 1.7), atomic spectrometric tests (see section 1.9) and also by chromatographic procedure see section 1.10).

The analysis of insoluble salts, alloys and rare elements are discussed in section 1.11, 1.12 and 1.13.

1.5 ANALYSIS OF CATIONS USING POTASSIUM THIOCARBONATE REAGENT

In the potassium thiocarbonate scheme for the analysis of cations, H_2S is generated *in-situ*. This scheme, commonly called PTC scheme was developed by Prof. K.N. Johri (K.N. Johri, Macro and Semi micro Analysis without H_2S using Potassiumthiocarbonate (PTC) Reagent 2nd Ed., Asia Publishing House, 1971). The outline of the scheme is given below.

Preparation of Solution of the Mixture

As already discussed (Section 1.4.1) the mixture (50-100 mg) is dissolved in one of the following solvents (first in cold and then in hot) in the order (*i*) water (*ii*) dil. HCl (*iii*) conc. HCl and (*iv*) conc. HNO_3.

The solution of the mixture in any of the above solvents is called the original solution (O.S.). In case the mixture is soluble in dil. HCl, the cations of 1st group are absent. If the mixture is soluble in water or HNO_3 then add 4-5 drops of conc. HCl. In case a white ppt. is obtained 1st group is present. The addition of HNO_3 should be avoided as far as possible, since it causes formation of colloidal sulphur on the addition of potassium thiocarbonate reagent for group II. In case, the original solution has been prepared in HNO_3, the solution is evaporated to almost dryness (to remove HNO_3) and the residue is dissolved in distilled water. The solution thus obtained is called the original solution.

In case the original solution gives a white ppt. on cooling (when the solution was made in hot HCl or water) the presence of $PbCl_2$ is indicated. However, in case a white ppt. is obtained on dilution of the solution prepared in HCl, it may be due to the oxychlorides of Bi, Sb or Sn.

In the PTC scheme also, the zero group cations (NH_4^+ and K^+) are tested individually in the mixture itself (*see Section 1.4.1.1*).

Analysis of group I cations (Pb^{2+}, Ag$^+$, Hg$_2^{2+}$)

In case 1st group is present, proceed as given in Scheme 1.2.

Analyses of group II cation

The filtrate obtained after separation of group I cations is used for testing group II cations. Before adding PTC to the whole of the centrifugate from group I, a small portion of the centrifugate is tested with freshly acidified PTC solution. If the solution on heating in a water bath for a minute gives a permanent coloured ppt., then group II is present. In case there is no ppt. formation, group II cations are absent and the test solution discarded and the remaining centrifugate used for group III. However, if group II is present (as indicated by the test given above), the total centrifugate (from group I) is treated with PTC as follows. The pH of the filtrate from group I is set using methyl violet test paper, a blue spot requires addition of more HCl and a yellow spot requires slight dilution, green spot shows the right pH. To this solution is added dropwise pre-acidified PTC reagent till further addition gives only a light coloured solution or a white ppt. The solution is heated for 1-2 min. in a water bath and centrifuged. The centrifugate is used for group III onwards. The residue is sulphides of group II (A) (Hg, Pb, Bi, Cu, Cd) and group II (B) (As, Sb, Sn, Co). The separation of group II (A) and group II (B) cations is achieved by treating the washed residue with hot 3 molar KOH solution, when group II (B) sulphides dissolve leaving behind those of group II (A) sulphides (Scheme 1.8).

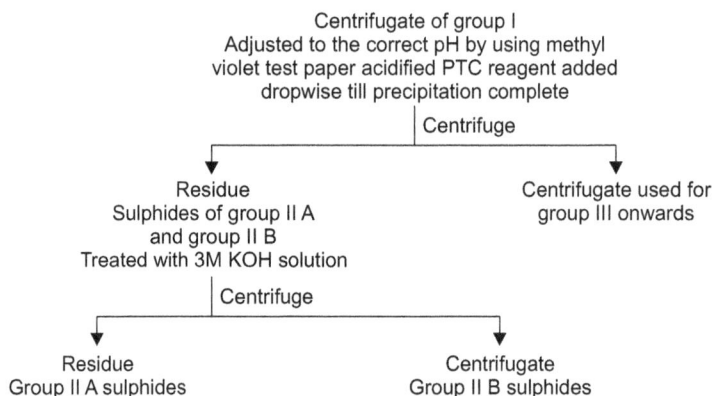

Scheme 1.8. Separation of group II cations into group II A and group II B cations.

Analysis of group II (A) cations

The residue obtained above (Scheme 1.8) contains group II (A) cations. It is washed with small amount of water (to remove excess KOH) and treated with dilute HNO_3 and the cations Hg, Cu, Cd, Pb and Bi analysed as given in Scheme 1.3.

Analysis of group II (B) cations

The centrifugate obtained after separation of group II A cations (Scheme 1.8) contains group II (B) cations (As, Sb, Sn, Co). These are analysed as given in Scheme 1.4.

Analysis of group III cations

The centrifugate obtained after separation of group II cations (Scheme 1.8) is used for the analysis of group III cations.

Removal of phosphate

Before proceeding for group III onwards PO_4^{-3} (if present) has to be eliminated since it interferes. This is done by zirconyl nitrate method which is carried out as follows:

The centrifugate from group II is boiled to expel H_2S and then conc. HNO_3 (1 mL) added. The solution is gently boiled (5 min), cooled and ammonium chloride (10 mg) added. This is followed by the dropwise addition of zirconyl nitrate reagent till the precipitation in complete. The mixture is warmed and centrifuged. The ppt. is discarded and to the centrifugate is added NH_4Cl (5 drops) and NH_4OH with stirring till the solution is ammonical (smell of ammonia). The mixture is centrifuged.

Analysis of group IV cations

The precipitate is analysed for group III cation as given in scheme 1.5 and the centrifugate analysed for group IV (as given in Scheme 1.6).

Analysis of group V cations

The centrifugate obtained after separation of group IV cations is analysed for group V cations as given in Scheme 1.7.

Analysis of group VI cations

The centrifugate after separation of group V cations (Scheme 1.7) is analysed for the analysis of group VI cations as given in Section 1.4.1.6.

1.6   GREENER ALTERNATIVE TO QUALITATIVE ANALYSIS OF CATIONS

The scheme developed by Indu Tucker Sindwani and Susmita Chowdry (J. Chem. Edn., 2008, 85(8), 1099–1101) is believed to be a greener alternative to qualitative analysis of cations. In this scheme, the analysis of arsenic and mercury compounds is not done since these are highly toxic and cannot be analysed by this scheme. The group reagents in the scheme are hydrochloric acid, sodium sulphate, sodium hydroxide and ammonia solution. The cations are confrmed by eco-friendly spot tests. The scheme, as developed has been carried in the absence of interferring cations like fluoride, oxalate, phosphate and borate.

Division of Cations in Different Groups

In the present scheme the separation of cations is based on common chemical properties requiring only few steps (Table 1.14).

Table 1.14. *Classification of cations on the basis of common chemical properties.*

Group	Cation	
Zero	NH_4^+, K^+	(Directly tested in the mixture or water extract)
I	Pb^{2+}, Ag^+	(Precipitated as chlorides)
II	Ba^{2+}, Sr^{2+}, Ca^{2+}, Pb^{2+}	(Precipitated as sulphates)
III A	Fe^{3+}, Mn^{2+}, Mg^{2+}	(Precipitated as hydroxides)
III B	Cu^{2+}, Cd^{2+}, Ni^{2+}, Co^{2+}	(Soluble amine complexes)
IV	Cr^{3+}, Al^{3+}, Zn^{2+}, Sn^{2+}	(Present as soluble hydroxo complex (Cr^{3+} as CrO_4^{2-})

Separation of cations in different groups

The separation of cations into various groups using appropriate group reagents is schematically represented in Scheme 1.9.

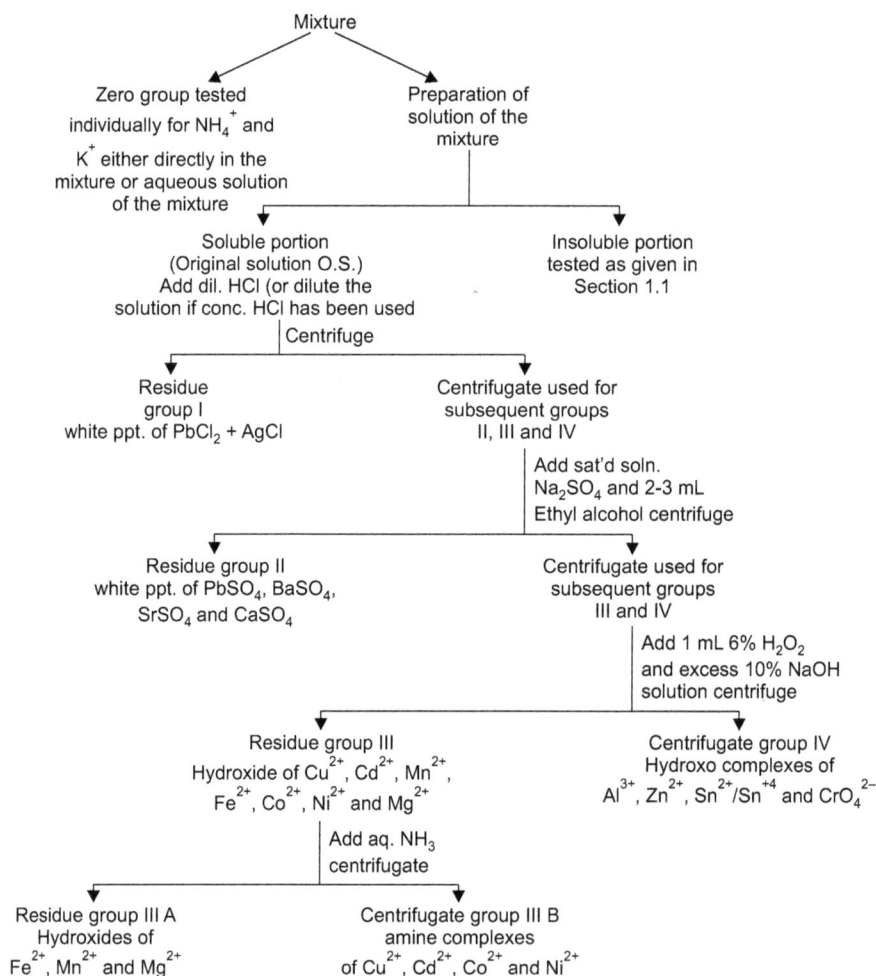

Scheme 1.9. Separation of mixture into various groups.

1.6.1 Analysis of Group Zero Cations

As already stated, the group zero cations are NH_4^+ and K^+. These are tested either directly in the mixture or in its aqueous solution.

Ammonium (NH₄⁺)

(*i*) The mixture on heating with NaOH solution gives smell of ammonia. A glass rod dipped in conc. HCl brought to the mouth of the test tube gives dense white fumes of NH_4Cl.

(*ii*) Addition of Nesslers reagent to aqueous solution of the mixture gives brown ppt. or colouration.

Potassium (K⁺)

(*i*) Flame test directly with the mixture.

(*ii*) To aqueous solution of the mixture is added dilute acetic acid (to acidity the solution) and 3-4 drops of freshly prepared solution of sodium cobaltinitrite is added. A yellow ppt. is obtained.

(*iii*) Potassium can be tested in presence of NH_4^+ as follows. To the water extract of the mixture is added few drops of phenolphthalein followed by addition of sodium carbonate solution. The solution is warmed and the ppt. (if formed) is rejected. To the remaining solution is added dil. CH_3COOH till phenolphthalein indicator in colourless. Finally potassium is tested by sodium cobaltinitrite solution as in test (*ii*).

Note: For all equations involved in the above tests see Section 1.4.1.1.

1.6.2 Analysis of Group I Cations

The mixture is dissolved in minimum volume of conc. HCl (heated if necessary) and the solution diluted with distilled water and cooled to room temperature. A white crystalline ppt. indicates the presence of group I. Centrifuge and analyse the ppt. for group I cations. The centrifugate is used for analysis of subsequent groups II, III and IV (Scheme 1.10).

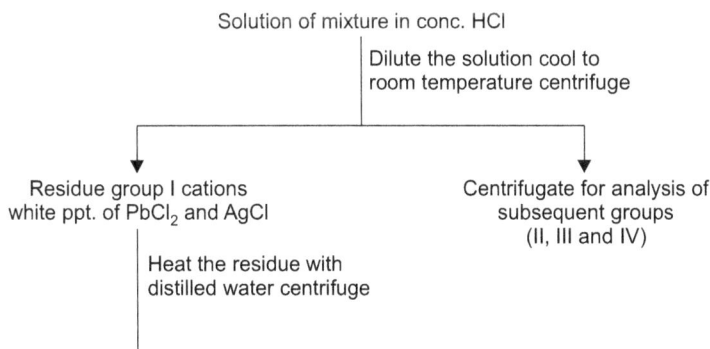

Solution of mixture in conc. HCl

Dilute the solution cool to room temperature centrifuge

Residue group I cations white ppt. of $PbCl_2$ and AgCl

Centrifugate for analysis of subsequent groups (II, III and IV)

Heat the residue with distilled water centrifuge

....*contd.*

Residue AgCl
Dissolve in NH_4OH and
divide in three parts
$AgCl + 2NH_3(aq) \rightarrow [Ag(NH_3)_2]Cl$
                                Soluble

(i) To one part is added few drops
of K_2CrO_4 solution. Formation

of red ppt. it Ag_2CrO_4 confirms Ag^+
$2Ag^+ + CrO_4^{2-} \rightarrow Ag_2CrO_4\downarrow$
                        red

(ii) To the second part is added a drop
of aqueous KI solution. A yellow
ppt. of AgI is formed
$Ag^+ + I^- \rightarrow AgI\downarrow$
                yellow

Centrifugate $PbCl_2$
Divide in two parts

(i) To one part is added CH_3COOH
and aq. $K_2Cr_2O_4$

– A yellow ppt. it $PbCrO_4$ confirms Pb^{2+}

(ii) To another part is added aq. KI
solution. A yellow ppt. of PbI_2
(Soluble on heating and

reappearing on cooling) confirms Pb^{2+}

(iii) Spot test
A scarlet spot with sodium
rhodizonate is seen.

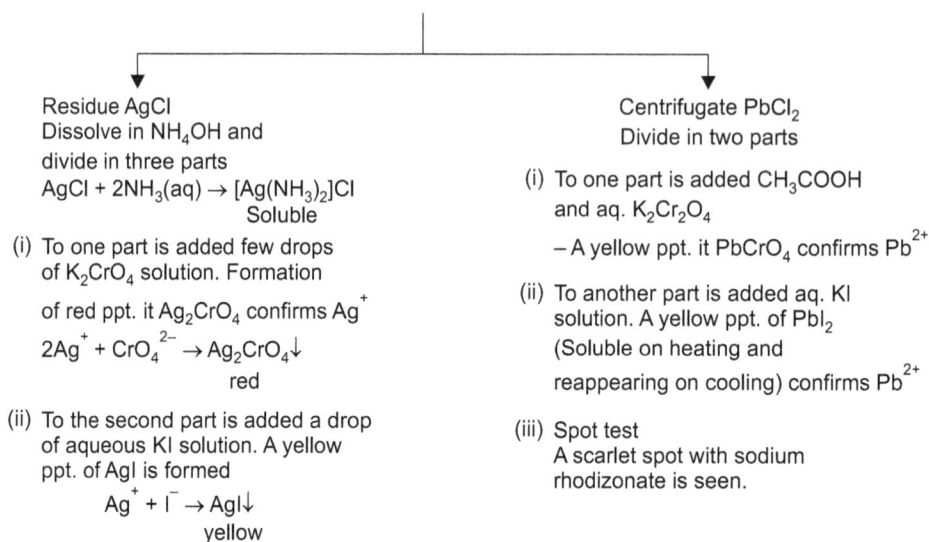

Scheme 1.10. Analysis of group I cations.

1.6.3   Analysis of Group II Cations

The centrifugate after separation of group I cations (Scheme 1.10) is treated with saturated solution of sodium sulphate and 2-3 mL of ethyl alcohol. A white ppt. indicates group II. The centrifugate is used for groups III and IV (Scheme 1.11).

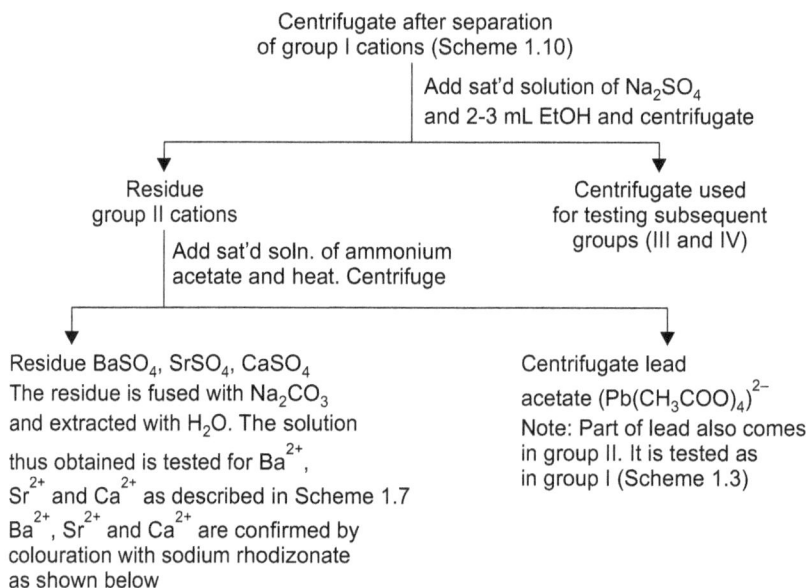

Centrifugate after separation
of group I cations (Scheme 1.10)

Add sat'd solution of Na_2SO_4
and 2-3 mL EtOH and centrifugate

Residue
group II cations

Add sat'd soln. of ammonium
acetate and heat. Centrifuge

Centrifugate used
for testing subsequent
groups (III and IV)

Residue $BaSO_4$, $SrSO_4$, $CaSO_4$
The residue is fused with Na_2CO_3
and extracted with H_2O. The solution

thus obtained is tested for Ba^{2+},
Sr^{2+} and Ca^{2+} as described in Scheme 1.7
Ba^{2+}, Sr^{2+} and Ca^{2+} are confirmed by
colouration with sodium rhodizonate
as shown below

Centrifugate lead
acetate $(Pb(CH_3COO)_4)^{2-}$
Note: Part of lead also comes
in group II. It is tested as
in group I (Scheme 1.3)

	Colouration with Sod. rhodizonate	Colouration with sod. rhodizonide and NaOH	Flame colour with residue
$BaSO_4$	—	—	Apple green
$SrSO_4$	Violet	—	Crimson
$CaSO_4$	—	Violet	Brick red

Scheme 1.11. Analysis of group II cations.

1.6.4   Analysis of Group III Cations

To the centrifugate obtained after removal of group II cations (Scheme 1.11) is added 6% H_2O_2 (1 mL) to oxidise Fe^{2+} to Fe^{3+} followed by excess of NaOH solution. Formation of a precipitate indicates the presence of group III cations. The centrifugate is used for group IV (Scheme 1.12).

Separation of group III Cations into Group IIIA and Group IIIB

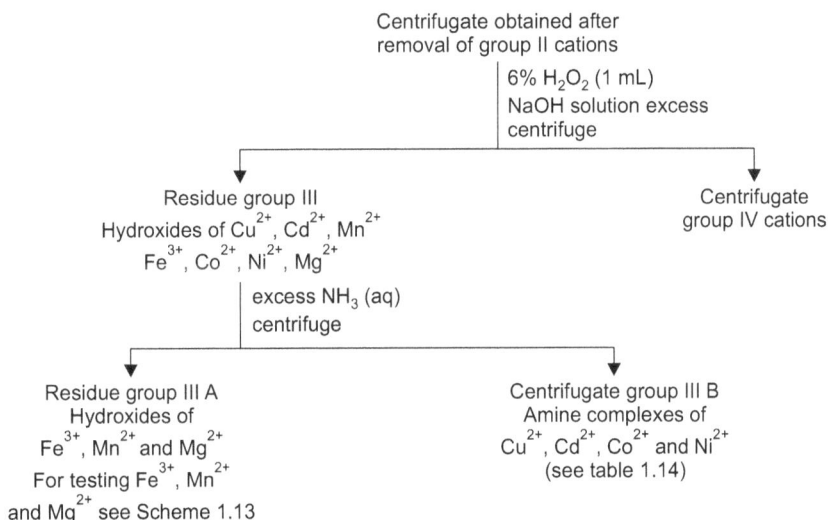

Scheme 1.12. Separation of group III cations into group III A and group III B cations.

Analysis of group III A cations

For the analysis of group III A cations, the ppt. of the hydroxides of Fe^{3+}, Mn^{2+} and Mg^{2+} is dissolved in dilute HCl and the solution thus obtained is used for testing for the presence of group III A cation (Scheme 1.13).

Residue obtained in Scheme 1.12

↓ Dissolved in dil. HCl

The solution is divided in three parts

for testing the presence of

Fe^{3+}, Mn^{2+} and Mg^{2+}

Tests for Fe³⁺

1. To one drop of the solution taken on a filter paper or groove tile is added a drop of KCNS solution. Red colouration confirms Fe^{3+}

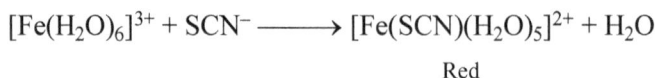

$$[Fe(H_2O)_6]^{3+} + SCN^- \longrightarrow [Fe(SCN)(H_2O)_5]^{2+} + H_2O$$

Red

2. In a similar way perform test with $K_4[Fe(CN)_6]$ reagent. A blue spot is obtained

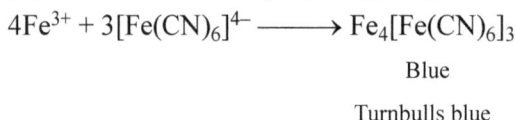

$$4Fe^{3+} + 3[Fe(CN)_6]^{4-} \longrightarrow Fe_4[Fe(CN)_6]_3$$

Blue

Turnbulls blue

Tests for Mn²⁺

To a drop of the solution taken on a filter paper or a groove tile is added one drop of conc. HNO_3 and 1 drop of sodium bismuthate. A purple spot confirms Mn^{2+}

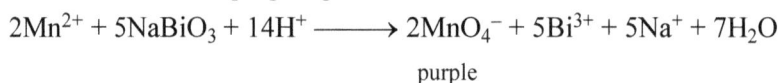

$$2Mn^{2+} + 5NaBiO_3 + 14H^+ \longrightarrow 2MnO_4^- + 5Bi^{3+} + 5Na^+ + 7H_2O$$

purple

Test for Mg²⁺

To 2 mL of the solution is added a pinch of solid NH_4Cl and 1 mL Na_2HPO_4 solution. The test tube on stratching the sides with a glass rod gives a white ppt. confirming Mg^{2+}

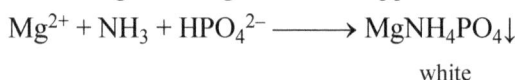

$$Mg^{2+} + NH_3 + HPO_4^{2-} \longrightarrow MgNH_4PO_4\downarrow$$

white

Scheme 1.13. Analysis of group III A cations.

Analysis of group III B cations

The centrifugate obtained in Scheme 1.12 contains group III B cations, which are present as amine complexes, $[Cu(NH_3)_4]^{2+}$, $[Cd(NH_3)_4]^{2+}$, $[Co(NH_3)_6]^{2+}$ and $[Ni(NH_3)_6]^{2+}$. The solution is divided into four parts for testing the presence of Cu^{2+}, Cd^{2+}, Co^{2+} and Ni^{2+}.

Tests for Cu²⁺

A drop of the test solution (acidified with acetic acid) is placed on a filter paper impregnated with rubeanic acid. Appearance of a black or olive green circle confirms Cu^{2+}. In case Co^{2+} and Ni^{2+} are present, their interference can be averted by masking with malonic acid.

Test for Cd²⁺

Treatment of a drop of the test solution with Fe (α, α′-diphy) I_2 gives a red spot. The reagent is prepared by dissolving 0.25 g of α, α′-dipyridyl and 0.15 g. $FeSO_4 . 7H_2O$ in 50 mL water and adding KI solution. The solution is filtered and is used as a reagent for testing Cd^{2+}.

Test for Co²⁺

(*i*) To the test solution is added acetic acid and sat'd solution of KNO_2. Formation of a yellow ppt. confirms Co^{2+}.

(*ii*) Treatment of the test solution with aqueous solution of $Na_2S_2O_3$ gives blue colour due to the formation of $[Co(S_2O_3)_2]^{2-}$. The colour becomes intense by addition of alcohol. The ions Zn^{2+}, Cd^{2+}, Cu^{2+}, Ni^{2+} if present do not cause any interference.

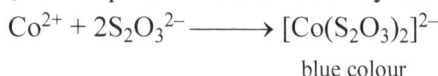

$$Co^{2+} + 2S_2O_3{}^{2-} \longrightarrow [Co(S_2O_3)_2]^{2-}$$
$$\text{blue colour}$$

Test for Ni²⁺

The test solution on treatment with dimethyl glyoxime and making the solution ammonical gives a red ppt. confirming Ni^{2+}. This test is specific for Ni^{2+}

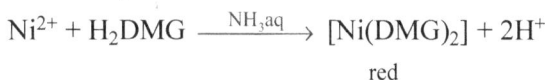

$$Ni^{2+} + H_2DMG \xrightarrow{NH_3aq} [Ni(DMG)_2] + 2H^+$$
$$\text{red}$$

Scheme 1.14. Analysis of group IIIB cations.

1.6.5 Analysis of Group IV Cations

The centrifugate obtained after separation of group III cations (Scheme 1.12) contains group IV cations as hydroxo complexes of Al^{3+}, Zn^{2+}, Sn^{2+}/Sn^{4+} and $CrO_4{}^{2-}$. The centrifugate is divided in four parts and the various cations are individually tested as given below.

Test for CrO₄²⁻

To the test solution is added acetic acid and lead acetate solution. Formation of yellow ppt. confirms $CrO_4{}^{2-}$

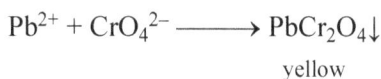

$$Pb^{2+} + CrO_4{}^{2-} \longrightarrow PbCr_2O_4\downarrow$$
$$\text{yellow}$$

Test for Sn²⁺/Sn⁴⁺

To the test solution is added a pinch of magnesium powder and few drops of dil HCl (to reduce Sn^{4+} to Sn^{2+}). A drop of this solution is placed on a filter paper impregnated with 5% phosphomolybdic acid. The filter paper on exposing to NH_3 gives a blue spot confirming Sn^{2+}.

Test for Al³⁺

A drop of the test solution is placed on a filter paper impregnated with $K_4[Fe(CN)_6]$ (to remove Zn^{2+} if present) and alizarin. This is followed by exposing the filter paper to NH_3. Development of red spot confirms Al^{3+}.

Test for Zn²

The test solution is acidified with acetic acid and treated with $K_4[Fe(CN_6)]$ solution. Appearance of a bluish white ppt. confirms Zn^{2+}

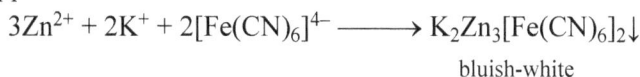

$$3Zn^{2+} + 2K^+ + 2[Fe(CN)_6]^{4-} \longrightarrow K_2Zn_3[Fe(CN)_6]_2\downarrow$$
$$\text{bluish-white}$$

1.7   SPOT TESTS FOR CATIONS

Identification and confirmation of cations can be easily done by spot tests. As the name implies, in spot tests development of a coloured spot occurs and very small amounts of the reagents are used. As such, the spot tests can be described as Eco-friendly tests.

Following are given spot tests of various cations in **alphabatical order**. Some spot tests which have been used in the confirmation of various cations have been described in earlier sections. However, in the present section all spot tests including those described earlier have also been included. The spot tests are performed with the fraction of the precipitate or solution which is supposed to contain a particular cation. This implies that the separation of the cations are first affected into various groups (as described in section 1.4 and or 1.5) and then spot tests performed on various fractions.

1. Aluminium (Al^{3+})

(*i*) **Alizarin reagent (saturated solution of alizarin in alcohol)**

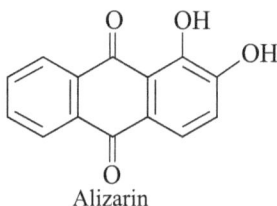

Alizarin

A drop of acidified test solution is placed on a filter paper imprignated with the reagent. The paper is held over ammonia fumes until violet colour (which is due to ammonium alizarinate) appears. On drying the paper at 100°C, the violet colour disappears (due to conversion into ammonia and alizarin) and red colour is visible.

(*ii*) **Quinalizarin reagent (0.05 percent in pyridine)**

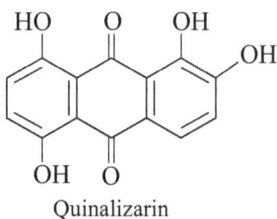

Quinalizarin

A drop of the test solution is placed on a filter paper impregnated with the reagents. It is then held over a bottle containing ammonia solution and then over glacial acetic acid till the blue colour (due to ammonium quinalizarinate) first formed disappears and the paper regains the brown colour of free quinalizarin. Finally a red-violet or red-spot is visible.

(*iii*) **Potassium ferrocyanide reagent**

A drop of test solution is placed on a filter paper impregnated with $K_4[Fe(CN)_6]$ (in order to remove Zn^{2+} if present) and alizarin. Finally the filter paper is exposed to ammonia vapours. A red spot appears.

2. Ammonium (NH_4^+)

(*i*) *p*-Nitrobenzene diazonium chloride reagent (prepared by diazolisation of *p*-nitro aniline with $NaNO_2$ and HCl)

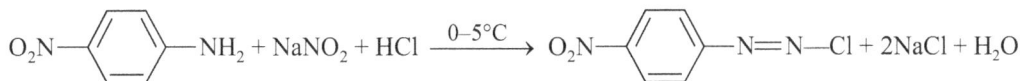

$$O_2N-\!\!\!\!\bigcirc\!\!\!\!-NH_2 + NaNO_2 + HCl \xrightarrow{0-5°C} O_2N-\!\!\!\!\bigcirc\!\!\!\!-N\!\!=\!\!N-Cl + 2NaCl + H_2O$$

A drop of the neutral or slightly acidic test solution is placed on a spot plate followed by a drop of the reagent and a pinch of CaO. A red zone is formed round the CaO.

(*ii*) **Tannic acid-silver nitrate test**

A mixture of 2 drops of 5% tannic acid solution and 2 drops of 20% $AgNO_3$ solution is placed on a reaction paper or cotton wool. The paper or the cotton wool is held in the vapour produced by heating an ammonium salt will NaOH solution. Formation of a black stain is observed in the filter paper or the cotton wool.

3. Antimony (Sb^{3+})

(*i*) **Rhodamine B reagent (0.01% aqueous solution)**

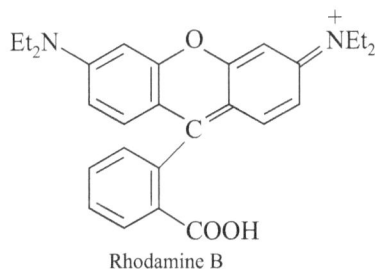

Rhodamine B

A drop of the test solution [which is rendered strongly acidified with HCl and oxidised by the addition of little $NaNO_2$ or KNO_2 [in order ot oxidise Sb(III) to Sb(V)] is placed on a spot plate and to this is added a drop of the reagent. A blue colour appears.

(*ii*) **Phosphomolybdic acid reagent ($H_3[PMo_{12}O_{40}]$)**

A drop of the test solution is placed on a filter paper impregnated with phosphomolybdic acid reagent. Finally, the paper is held in steam. A blue colour appears is 2-3 minutes.

4. Arsenic (As^{3+})

Mercuric bromide reagent. To pure zinc and dilute H_2SO_4 taken in a test tube is added arsenic compound. A filter paper soaked in $HgBr_2$ solution is held over the mouth of the test tube. The paper gets yellow or orange.

5. Barium (Ba^{2+})

Sodium rhodizonate reagent

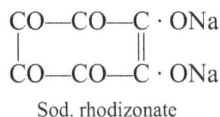

$$\begin{array}{l} CO\!-\!CO\!-\!C \cdot ONa \\ \;|\qquad\qquad\;\| \\ CO\!-\!CO\!-\!C \cdot ONa \end{array}$$

Sod. rhodizonate

A drop of neutral or faintly acidic test solution is placed on a reaction paper. Addition of a drop of the reagent gives a brown or reddish-brown spot. To the reddish-brown spot is added a drop of 0.5 M HCl. A bright-red stain in formed.

6. Bismuth (Bi^{3+})

(*i*) **Cinchonine-potassium iodide reagent (1 percent).** A drop of the slightly acidic test solution is placed on a moistened piece of drop reaction paper with the reagent. An orange red spot is obtained.

(*ii*) **Thiourea reagent (10 percent aqueous solution).** To a drop of the test solution taken on a spot plate is added one drop of HNO_3 followed by a drop of the thiourea reagent solution. A yellow colour is obtained.

7. Cadmium (Cd^{2+})

(*i*) Diphenyl carbazide reagent (0.1 percent alcoholic solution).

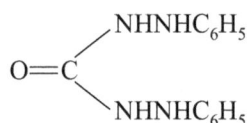

$$O\!\!=\!\!C\underset{\diagdown NHNHC_6H_5}{\overset{\diagup NHNHC_6H_5}{}}$$

To a drop of the test solution taken on a filter paper is added a drop of the reagent. The filter paper on exposing to ammonia gives a red spot.

(*ii*) Fe (α, α'-dipyridyl) I_2 reagent (obtained by dissolving 0.25 g of α, α'-dipyridyl and 0.15 g $FeSO_4$. $7H_2O$ in water (50 mL) and adding KI solution. The deep red solution is filtered if necessary).

To a drop of the test solution taken on a filter paper is added a drop of the reagent. A red spot is developed. Cu^{2+}, Co^{2+} and Ni^{2+} do not interfere is ammonical medium.

(*iii*) 4-Nitronaphthalene-diazmino-azobenzene reagent (0.02 % in alcohol).

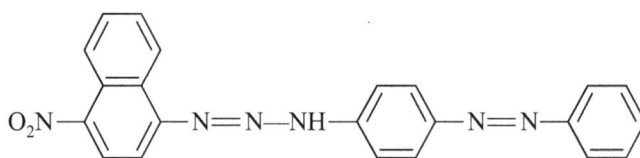

To a drop of the reagent placed on a drop-reaction paper is added one drop of the test solution (which should be slightly acidic (2M CH_3COOH) containing a little sodium potassium tartarate). This is followed by the addition of one drop of 2 M KOH solution. A bright-pink spot, surrounded by a blue circle is produced.

8. Calcium (Ca^{2+})

(*i*) Sodium dihydroxy tartarate osazone [sodium salt of succinic acid-dioxo-2, 3-bis (phenyl hydrazone)] reagent.

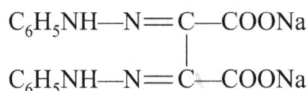

$$C_6H_5NH\!\!-\!\!N\!\!=\!\!C\!\!-\!\!COONa$$
$$C_6H_5NH\!\!-\!\!N\!\!=\!\!C\!\!-\!\!COONa$$

To a drop of the neutral test solution placed on a black watch glass is added a minute fragment of the solid reagent. A white film is formed over the surface of the liquid which separation as a precipitate.

(*ii*) Picrolonic acid (1-*p*-nitrophenyl-3-methyl-4-nitro-5-pyrazoline) reagent.

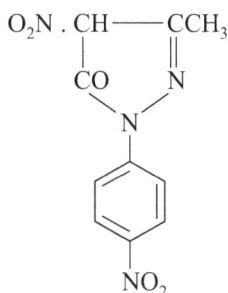

Picrolonic acid

A neutral or acidic test solution is placed on a warm spot plates and a drop of saturated solution of picrolonic acid added. Characteristic rectangular crystals are produced.

9. Chromium (Cr^{3+})

(*i*) **1, 5-Diphenyl carbazide reagent.** To a drop of the acidified test solution (with mineral acid) placed on a spot plate is added 2 drops of 0.1 M potassium peroxodisulphate solution and one drop of 0.1 M AgNO$_3$ solution. After 2-3 min a drop of the reagent is added. A violet or red colour is produced due to the formation of a complex (as shown below):

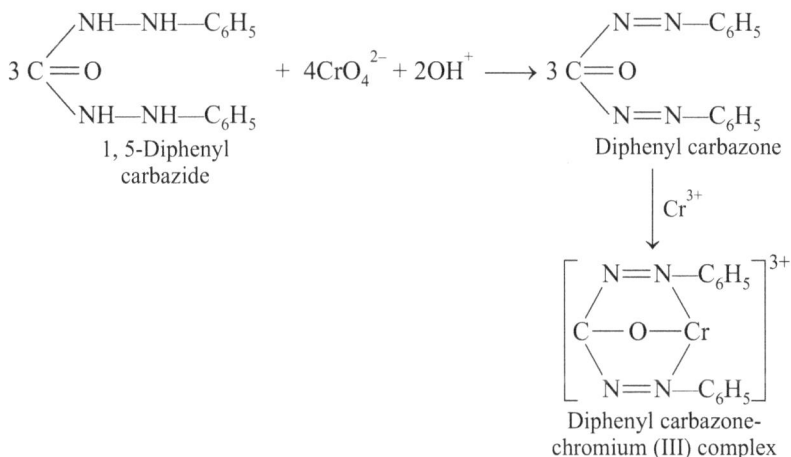

1, 5-Diphenyl carbazide

Diphenyl carbazone

Diphenyl carbazone-chromium (III) complex

(*ii*) Chromatropic acid reagent (sodium salt of 1,8-dihydroxy naphthalene-3,6-disulphonic acid)

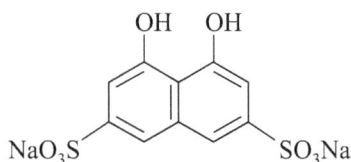

To a drop of the test solution, placed in a semimicro test tube is added a drop of the reagent, a drop of dil HNO_3 (8 M). The resultant solution on dilution with water (2 mL) gives a red colouration (observed against a white background).

10. Cobalt (Co^{2+})

(*i*) **Ammonium thiocyanate reagent.** To a drop of the test solution placed on a spot plate is added 4-5 drops of the reagent in acetone. A green colouration appears.

(*ii*) **1-Nitroso-2-naphthol reagent.** To a drop of the acidic test solution placed on a spot plate is added a drops of the reagent. A brown spot is produced.

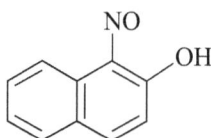

1-Nitroso-2-naphthol

(*iii*) **Sodium 1-nitroso-2-hydroxynaphthalene-3, 6-disulphonate reagent.**

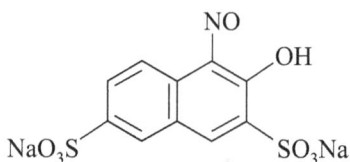

A drop of the neutral test solution (buffered with sodium acetate) is placed on a spot plate. Subsequent addition of 2-3 drops of the reagent gives a red colour. In this test tin and iron interfere and should be removed before carrying out the spot test.

(*iv*) **Rubeanic acid reagent (1 percent solution in ethanol)**

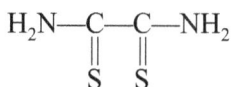

The filter paper having a drop of the test solution is exposed to ammonia vapour. This is followed by adding a drop of the reagent solution. A brown spot is produced.

11. Copper (Cu^{2+})

(*i*) **Benzoin-α-oxime (or cupron) reagent (5 percent in alcohol)**

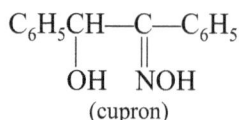

(cupron)

The filter paper having a drop of the weakly acidic test solution and a drop of the reagent on exposing to ammonia gives a green colouration.

In case, ions which can be precipitated with ammonia are present, a drop of 10 percent potassium sodium tartarate solution is placed on the filter paper before the reagent is added.

(*ii*) **Salicylaldehyde oxime (salicylaldoxime) (1 percent solution in alcohol)**

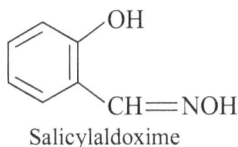
Salicylaldoxime

To a drop of the neutral test solution is added 1 drop of dilute acetic acid followed by addition of a drop of the reagent. A yellow-green colour is produced.

(*iii*) **Dithio-oxamide (or rubeanic acid) (0.5% solution)**

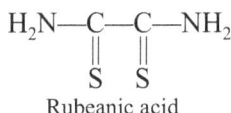
Rubeanic acid

A drop of the neutral test solution placed on a drop-reaction paper is exposed to ammonia vapour. Subsequent addition of a drop of the reagent gives a black or greenish spot.

Alternatively, a drop of the test solution acidified with CH_3COOH is placed on a filter paper impregnated with rubeanic acid. A black or olive green circle confirms Cu^{2+}. Co^{2+} and Ni^{2+} if present interfere. If present, the interference is averted by masking with malonic acid.

12. Iron (Fe^{3+})

(*i*) **2, 2′-Bipyridyl reagent**

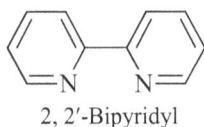
2, 2′-Bipyridyl

A drop of the test solution (faintey acidic) is placed on a filter paper which has been impregnated with the reagent (and dried). A pink or red spot is produced.

(*ii*) **1, 10-Phenanthroline reagent**

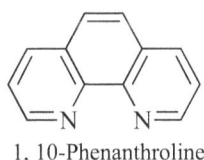
1, 10-Phenanthroline

To a drop of the test solution placed on a spot plate is added is drop of the reagent. A red colour is obtained.

(*iii*) **Ammonium thiocyanate solution**

A drop of the test solution is put on a spot plate and a drop of 0.1 M ammonium thiocyanate solution added. A deep-red colour appears.

(*iv*) **8-Hydroxy-7-iodoquinoline-5-sulphonic acid (Ferron) reagent**

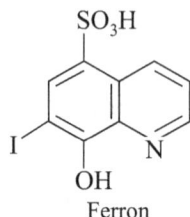
Ferron

A drop of the slightly acidic test solution is placed on a groove tile. Addition of a drop of the reagent gives a green colour.

(*v*) **Potassium ferrocyanide, $K_4[Fe(CN)_6]$**

A drop of the test solution placed on a groove tile on treatment with a drop of the solution of potassium ferrocynamide produces blue colour.

13. Lead (Pb^{2+})

(*i*) **Gallocyanine reagent (1 percent)**

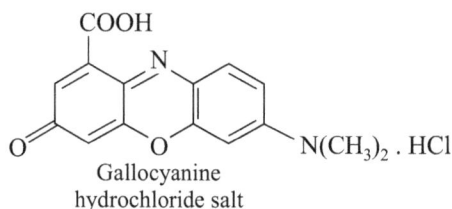
Gallocyanine
hydrochloride salt

To a drop of the test solution placed on a drop reaction paper is added 1 percent aqueous pyridine and 1 percent gallocyanine reagent (blue). Another filter paper is placed below the drop reaction paper and 2-3 drops of pyridine solution added. The blue colour of the reagent percolates to the filter paper kept below. A deep violet spot appears on the drop reaction paper.

Alternatively, if silver, bismuth, cadmium or copper is present the test is performed by transferring a drop of the test solution to a drop reaction paper and adding a drop of 1M H_2SO_4 in order to fix lead as $PbSO_4$. The paper is washed again with 3-4 drops of 1M H_2SO_4 followed by a little 95% ethanol. The drop reaction paper is dried and then the gallocyanine test performed as given above.

(*ii*) **Diphenylcarbazone (or Dithiazone) reagent (0.005 percent)**

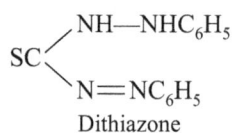
Dithiazone

To 0.5 mL of the neutral or faintey alkaline test solution taken on a groove tile is added 1-2 crystals of KCN (deadly poison) and then 2 drops of the reagent. The green colour of the reagent changes to red in 1-2 minutes (the mixture is throughly mixed with a glass rod).

(*iii*) **Benzidine reagent (0.05% solution in 10% acetic acid).**

To a drop of the test solution taken on a filter paper is added 2 drops of dilute NaOH solution followed by a drop of bromine water and 2 drops of dilute ammonia (1 : 1). The spot paper is dried (in air) and 2 drops of the reagent solution added. Blue colour confirms Pb^{2+}.

(*iv*) **Rhodizonate solution**

A drop of the test solution is placed on a filter paper followed by addition of freshly prepared sodium rhodizonate solution on the moist fleck. A scarlet spot is obtained.

14. Magnesium (Mg^{2+})

(*i*) **4-(4-Nitrophenylazo)-2-resorcinol (Magneson I) reagent (0.5 percent).**

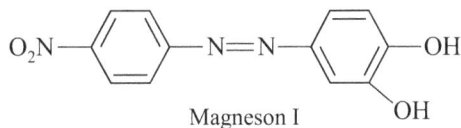

Magneson I

To a drop of the strongly alkaline test solution (by addition of 1 drop of 2 M NaOH solution) is added 1-2 drops of the reagent. The reddish-violet reagent attains a blue colour.

In an alternative method, instead of magneson I, magneson (II) (4-nitrobenzeneazo-1-naphthol) is used.

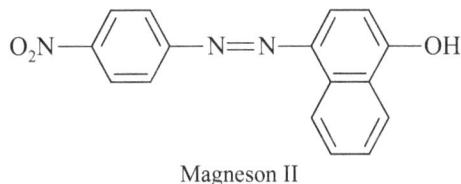

Magneson II

The procedure used is similar to that used for magneson I.

(*ii*) **Titan yellow reagent (also known as clayton yellow reagent)**

To a drop of the test solution placed on a spot plate is added a drop of the reagent (aqueous solution) and a drop of 2 M NaOH. A red colour is produced.

15. Manganese (Mn^{2+})

Sodium bismutate ($NaBiO_3$). To a drop of the test solution placed on a spot plate is added a drop of conc. HNO_3 followed by a drop of sodium bismutate solution. Purple colour of permanganate appears.

16. Mercury (Hg_2^{2+})

(*i*) **1, 5-Diphenyl carbazide (1% in alcohol).** To a filter paper impregnated with freshly prepared reagent is added a drop of 0.4 M HNO_3 followed by a drop of the test solution. A violet colour is observed.

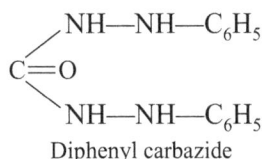

Diphenyl carbazide

(*ii*) **Cobalt (II) thiocyanate.** To a drop of the test solution placed on a spot plate is added a small crystal of ammonium thiocyanate followed by a little of solid cobalt (II) acetate. A blue colour is produced.

The presence of mercuric (Hg^{2+}) is detected by taking 2 drops of the test solution on a filter paper followed by the addition of a drop of dil. HNO_3 and a drop of diphenyl carbazide reagent (a cold saturated solution of diphenyl carbazide and a few crystals of KI). A violet blue colour is produced.

17. Nickel (Ni^{2+})

(*i*) **Dimethyl glyoxime reagent.** To a drop of the test solution placed on a spot plate is added a drop of the reagent followed by addition of a drop of dilute ammonia solution. A red colouration is produced.

Red

(*ii*) **Rubeanic acid reagent (1 percent solution in alcohol).** A drop of the test solution is placed on a filter paper and exposed to ammonia vapour. This is followed by addition of a drop of the reagent. Blue or blue-violet spot is produced.

18. Potassium

Sodium cobaltinitrite-silver nitrate reagent. To 2 drops of the test solution placed on a spot plate is added a drops of CH_3COOH followed by addition of a drop of $AgNO_3$ solution. To the resulting spot is added a tiny amount of powdered sodium cobaltinitrite. A yellow colour is observed.

In case NH_4^+ is also present a few drops of phenolphthalein indicator is added to the test solution, NH_4^+ is decomposed by adding few drops of formaldehyde (40%) followed by the additional Na_2CO_3 solution. The solution is warmed and the ppt. is rejected. To the filtrate is added till the indicator is colourless. Finally test for K^+ with $Na_3[Co(NO_2)_6]$ as described above.

19. Silver (Ag^+)

(*i*) **Potassium chromate in neutral solution.** To a drop of the test solution placed on a spot plate is added a drop of ammonium carbonate solution and mixed, this renders any Hg(I) or Pb^{2+} ions (if present unreactive by precipitation as highly insoluble carbonates). To a drop of the clear liquid is added a drop of potassium chromate. A red ring of silver chromate is observed.

(*ii*) **Manganese nitrate reagent.** To a drop of 0.1 N HCl placed on a filter paper is added a drop of the test solution. This is followed by addition of one more drop of HCl. Subsequent addition of a drop of 0.1 N $Mn(NO_3)_2$ solution and a drop of 0.1 N NaOH produces blackening of the fleck.

(*iii*) **4-Dimethylaminobenzylidine rhodanine (0.3 percent solution in acetone)**

To a drop of the test solution taken in a spot plate is added a drop of 2 M HNO_3 followed by one drop of the reagent. A reddish-violet colour appears.

20. Strontium (Sr^{2+})

Sodium rhodizonate reagent. To a drop of the neutral test solution placed on a drop reaction paper or an a spot plate is added a drop of the reagent. A brownish-red colouration is produced. This test is possible only if Ba^{2+} is absent.

In case Ba^{2+} is present, the spot test is conducted as follows. On the impregnated filter paper with sat'd K_2CrO_4 solution (and dried), a drop of the test solution is placed. After a minutes a drop of the reagents placed on the moistened spot. A brownish-red spot or ring becomes visible.

21. Tin (Sn^{2+})

(*i*) **Phosphomolybdic acid reagent (5% solution)**

To the test solution (1 mL) is added a pinch of Mg powder a few drops of dil HCl in order to reduce Sn^{4+} to Sn^{2+}. A drop of this solution is placed on a filter paper impregnated with 5 percent phosphomolybdic acid. This is followed by exposing the filter paper to ammonia. A blue spot confirms Sn^{2+}.

(*ii*) **Cacotheline reagent (nitro derivative of brucine, $C_{21}H_{21}O_7N_3$)**

To a filter paper impregnated with the reagent (and dried) is added a drop of the test solution. A violet colour surrounded by a less coloured zone in observed (**Caution:** The reagent is highly toxic).

(*iii*) Diazine green reagent (dyestuff formed by the coupling diazotized safranin with dimethyl aniline).

A drop of the test solution is placed on a spot plate and mixed with 1 mL of the reagent. The colour changes from blue to violet.

(*iv*) **Dimethyl glyoxime reagent (1 percent in EtOH).** To a drops of the test solution placed on a groove tile is added dilute HCl followed by addition of 3 drops of 5% tartaric acid and 3 drops of the reagent solution and 4-5 drops of ammonia solution. A red colour is seen.

22. Zinc (Zn^{2+})

(*i*) Ammonium tetracyanatomercurate (II)–copper sulphate reagent.

To a drop of the test solution (acidified with dil. H_2SO_4) placed on a spot plate is added a drop of 0.25 M $CuSO_4$ solution and one drop of ammonium tetracyanatomercurate (II) reagent. A violet (or blackish-purple) colour appears.

(*ii*) Potassium cobalticyanide reagent (Rinmannas reagent) (Prepared by dissolving 2g of potassium cobalticyanide and 0.5 g $KClO_3$ in 50 mL distilled water)

To a filter paper imprignated with the test reagent (and dried at room temperature) is placed 3-4 drops of the neutral test solution. The filter paper is dried (over a flame) and ignited in a crucible. A green coloured ash (observed against a white background) is obtained.

1.8   DETECTION OF METALS (CATIONS) BY FLAME PHOTOMETRY

Introduction. Most of the laboratories are equipped with modern instruments for flame photometry, also known as F**flame Atomic Emission Spectrometry (FAES).** Using this technique, detection of metals (cations) can be carried out in much shorter time. The only draw back is that by this technique no information can be obtained about the oxidation state of the metals.

The flame atomic emission spectrometry (FAES) is based on the emission of radiation when a metal is aspirated into a flame. Each metal emits characteristic radiation. As will be discussed subsequently, the intensity of radiation is also helpful in knowing the concentration of the metal aspirated into the flame.

Principle of FAES

In FAES, the solution of the metallic salt is aspirated into the flame. Following sequence of events take place.

- The solvent in the sample gets evaporated giving a residue.
- The residue gets decomposed into neutral atoms.
- The neutral atoms get excited to form excited atoms, which are unstable.
- The unstable excited atoms revert back to the ground state; in this process emission of radiation takes place.
- The wavelength (and also the intensity) of the emission radiation (which is characteristic of the metal) is measured.

The radiation emitted ($h\nu$) is given by the expression

$$h\nu = E_2 - E_1 \qquad \qquad ...(i)$$

where E_2 is the energy of higher energy level (excited state) and E_1 is the lower energy level (ground state).

h is Plancks constant and ν is the frequency of the light emitted, which is given by the expression

$$\nu = \frac{c}{\lambda} \qquad \qquad ...(ii)$$

The equations (i) and (ii) on combining gives

$$E_2 - E_1 = \frac{hc}{\lambda}$$

$$\therefore \qquad \qquad \lambda = \frac{hc}{E_2 - E_1} \qquad \qquad ...(iii)$$

The above equation (iii) is helpful to calculate the wavelength of the emitted radiation which is characteristic of the atoms of an element. The intensity of the emitted radiation helps to find the amount or the concentration of the metal present. This aspect forms the basis of subsequent discussion (*see Chapter 6*).

It is found that a part of the neutral atom (or radical) formed in the flame may combine resulting in the formation of a new gaseous compound.

The formation of new gaseous product results in the decrease in the concentration of neutral atoms thereby reducing the intensity of the emitted radiation. This is called **chemical interference** in FAES.

The fraction of the free atoms that are thermally excited is given by Boltzmann distribution as given below:

$$\frac{N^*}{N} = Ae^{-DE/kT}$$

...(iv)

where N^* and N are the number of excited atoms and the number of atoms remaining is the ground state respectively. A is a constant for a particular element, ΔE is the difference in energies of the two levels (excited state and ground state), K is Boltzmann constant and T is the temperature of the flame.

It is thus, evident [equation (iv)] that the fraction of the excited atoms is dependent on the temperature of the flame, which depends on the type of the fuel and the oxidant used (Table 1.15).

Table 1.15. *Flame temperature using a combination of fuel and oxidant*

Fuel	Oxidant	Flame temp. (°C)
Hydrogen	Argon (Ar)	1600
	Air	2100
	O_2	2800
Acetylene	Air	2200
	O_2	3000
	N_2O	3000
Propane	Air	1900
	O_2	2800

Instrumentation

Different types of FAES instruments are in use. Two of such instruments are simple flame photometer and internal standard flame photometer. For finding the metal (or cation) present, the former type, *viz.*, simple flame photometer is used. The second type (*viz.* internal standard flame photometer) as we will discuss subsequently (Chapter 6) is useful for determining the concentration or the amount of metal (cation) present.

Simple FAES instrument

A diagrammatic representation of a simple FAES instrument is shown in the (Fig. 1.1).

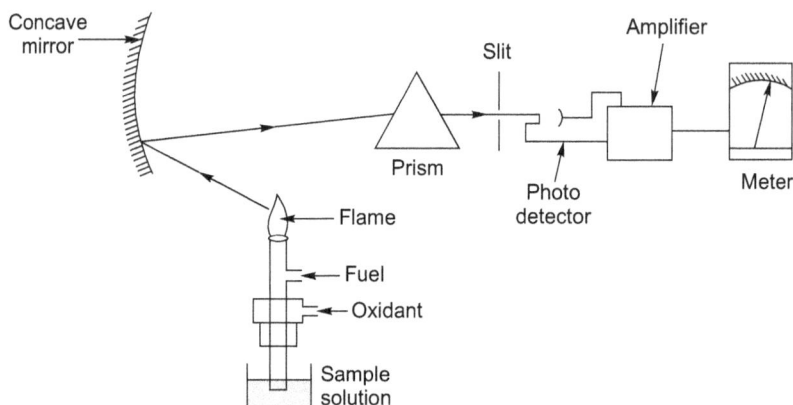

Fig. 1.1. Diagramatic representation of a simple FAES instrument.

As seen the FAES instrument consists of a burner which is a total consumption type (however, other types of burners can also be used), a collimating concave mirror (used behind the flame in order to increase the intensity of emission). The solution of the sample to be analysed is sucked by an automiser which is operated by one of the flame producing gases (fuel or the oxidant) and is aspirated into the flame as a fine spray. During the process of combustion in the flame spectral emission comes from the excited atoms formed. The radiation emitted from the flame is collected by the collimating concave mirror and this passes through a prism and a slit. The radiation of the selected wavelength on striking a photodetector produce electrical signal which is noted from the meter.

Following are given the wavelength of the radiation (which are characteristic of a particular element and the flame composition used).

Table 1.16 *Wavelength of emitted radiation*

Metal	Wavelength (nm) of emitted radiation	Fuel-mixture used for the flame
1. Aluminium[a]	396.15	acetylene-Nitrous oxide
2. Antimony	607.90	acetylene-air
3. Ammonium	—	—
4. Arsenic	—	—
5. Barium[b]	455.4	acetylene-air
6. Bismuth[c]	223.06	acetylene-air
7. Cadmium[b]	455.4	acetylene-air

8. Calcium	422.67	acetylene-air
9. Chromium[d]	425.44	acetylene-air
10. Cobalt	345.35	acetylene-nitrous oxide
11. Copper[e]	324.75	acetylene-air
12. Iron[e]	371.99	acetylene-air
13. Lead	405.78	acetylene-nitrous oxide
14. Magnesium	285.21	acetylene-air
15. Manganese[d]	403.08	acetylene-air
16. Mercury	—	—
17. Nickel	341.48	acetylene-air
18. Potassium	766.49	acetylene-air
19. Silver[f]	328.07	acetylene-nitrous oxide
20. Strontium	460.73	acetylene-air
21. Tin	284.00	acetylene-nitrous oxide
22. Zinc	213.86	acetylene-oxygen
23. Sodium[g]	589.00 589.59	acetylene-oxygen

Notes:

(a) The use of acetylene-air flame reduces the sensitivity. The presence of some metals like iron interfere.

(b) By using acetylene-air flame, some ions like PO_4^{3-}, SO_4^{2-}, Al^{3+} result in decrease of the emission and absorption signal. This interference can be overcome by addition of $LaCl_3.7H_2O$ to the solution to be tested.

(c) The test is comparatively more sensitive if acetylene nitrous oxide flame is used.

(d) Use of acetylene-nitrous oxide flames increases the sensitivity ten fold.

(e) The test is more sensitive using acetylene-nitrous oxide flame.

(f) Silver cannot be easily detected using acetylene-air flame.

(g) Sodium is generally not given in the mixture to the analysed. However, the absorption of emitted radiation in FAES is given.

1.9 DETECTION OF METALS (CATIONS) BY ATOMIC ABSORPTION SPECTROSCOPY

Atomic Absorption Spectroscopy (AAS) is useful for the detection (and also for the estimation) of metallic elements in solution. The procedure involves conversion of the sample to the atomic vapour followed by the measurement of absorption of atomic vapour at some selected wavelength, which is characteristic of each individual element. In this technique the element to be determined is first reduced to the elemental state followed by vapourisation and imposing the vapourised metal in the beam from the source. The procedure is specific as the atoms of only a particular element can absorb radiation of its characteristic wavelength. Thus, a metal can be detected (and also estimated) in presence of other metals. The method is very sensitive and even trace elements can be determined in as low as 1 ppm. However, this method cannot be used for elements like Ti, W, Mo, V etc. as oxides of these metals are formed is the flame.

When light of a particular wavelength (which is specific to the element under study) passes through an atom (in the flame), part of the energy is absorbed. In this process, the atom changes from low energy state to higher energy state.

The total amount of energy absorbed by the atoms is given by the equation

$$ n = \frac{pe^2}{mc} \times Nf $$

where e is the electronic charge, m is the mass of the electron, c is the velocity of light, N is the total number of atoms that absorb a frequency v in the light and f is the oscillator strength or ability of each atom to absorb at frequency v. As π, e, m and c are constants, the above equation can be expressed in a simple form.

Total amount of energy absorbed = constant $\times N \times f$.

As seen from the above simple expression, the absorption by an atom is independent of the wavelength of absorption and temperature of the atom. It is due to this reason, the AAS has an advantage over flame photometry (Section 1.8).

AAS Instrument

The AAS instrument consists of a glass tube containing an inert gas 'Argon' at several mm pressure, an anode and a hollow cathode. The inside surface of the hollow cathode is coated with the metal under investigation (Fig. 1.2.).

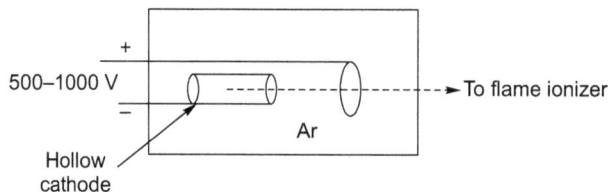

Fig. 1.2. Anode and cathode in a glass tube.

High potential is applied across the electrodes resulting in current of several milliamperes flowing through the lamp. The argon gas is converted into Ar^+ ions (inside the tube) which impringe on the cathode with high energy, leading to sputtering of metal atoms from the inner surface of the cathode. These energised metal atoms emit radiations having a very narrow wavelength, whch are characteristic of the metal. Finally, the radiations from the hollow cathode lamp pass through a flame automizer (Fig. 1.3)

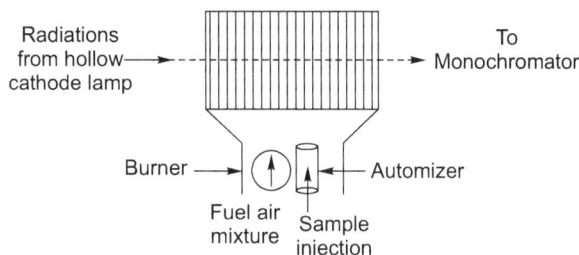

Fig. 1.3. Flame automizer.

In the flame automizer, the metal (to be analysed) in the form of solution is aspirated. The solution gets converted into the metal, which gets decomposed in the flame and is reduced to the elemental state forming cloy of atoms, which in turn absorb a fraction of the radiation in the flame. In case, a flame automizer is used, the emission by the metal in the flame is at the same wavelength as the absorption wavelength of the metal. This is because same electronic transitions is involved. The resultant energy is passed through a monochromator (which can be a prism or a gratings) to eliminate any extraneous light resulting from the flame) and finally to the detector and recorder (commonly used detector is photomultiplier type). A diagrammatic representation of an atomic absorption spectrometer is shown below (Fig. 1.4).

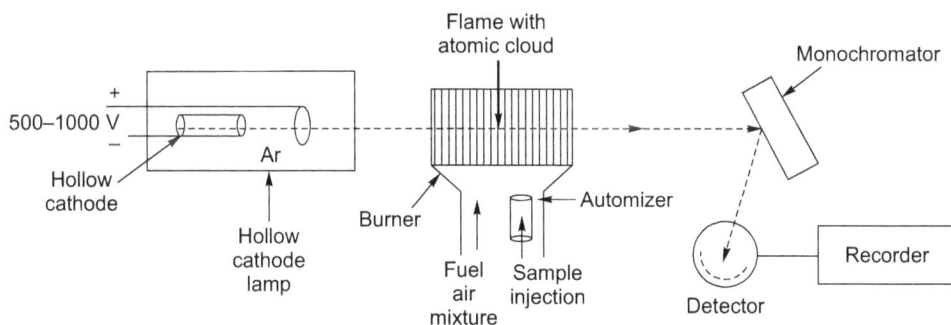

Fig. 1.4. Diagramatic representation of AAS showing various components.

The common fuel used in the burner of an AAS instrument are acetylene-air (temp. 2800°C), acetylene-oxygen or nitrous oxide (temp. 3000°C) and acetylene-air (temp. 2200°C). Depending on the metal to be analysed (or estimated) the temperature in the owen is adjusted accordingly.

It has been found that much better results are obtained by using a graphite furnace in place of a flame automizer and the sample holder. The graphite furnace has a hallow graphite cylinder, which is placed is such a way that the beam of light from the hallow cathode lamp passes through it. The sample (small amount) is injected in the tube through a small hole at the top.

When an electric current is passed through the tube the metal sample is obtained in dry state and the absorption of the metal recorded as a single peak. The (Fig. 1.5) below shows a graphite furnace and other components of the spectrometer.

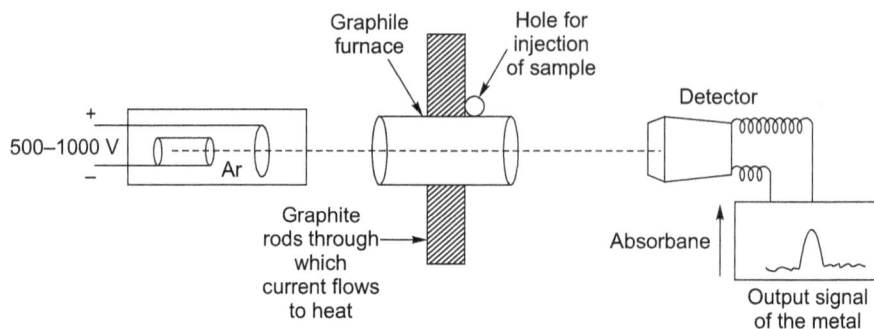

Fig. 1.5. Atomic absorption spectrometer using graphite furnace.

The graphite furnace is much superior compared to the flame detector and can detect various elements even in minute amounts.

Following are given the wavelengths of emitted radiation of various metals (in alphabatical order).

Table 1.17. *Wavelengths of emitted radiation in AAS*

Metal	Coating in the inside of the hollow cathode lamp	Flame used	Wavelength (nm) of emitted radiation
1. Aluminium	Aluminium	acetylene-nitrous oxide	396.15
2. Ammonium	—	—	—
3. Antimony	Antimony	acetylene-air	200.83
4. Arsenic	—	—	—
5. Barium	Barium	acetylene-air	553.55
6. Bismuth	Bismuth	acetylene-air	223.06
7. Cadmium	Cadmium	acetylene-air	228.80
8. Calcium	Calcium	acetylene-air	422.67
9. Chromium[a]	Chromium	acetylene-air	357.87
10. Cobalt	Cobalt	acetylene-air	240.73
11. Copper[b]	Copper or brass	acetylene-air	324.75
12. Iron	Iron	acetylene-air	248.33
13. Lead	Lead	acetylene-air	283.31
14. Magnesium	Magnesium	acetylene-air	285.21
15. Manganese	Manganese	acetylene-air	279.48
16. Mercury[c]	Mercury	see note C	253.7
17. Nickel	Nickel	acetylene-air	341.48
18. Potassium	Potassium	acetylene-air	766.49
19. Silver	Silver	acetylene-air	328.07
20. Strontium	Strontium	acetylene-air	460.73
21. Tin	Tin	acetylene-nitrous oxide	224.61
22. Zinc	Zinc or brass	acetylene-air	213.86

Notes:

(*a*) Better sensitivity is obtained by using acetylene-nitrous oxide flame and the signal is less prone to interference by phosphate and transition metals.

(*b*) The test is more sensitive in the absence of transition metals.

(*c*) Mercury an extremely toxic metal.

Its intake is responsible for various neurological disorders ultimately leading to death. Following steps are involved in testing the presence of mercury in water:

(*i*) To water sample (100 mL) taken in a 500 mL distillation flask is added 5-10 mL 5 percent $KMnO_4$ solution. The solution is refluxed (4 hr.) using cold water for circulation in reflux condenser. The solution is cooled and 4-5 mL of 30 percent H_2O_2 added and solution boiled to remove excess H_2O_2. The volume of the solution is made upto 250 mL.

(*ii*) In the mean time 20 percent $SnCl_2$ in 10 mL conc. HCl and 88 mL 10% HNO_3 is added to the reaction vessel (*B*). The solution is stirred using a magnetic stirrer. The Hg vapour in the reagent is pumped through 3 mL 20% NaOH and 50% H_2SO_4 (vessels *C* and *D* respectively).

(*iii*) Air is bubbled through vessel A and this pumps Hg vapour through 20% NaOH (vessel *C*), 50% H_2SO_4 (vessel *D*) and then to the absorption cell.

(*iv*) The absorbance is measured at 253.7 nm using a mercury hollow cathode lamp as the light source.

(*v*) The resulting Hg vapours are absorbed in 1% $KMnO_4$ – 10% H_2SO_4 in vessel *E*.

All the above steps are shown in (Fig. 1.6) below.

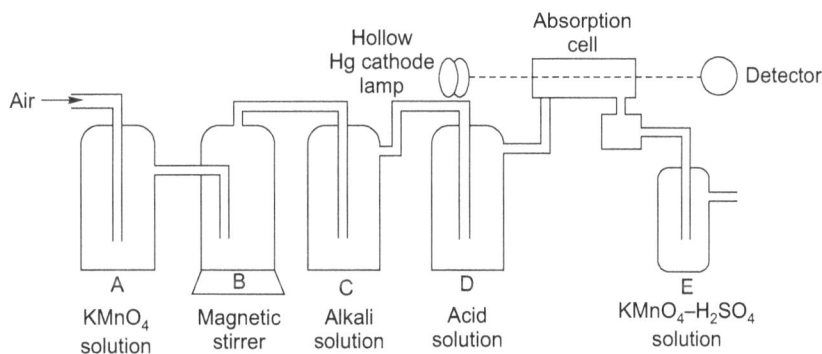

Fig. 1.6. Set up the detection of mercury.

1.10 CHROMATOGRAPHIC SEPARATION AND IDENTIFICATION OF CATIONS

The term chromatography was coined by a Russian scientist Tswett (1966), who successfully separated different coloured constituents of leaves by extraction using a suitable solvent and passing through a column of calcium carbonate, alumina and sucrose. The name chromatography was derived from the greek word chromon meaning colour. As per the International Union of Pure and

Applied Chemistry (IUPAC), Chromatography is defined as a physical method of separation of components of a mixture by distribution between two phases, *viz.*, a stationary phase and a mobile phase.

Chromatography is used for the purification of compounds, identification of the components of a mixture and their separation. A special advantage of chromatography is that a very small amount (few milligram) of a mixture can be separated which is not possible by other methods.

Principle of Chromatographic Separation

A mixture is separated into its components by using a stationary phase (which may be a solid or liquid), which adsorbs the mixture and a mobile phase (which may be a liquid or gas), which on passing through the stationary phase transports the components of the mixture on to it. The components of the mixture are transported at different rates. The strongly adsorbed component of the mixture moves slowly with the moving phase while the weakly adsorbed components move faster along the moving phase. As the mobile phase passes through the stationary phase, differently adsorbed components are separated.

Types of Chromatography

Chromatography are of different types depending on the process of either adsorption or partition. Different type of chromatography include partition chromatography, adsorption chromatography, exclusion chromatography and ion-exchange chromatography.

In the present section, we are mainly concerned with partition chromatography which has been used for the separation and identification of various cations. In partition chromatography, the mixture is separated into the constitutes by partition between a stationary phase and a moving solvent (mobile phase). If the moving phase is a liquid, the chromatography is called liquid-liquid chromatography. On the other hand, if the moving phase is a gas, the chromatgraphy is called gas-liquid chromatography. We are at present mainly concerned with liquid-liquid chromatography, which is paper chromatography, solid-liquid chromatography like thin layer chromatography (TLC) or column chromatography (see Table 1.18).

Table 1.18. *Different types of partition chromatography*

Type of chromatography	Stationary phase	Mobile phase	Principle
Paper chromatography	Liquid	Liquid	Partition
Thin layer chromatography (TLC)	Solid	Liquid	Adsorption
Column chromatography	Solid	Liquid	Partition or adsorption

Paper Chromatography

Paper chromatography is the simplest and most widely used technique for the separation and identification of various cations. In paper chromatography, the stationary phase is liquid (*i.e.*, water adsorbed on cellulose of Whatman filter paper) and the moving phase (known as irrigant) is also a liquid, which may be single solvent or a mixture of solvents. Paper chromatography as has already been stated is a type of partition chromatography (partition between two liquid phases). The components of the mixture to be separated travel at different rates and appear as spots at different points on the Whatmann filter paper. The paper chromatography is of three types. These are circular (or radial) paper chromatography, ascending paper chromatography and descending paper chromatography.

In this section, we are concerned mainly with ascending paper chromatography which is used for the separation and identification of cations.

Ascending Paper Chromatography

In ascending paper chromatography, a strip (3 × 20 cm) of Whattmann filter paper is taken. Small spots of the solution of the mixture are put at a distance of about 2 cm above the edge of the paper (using a capilliary tube) [Fig. 1.7(a)].

A suitably selected solvent (irrigant) is placed in a glass jar, which has a small hook in the lid. The spotted filter paper is dried in air and suspended in the glass jar with the help of the hook. Only the lower edge of the strip of paper dips in the solvent but the spots of the mixture remain above the solvent [Fig. 1.7(b)]. The solvent rises up by capillary action on the filter paper strip and during this process the components of the mixture move up at different rates. When the solvent front has reached near the top of the strip, the strip is taken out, solvent front marked with a pencil and the paper strip air dried.

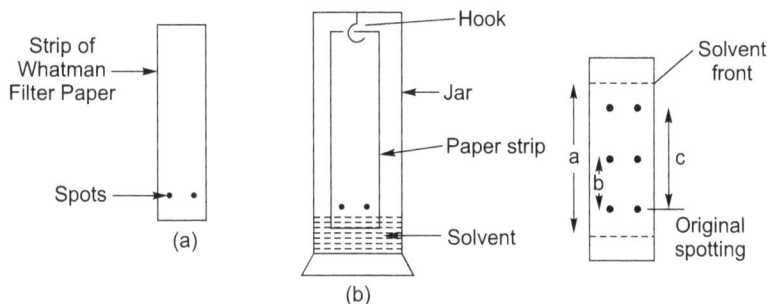

Fig. 1.7. Ascending paper chromatography.

Finally, the air-dried paper strip is developed using a suitable developer. The developer to be used depends on the nature of the cation (the developer should be such that it produces a coloured spot of different cations).

The R_f value of each spot in the chromatogram is calculated by measuring the distance travelled by the cation and the solvent from the origin [see Fig. 1.17(c)].

$$R_f = \frac{\text{Distance travelled by the cation}}{\text{Distance travelled by the solvent}}$$

The R_f value is less than 1.

Thin Layer Chromatography

Besides paper chromatography, thin layer chromatography is also useful for the separation and identification of cations. Thin layer chromatography is an adsorption type of chromatography. The stationary phase or the adsorbent in TLC is a thin layer of an adsorbent (mostly silica gel) coated on a glass plate. The strength of silica gel layer is increased by addition of a small amount of $CaSO_4$. The thickness of the adsorbent is mostly in the range 0.15 to 2.0 mm.

Principle of TLC Separation

The mixture to be separated is adsorbed on to the adsorbent. The mobile phase, as it rises above on the TLC plate transports the components of the mixture on the TLC plate. The components of the mixture are transported at different rates. The weakly adsorbed component (as expected) moves faster with the moving phase and the strongly adsorbed component moves slowly. So the separation of the mixture depends on the selective adsorption on to the stationary phase.

Preparation of TLC Plate

Preparation of TLC plate is the most important part of TLC separation technique. The separation of the components depends on the quality of the TLC plate. Normally 20 × 5 cm size glass plates are used. These are properly cleaned and a layer of silica gel spread on it by using a easily assembled apparatus as shown in the (Fig. 1.8) below.

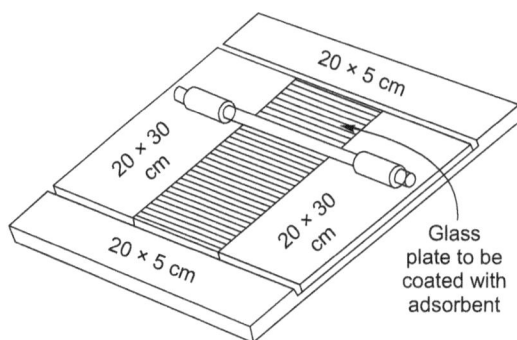

Fig. 1.8. Apparatus for making TLC Plates.

Procedure for TLC Separation

The separation of a mixture into its components involves the following steps:

(*i*) The mixture to be separated into its components is dissolved in a suitable solvent (2 mg in 1-2 mL solvent, *e.g.*, 1-2%). It is spotted on the plate about 2 cm above the lower edge of the plate. The spots must be as small as possible and at the same distance from the lower edge of the plate.

(*ii*) The spotted TLC plate is placed in a glass Jar [Fig. 1.9(*b*)] containing the moving phase (20–25 mL) such that the lower end of the adsorbent layer of the TLC plate is under the solvent but the spots of the applied mixture are above the solvent [Fig. 1.9(*c*)].

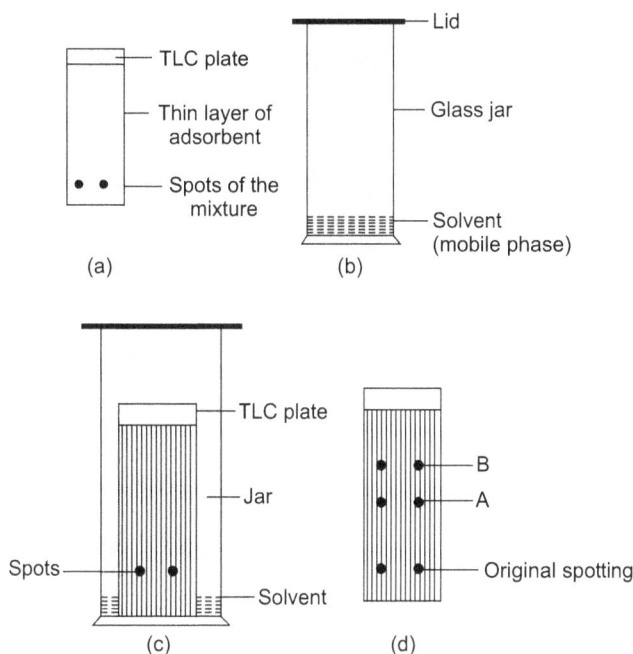

Fig. 1.9.

(*iii*) **The glass jar is covered.** The moving phase moves up the adsorbent carrying different components of the mixture at different rates. When the solvent has moves almost to the top of the adsorbent, the TLC plate is taken out of the Jar and the solvent front is marked with a pencil and the solvent allowed to evaporate (as in the case of paper chromatography).

In case, the components of the mixture are coloured, the coloured spots are visible on the plate. The number of spots is equal to the number of components in the mixture [Fig. 1.10(*d*)].

(*iv*) **Development of the TLC plate.** If the components are not coloured, the location and the number of components can be found by spraying with a suitable reagent.

The TLC plate may be viewed under UV light. Some components which are fluorescent in nature can be seen in the UV light.

(*v*) **Calculation of R_f values:** The R_f values (known as the retention factor or ratio of fronts) provide a useful index for comparing two compounds but these values are not always reprodusable.

$$R_f = \frac{\text{Distance travelled by the compound}}{\text{Distance travelled by the solvent}}$$

R_f of compound $A = x/z$, R_f of $B = y/z$. (Fig. 1.10)

x = Distance moved by compound A

y = Distance travelled by B,

z = Distance travelled by the solvent.

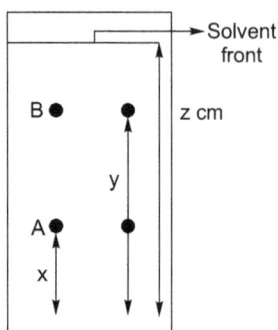

Fig. 1.10. Calculatin of R_f.

The identity of the components of the cation mixture can be established by putting simultaneously spots of authentic samples of cation on the same plate and comparing their R_f values.

Column Chromatography

Besides paper chromatography and TLC, column chromatography is also useful for the separation of cation mixtures and identification of the components.

In column chromatography, the stationary phase is a solid (the common adsorbents are alumina and silica gel) and the mobile phase is a liquid.

Principle of Column Chromatography

The principle of column chromatography is selective adsorption followed by elution. The mixture of cations to be separated is dissolved in minimum amount of a suitable solvent and allowed to pass through a glass column containing the adsorbent. The various components of the mixture are adsorbed in a particular order. The component which is most strongly adsorbed remains on the upper end of the column followed by the component which is adsorbed less strongly and so on. The overall result is that the components are separated in various parts of the column. As the column is eluted with a suitable solvent, the various bands in the column become more clear in different zones.

The different zones can be cut (after passing the adsorbent) and the bands extracted with suitable solvent. These can be finally characterised by comparing with authentic samples by paper chromatography or TLC.

Procedure of Column Chromatography

The procedure for column chromatography involves following steps:

(*i*) **Packing the column with an adsorbent.** The size of the glass column [Fig. 1.11(*a*)] to be used depends on the amount of the mixture to be separated. Long columns give better separations. A small amount of the eluting solvent is poured into the dry, clear column after inserting a ball of cotton or glass wool at the bottom end of the column (using a glass rod). The column is clamped, stopper opened (to allow the flow of solvent) into a flask placed below the column [Fig. 1.11(*b*)]. A slurry of the adsorbent in least polar solvent is continuously added from the top of the column [Fig. 1.11(*c*)]. The column is frequently tappd (with a rubber piece) during the addition of slurry (to remove the air bubbles).

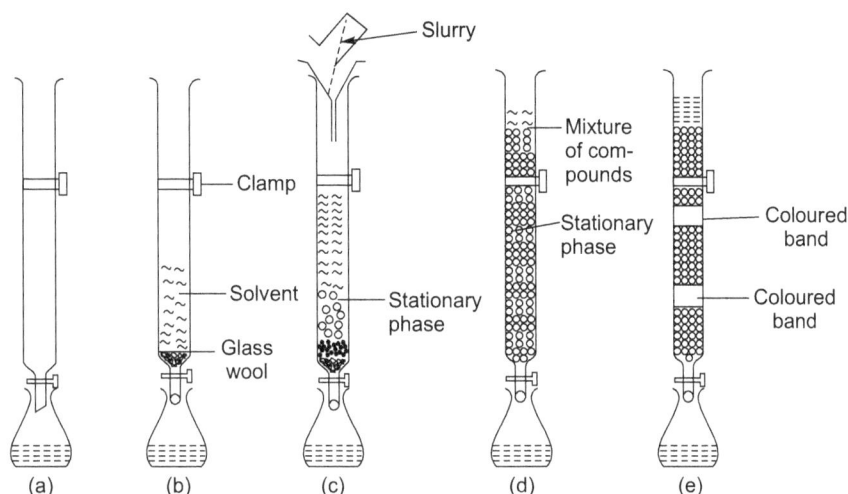

Fig. 1.11. Packing the column with the adsorbent.

The stopper is finally closed and the flow of the solvent stopped.

(*ii*) **Addition of mixture (to be separated) into the column.** A solution of the mixture is added carefully onto top of the adsorbent without disturbing the adsorbent. Excess of the solvent is allowed to drain out till the level of the liquid is just above the adsorbent [Fig. 1.11(*d*)].

(*iii*) **Elution of the column.** The column is eluted with a suitable solvent. In case the components of the mixture are coloured, the coloured bands separate out (as the solvent moves) and different zones are obtained [Fig. 1.11(*e*)].

1.10.1 Separation and Identification of Group I Cations (Pb⁺, Ag⁺, Hg⁺) by Paper Chromatography

Following procedure is followed:

(*i*) On a Whatman filter paper strip (3×20 cm) is placed a spot of the mixture (of group I cations) and one spot of each of the three cations (Pb⁺, Ag⁺ and Hg⁺) near the end of the strip. All the spots be at the same distance (2 cm above the end) from the end of the strip.

(*ii*) The spotted strip is air-dried and suspended into a cylinder containing 25 mL of the solvent (distilled water or Ethyl alcohol).

(*iii*) The chromatogram is run till the solvent front reaches just below the upper end of the paper strip (1.5 to 2 hrs).

(*iv*) The filter paper strip is taken out, solvent front marked, dried (in air) and sprayed with 0.25 M potassium chromate solution.

(*v*) The mixture shows three spots corresponding to Pb⁺, Ag⁺ and Hg⁺ (if all the three cations are present) which appear as yellow, orange-red and orange respectively.

(*vi*) The spots due to Pb⁺, Ag⁺ and Hg⁺ in the mixture corresponded to the spots due to pure specimen of Pb⁺, Ag⁺ and Hg⁺.

(*vii*) The R_f value of each of the spot is calculated and recorded.

1.10.2 Separation and Identification of Group II A Cations (Hg^{2+}, Cu^{2+}, Cd^{2+} and Bi^{3+}) by Paper Chromatography

The procedure involves following steps:

(*i*) On a Whatman filter paper strip (3 × 20 cm) is placed a spot of the mixture (of group II A cations) and one spot each of the four cations (Hg^{2+}, Cu^{2+}, Cd^{2+} and Bi^{2+}). For these cations solution of $HgCl_2$, $CuCl_2$, $CdCl_2$ and $BiCl_3$ respectively is used. All the five spots be at the same distance (2 cm above the end) from the end of the strip.

(*ii*) The chromatogram is run by using 25 mL of the solvent (Ethanol-5 N HCl; 9 : 1) taken in a cylinder till the solvent front reaches just below the upper end of the paper strip (about 1 hr.).

(*iii*) The filter paper strip is taken out, solvent front marked, dried (in air) and sprayed with a solution of dithiazone in chloroform. Following spots become visible:

$$Cu^{2+} - \text{Brown}$$
$$Hg^{2+} - \text{Pink}$$
$$Cd^{2+} - \text{Purple}$$
$$Bi^{3+} - \text{Brown}$$

The original spot of the mixture is also resolved into four spots (as stated above).

(*iv*) The R_f values of each of the spots is calculated and recorded.

Note: The separation and identification of Hg^{2+}, Cu^{2+}, Cd^{2+} and Bi^{3+} can also be affected by using column chromatography.

1.10.3 Separation and Identification of Group II A Cations (Hg^{2+}, Cu^{2+}, Cd^{2+} and Bi^{3+}) by Column Chromatography

Following steps are followed:

(*i*) **Preparation of the Column.** In a glass tube column (2 cm diameter and 20 cm long), is placed a filter disc and glass wool at the bottom (of the tube). Pour a little distilled water into the column and let some of it pass through the glass wool by opening the stopper.

Now prepare a slurry of 12 g powdered alumina with 50 mL distilled water (in a beaker) and pour the stirred slurry into the column until the column is 2/3 full. Pour some distilled water at the top. The column is ready for use in step (*iii*). The stopper should be kept open for the solution to pass through.

(*ii*) The precipitate of the sulphides of Hg^{2+}, Cu^{2+}, Cd^{2+} and Bi^{3+} (cations of group II A) is washed with water and dissolved in dil. HNO_3 (1 : 4), neutralised with NH_4OH (1 : 2) (using litmus paper) and finally made slightly acidic with dil. HNO_3.

(*iii*) Pour slurry of the above solution [obtained in step (*ii*)] in the column keeping the stopper slightly open. This is followed by addition of 1 mL H_2O, 5 drops of KI reagent, 2 mL H_2O,

1 mL $K_4[Fe(CN)_6]$ reagent, 2 mL H_2O, 5 drops of $(NH_4)_2S$ reagent, 5 mL H_2O and finally 3 mL, 0.1 N HCl. Following bands become visible.

Bi – Lemon coloured band

Pb – Lemon coloured band

Cu – Bluish zone

Cd – Black band.

1.10.4 Separation and Identification of Cu^{2+} and Cd^{2+} by paper Chromatography

Following procedure is followed:

(*i*) On a Whatman filter paper strip (3 × 20 cm) is placed a spot of the mixture (of Cu^{2+} and Cd^{2+}) and one spot of each of $CuSO_4$ and $CdCl_2$ solution. All the three spots be at the same distance (2 cm above the end) from the end of the strip.

(*ii*) The chromatogram is run using a solvent (Ethanol-5 N HCl; 9 : 1) taken in a cylinder till the solvent front reaches near the top of the filter paper strip.

(*iii*) The filter paper strip is taken out, solvent front marked and dried (in air).

(*iv*) The chromatogram is developed with colourless ammonium sulphide solution (obtained by bubbling H_2S in dilute NH_4OH solution). Both Cu^{2+} and Cd^{2+} appear as black spots. The mixture also shows two black spots.

(*v*) The R_f of Cu^{2+} and Cd^{2+} is calculated and recorded ($Cd^{2+} > Cu^{2+}$).

In place of Ethanol-5 N HCl (9 : 1) as mobile phase, a saturated solution of *n*-butanol in 3 N HCl can also be used.

1.10.5 Separation and Identification of Group III Cations (Fe^{3+}, Al^{3+}, Cr^{3+}) by Paper Chromatography

Following procedure is followed:

(*i*) On a Whatman filter paper strip (3 × 20 cm) is placed a spot of the mixture (of Fe^{3+}, Al^{3+} and Cr^{3+}) and one spot each of Fe^{3+}, Al^{3+} and Cr^{3+} salts in water. All spots be at the same distance (2 cm above the end) from the end of the strip.

(*ii*) The spotted filter paper strip is dried (air) and solvent allowed to rise (Ethanol-isopropyl alcohol- 5 N HCl; 9 : 9 : 2 or glacial acidic acid-dry methanol; 3 : 1).

(*iii*) The filter paper strip is taken out, solvent front marked and dried (in air).

(*iv*) The filter paper strip is covered with paper in such a way that the spot due to Cr^{3+} is not exposed. The rest of the strip (having Fe^{3+} and Al^{3+}) is developed with 1 percent alcoholic solution of alizarin, exposed to ammonia vapour and the filter paper strip warmed. Two spots are visible

red spot – Al^{3+}

purple spot – Fe^{3+}

Finally, the filter paper strip is sprayed with 5% solution of sodium peroxide followed by 0.05% benzidine in acetic acid. A blue colour (Cr^{3+}) develops after warming the strip in an oven.

(*v*) Calculate the R_f of each of the cations and record.

1.10.6 Separation and Identification of Fe^{3+} and Al^{3+} by Paper Chromatography

The procedure is the same as that used for the separation of Fe^{3+}, Al^{3+}, Cr^{3+}.

1.10.7 Separation and Identification of Group IV Cations (Ni^{2+}, Co^{2+}, Mn^{2+} and Zn^{2+}) by paper Chromatography

Following procedure is followed:

(*i*) On a Whatman filter paper strip (3 × 20 cm) is placed a spot of the mixture containing Ni^{2+}, Co^{2+}, Mn^{2+} and Zn^{2+} and one spot each of Ni^{2+}, Co^{2+}, Mn^{2+} and Zn^{2+} salts (chlorides) in water. All spots be at the same distance (2 cm above the end) from the end of the strip.

(*ii*) The spotted filter paper strip is dried (in air) and the chromatogram developed in solution [acetone + water + conc. HCl (44 mL + 3 mL + 4 mL)].

(*iii*) The chromatography is continued till the solvent front reaches almost the top of the strip. The strip is taken out, solvent front marked, and dried (in air).

(*iv*) The chromatogram is exposed to ammonia and then sprayed with a mixture of saturated alcoholic alizarin having 0.1 percent rubeanic acid and 0.1 percent salicylaldoxime. Finally, the filter paper strip is warmed in an oven. Following spots are observed.

$$Ni^{2+} - \text{blue colour}$$
$$Co^{2+} - \text{brown colour}$$
$$Mn^{2+} - \text{brown colour}$$
$$Zn^{2+} - \text{purple colour}$$

(*v*) The R_f value of each spot is calculated and recorded.

1.10.8 Separation and Identification of Ni^{2+} and Co^{2+} by Paper Chromatography

Following procedure is followed:

(*i*) On a Whatman filter paper strip (3 × 20 cm) is placed a spot of the mixture of Ni^{2+} and Co^{2+} and one spot each of Ni^{2+} and Co^{2+} as their salts (chlorides) in water. All the spots be at the same distance (2 cm above the end) from the end of the strip.

(*ii*) The spotted filter paper strip is dried (in air) and developed in solution [acetone + water + conc. HCl (44 mL + 3 mL + 4 mL)].

(*iii*) The chromatogram is developed till the solvent front reaches almost the top of the strip. The strip is taken out, solvent front marked and dried (in air).

(*ii*) The chromatogram is exposed to ammonia vapours and then sprayed with a mixture of saturated alcoholic alizarin having 0.1% rubeanic acid and 0.1% salicylaldoxime. The chromatogram is finally dried in an oven. The Ni^{2+} spot shows blue colour and Co^{2+} spot shows brown colour.

(*iii*) The R_f values of the spots is calculated and recorded.

The separation of Ni^{2+} and Co^{2+} can also be affected by using silica gel plate (TLC) using acetone : water : conc. HCl (44 : 3 : 4) as the mobile phase.

The chromatogram is developed as in the case of paper chromatography.

1.10.9 Separation and Identification of Cations by Thin Layer Chromatography (TLC)

A number of cations can be separated and identified by TLC. The procedure used is similar to that described on page 94. The TLC plate used is coated with silica gel.

The silica gel TLC plate is spotted with cation (mixture) and also of authentic cations. The dried TLC plate is dipped in the developing solvent. The various cations move with the solvent system at different speeds. When the solvent front has reached almost the top of the TLC plate. The plate is taken out, solvent front marked and dried (in air). Finally, the TLC plate is developed using a suitable visualising reagent. The R_f of various cations is calculated (as in the case of paper chromatography) and recorded.

Following Table 1.19 gives the various combinations of cations which can be successfully separated and identified by TLC.

Table 1.19. *Separation and Identification of various Combination of Cations by TLC*

Cation mixture	Solvent for irrigation	Reagent for development	Colour of spots
1. Ag^+, Hg^+	*n*-Butanol-dil. HCl-3 M tartaric acid (10 : 1 : 1)	PTC	Ag^+ – light pink, Hg^+ – black pinkish red
2. Cu^{2+}, Pb^{2+}, Bi^{3+}, Ca^{2+}, Hg^{2+}	*n*-Butanol-3 M HCl-Me-Et ketone (75 : 15 : 2)	PTC	Cu^{2+} – brown, Pb^{2+} – pinkish red, Bi^{3+} – brown, Cd^{2+} – yellow black, Hg^{2+} – black
3. As^{3+}, Sb^{3+}, Sn^{2+}	n-BuOH-3 M HCl-1 M tartaric acid (10 : 1 : 1)	PTC	As^{3+} – yellow, Sb^{3+} – orange, Sn^{2+} – brown
4. Fe^{3+}, Cr^{+6}, Al^{3+}	Acetone-3 M HCl–methyl isobutyl ketone (100 : 3 : 4)	Aluminon	Fe^{3+} – black, Cr^{6+} – green, Al^{3+} – red
5. Zn^{2+}, Co^{2+}, Ni^{2+}	–do–	Dithizone PTC	Zn^{2+} – cream, Co^{2+} – pink, Ni^{2+} – black-yellow
6. Cu^{2+}, Ni^{2+}, Co^{2+}	Acetone-methyl isobutyl ketone (3 : 1)	PTC	Cu^{2+} – brown, Ni^{2+} – yellowish orange, Co^{2+} – green
7. Pt^{4+}, V^{+5}, Rh^{2+}, Hg^{2+}, Au^{3+}, Ag^{+1}	Isoamyl acetate-conc. HCl (30 : 2)	PTC	Pt^{+4} – light brown V^{+5} – dark brown, Rh^{2+} – pink, Hg^{2+} – grey Au^{3+} – golden yellow, Ag^{+1} – light pink

...contd.

8. Mo^{+6}, Se^{+4}, V^{+6}, Te^{+4}	n-Butyl acetate-conc. HCl (40 : 6)	PTC	Mo^{6+} – brown, Se^{+4} – orange, yelllow, V^{+6} – light brown, Te^{+4} – dark brown
9. Se^{+4}, V^{+5}, Mo^{+6}, Te^{+4}	Diethoxyoxalate-conc. HCl (60 : 1)	PTC	Se^{+4} – orange, V^{+5} – yellow, Mo^+ – light brown, Te^{+4} – dark brown
10. Zn^{+2}, Cd^{+2}, Hg^{+2}	Isobutanol-acetic acid (3 : 1)	Dithiozone PTC	Zn^{2+} – pink, Cd^{2+} – yellow, Hg^{2+} – brownish black

1.11   ANALYSIS OF INSOLUBLES

In a mixture, the portion that remains insoluble in conc. HNO_3 or aqua-regia is described as an insoluble substance. [(see Scheme 1.9], page 68). Some of the common insoluble substances are:

>   Sulphates of Sr, Ba and Pb
>
>   Chlorides, bromides and iodides of Ag
>
>   Oxides of Al, Fe, Cr, Sn and Sb
>
>   Silica (SiO_2) and silicates
>
> CaF_2, mosaic gold (SnS_2)
>
>   Chromates (minerals) like $PbCrO_4$ and $FeCrO_4$
>
>   Hexacyanoferrates like $Cu_2[Fe(CN)_6]$, $Zn[Fe(CN)_6]$

The insoluble portion of the mixture [Scheme 1.9] is washed with water (in order to remove any adhering soluble salts). The insoluble portion may be a single insoluble salt or a mixture of insoluble salts.

Analysis of Single Salt

In case the insoluble portion of the mixture is a single salt, following procedure is followed for its analysis.

(*i*) **Colour of the salt.** The colour of some of the common insoluble substances and given below:

> White   : $PbSO_4$, $SrSO_4$, $BaSO_4$, AgCl, CaF_2, SnO_2, Al_2O_3
>
> Yellow   :   AgI, AgBr
>
> Green   : Cr_2O_3
>
> Brown   : Fe_2O_3, $PbCrO_4$
>
> Drark brown : $Cu_2[Fe(CN)_6]$
>
> Dark grey   : $FeCr_2O_4$
>
> Bronze   : SnS_2

(*ii*) **Charcoal cavity test:** A small amount (0.1 *g*) of the insoluble component is heated with twice its weight of Na_2CO_3 in a charcoal cavity using a blow pipe.

In case a bead is formed, it is dissolved in HNO_3 and the solution divided into two parts. To one part of the solution is added HCl. Formation of a white ppt., which is soluble in NH_4OH indicates the presence of Ag^+. To the second part of the solution (in case Ag^+ is absent) is added Hg_2Cl_2 solution. Formation of a white ppt. indicates the presence of Sn^{4+}.

However, if in the charcoal cavity test, no bead formation takes place, then to the residue is added a drop of cobalt nitrate solution and the residue again heated. Formation of a blue mass indicates the presence of Al^{3+}.

In case a bead is not formed, and Al^{3+} is absent, the charcoal cavity test is repeated with fresh insoluble salt and the residue is moistened with 2-3 drops of dilute HCl and placed in contact with lead acetate paper. A black stain on the paper indicates the presence of sulphide and therefore, the presence of SO_4^{2-} in the insoluble salt.

(iii) **Action of conc. H_2SO_4.** A small amount (0.1 g) of the insoluble substance is heated with conc. H_2SO_4 (1 mL).

   (a) The gas evolved turns a drop of water (on a glass rod) turbid, indicating the presence of fluoride.

   (b) The gas evolving contains CO which burns with a blue flame indicating the presence of ferrocyanide.

(iv) **Fusion with Na_2CO_3 and KNO_3.** The insoluble salt (0.1 g) is heated with 1:1 mixture of Na_2CO_3 and KNO_3 (0.1 g) in a porcelain dish. Formation of yellow melt indicates the presence of Cr^{3+}. The cooled melt is dissolved in water and the solution acidified with CH_3COOH. Addition of $AgNO_3$ solution to the clear solution giving a reddish brown precipitate indicates the presence of Cr^{3+}. Alternatively, if addition of lead acetate solution to the above clear solution gives a yellow ppt. presence of Cr^{3+} is confirmed.

(v) **Heating with NaOH solution.** The insoluble salt (0.1 g) is boiled with NaOH solution. Following may be observed.

   (a) If Al^{3+} or Si^{+4} is present, a clear solution of sodium aluminate or sodium silicate is obtained. The clear solution is acidified with HCl and NH_4OH added till the solution is alkaline. The solution is heated to boiling. Formation of white gelatinous ppt. shows the presence of Al^{3+} or Si^{4+}.

   (b) The presence of $PbCrO_4$ is indicated by formation of yellow coloured solution of Na_2CrO_4.

$$PbCrO_4 + 4NaOH \longrightarrow Na_2CrO_4 + Na2PbO_2 + 2H_2O$$

   (c) The presence of $Fe_4[Fe(CN)_6]_3$ is indicated by formation of ferric hydroxide.

$$Fe_4[Fe(CN)_6]_3 + 12NaOH \longrightarrow 4Fe(OH)_3 + 3Na_4Fe(CN)_6$$

(vi) **Action of hydriodic acid.** The powdered insoluble substance (0.2 g) is heated to just below the b.p. with hydriodic acid (sp. gr. 1.7, 2 mL). Following observation may be there:

   (a) The presence of stannic oxide is indicated by forming a clear solution of stannic iodide

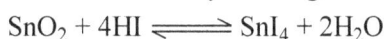

$$SnO_2 + 4HI \rightleftharpoons SnI_4 + 2H_2O$$

When H_2S passed into the clear solution, tin is precipitated as tin sulphide.

(b) The presence of $PbSO_4$, $BaSO_4$ or $SrSO_4$ is indicated by evolution of H_2S (arising due to decomposition of the sulphates). The presence of H_2S is indicated by lead acetate paper turning black.

$$PbSO_4 + 11HI \longrightarrow H[PbI_3] + 4I_2 + H_2S\uparrow + 4H_2O$$
$$BaSO_4 + 10HI \longrightarrow BaI_2 + 4I_2 + H_2S\uparrow + 4H_2O$$
$$SrSO_4 + 10HI \longrightarrow SrI_2 + 4I_2 + H_2S\uparrow + 4H_2O$$

The solution on filtration (through sintered glass crucible) and subsequent dilution gives golden yellow lead iodide.

Barium can be detected by adding dilute H_2SO_4. White ppt. of $BaSO_4$ is obtained. Both barium and strontium can be detected by flame test.

(c) Silver halides, if present dissolved in cold HI due to the formation of complex.

$$3AgI + HI \equiv H[Ag_3I_4]$$
$$3AgX + 4HI \equiv H[Ag_3I_4] + 3HX$$

On heating, hydrogen halides are expelled and on dilution AgI gets precipitated. If $PbSO_4$ is present, it forms PbI_2, which is removed by extraction with ammonium acetate solution.

(d) Calcium fluoride is attacked by HI in hot. The HF evolved etches the glass

$$CaF_2 + 2HI \longrightarrow CaI_2 + 2HF$$

The solution on dilution followed by neutralisation with NH_4OH and addition of ammonium oxalate solution gives a white ppt. of calcium oxalate.

(e) **Silver bromide and silver iodide.** The insoluble substance is treated with Zn and dil. H_2SO_4 for 2-3 min. and filtered (Filtrate X). The residue (y) contains metallic Ag. It is washed with water and dissolved in HNO_3 and HCl added. A curdy white ppt. soluble in NH_4OH indicates the presence of Ag. The filtrate (X) contains $ZnBr_2$ or ZnI_2. Br^- and I^- are tested as described on Section 1.3.6.

(vii) **Sodium carbonate extract.** The insoluble fraction (0.2 g), sodium carbonate (1 g) and water (about 10 mL) is heated till the volume is reduced to about half. The mixture is filtered. The residue is used for the analysis of some cations and the filtrate is used for finding the presence of anions as given below:

(a) The filtrate (1 mL) is acidified with dil. HCl (till effervescence ceases) and $BaCl_2$ solution added. Formation to a white ppt. insoluble in conc. HCl confirms SO_4^{2-}. If SO_4^{2+} is absent it is not necessary to test Ba^{2+}, Pb^{2+} and Sr^{2+}. In case SO_4^{2-} is present, and the insoluble substance is white in colour, it may be $PbSO_4$, $BaSO_4$ or $SrSO_4$.

(b) A part of the residue (y) is dissolved in hot ammonium acetate solution. To the solution is added acetic acid and K_2CrO_4 solution. Formation of yellow ppt. (of $PbCrO_4$) confirms Pb^{2+}.

(c) In case Pb^{2+} is absent, perform the flame test with the residue (y) (see Table 1.3). A green flame indicates Ba^{2+}, and can be confirmed by dissolving the residue (y) in CH_3COOH and adding K_2CrO_4 solution. A yellow ppt. is obtaned.

(d) In case no ppt. is obtained with K_2CrO_4 (test above), the insoluble substance may be $SrSO_4$. Dissolve a part of the residue (y) in acetic acid and add ammonium sulphate solution. A white ppt. confirms Sr^{2+}.

(viii) If no anion is present (as indicated by tests given above), it may be one of the following oxides.

Al_2O_3 : White in colour. Perform charcoal cavity test. Also perform the test (Test V, Page 103).

SnO_2 : White in colour. Test as given in Page 103.

Sb_2O_4 : White in colour. Test is given in Table 1.19.

Cr_2O_3 : Green in colour. Test as given in test (iv), Page 103.

Analysis of a Mixture of Insoluble Salts

For analysis of a mixture of insoluble salts, proceed as follows:

1. Test for the presence of Pb^{2+} and Ag^{2+} as given is analysis of single salts.

2. If lead and silver are absent, the mixture of insoluble salts is fused with 3-4 times its weight of a mixture Na_2CO_3 and K_2CO_3 (1:1) in a nickel crucible. The mixture is cooled and extracted with hot water and filtered (Scheme 1.14)

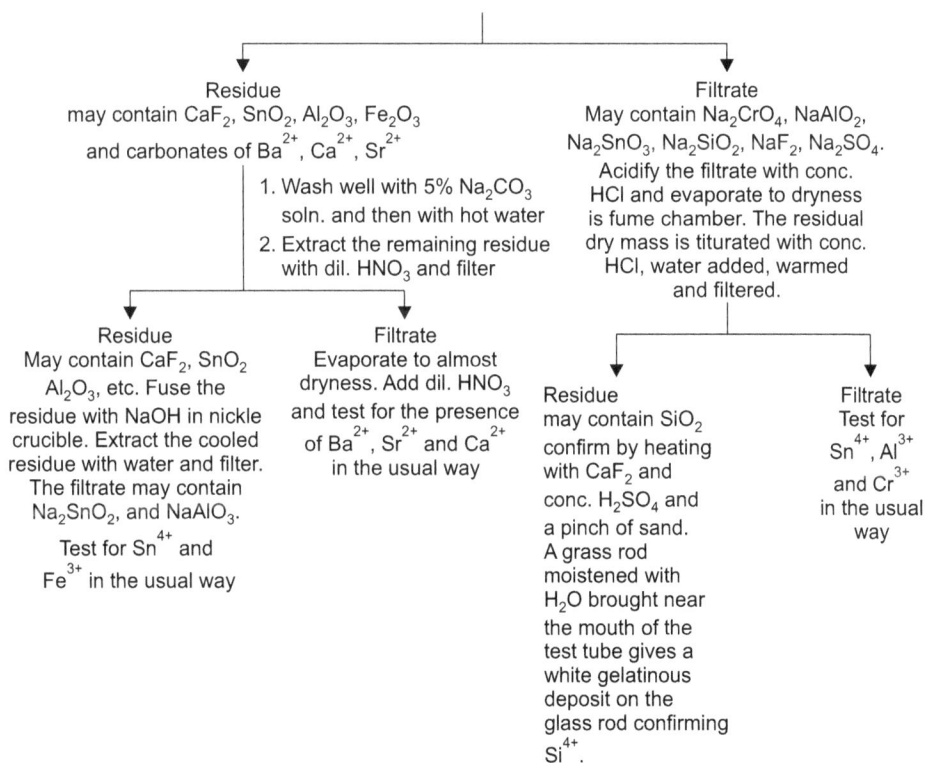

Scheme 1.14: Analysis of insolubles.

The presence or absence of individual cations in insoluble substance (or a mixture) can also be confirmed by flame photometry (Section 1.8) and atomic absorption spectroscopy (Section 1.9).

1.12   ANALYSIS OF A METAL OR AN ALLOY

In case of a metal or an alloy only the basic radicals (or cations) are analysed. Some alloys, however, contain small amounts of P, Si, C and S which are converted into phosphate, silicate etc. during the process of making these alloys.

Various steps involved in the analysis of metal or an alloy are:

(i) The metal or the allow is converted into shavings. About 0.5-1 g of the shavings are treated with conc. HNO_3 (about 10 mL) in a porcelain dish or silica dish in a fume cupboard. The mixture is gently warmed till the evolution of red fumes of oxides of nitrogen ceases and the solution is evaporated almost to dryness. During the process P is oxidised to H_3PO_4. As to H_3AsO_2 or H_3AsO_4, Sb to $Sb_2O_5 \cdot xH_2O$, Sn to SnO_2 and Si to gelatinous silicic acid. Some more HNO_3 (about 10 mL) is added to the residue and the mixture heated for 2-3 minutes and filtered if necessary.

(ii) In case, a clear solution is obtained, the cations are analysed as already discussed (Section 1.4) after testing for phosphate.

(iii) If the metal or the alloy does not dissolve completely, the mixture is filtered (using a sintered funnel) and the filtrate (A) is examined as in step (ii).

(iv) The residue left after treatment of the alloy with conc. HNO_3 is washed with water and dried by heating in a silica dish. To the dried residue is added about 5-6 times of its weight a mixture of equal parts of S and anhyd. Na_2CO_3 and the total mixture is heated on a small flame till all the S is burned off (20-30 min.). The mixture is cooled, extracted with hot water and filtered. Subsequent steps are given below (Scheme 1.15).

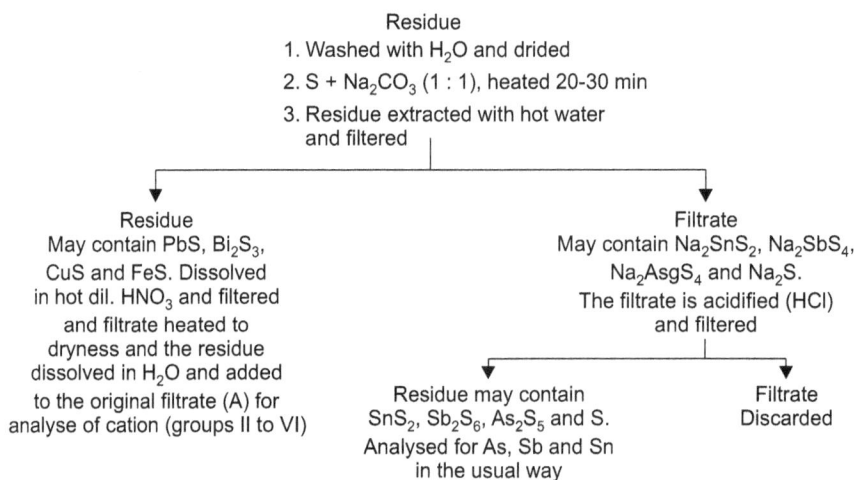

Residue
1. Washed with H_2O and drided
2. $S + Na_2CO_3$ (1 : 1), heated 20-30 min
3. Residue extracted with hot water and filtered

Residue
May contain PbS, Bi_2S_3, CuS and FeS. Dissolved in hot dil. HNO_3 and filtered and filtrate heated to dryness and the residue dissolved in H_2O and added to the original filtrate (A) for analyse of cation (groups II to VI)

Filtrate
May contain Na_2SnS_2, Na_2SbS_4, Na_2AsgS_4 and Na_2S. The filtrate is acidified (HCl) and filtered

Residue may contain SnS_2, Sb_2S_6, As_2S_5 and S. Analysed for As, Sb and Sn in the usual way

Filtrate
Discarded

Scheme 1.15: Analysis of a metal or alloy.

(*iv*) In case the metal or the alloy does not react with HNO_3 (1 : 1), it is separately treated with aqua-regia in a porcelein dish. The mixture is gently heated in a fume cupboard till the volume is reduced to about one third. Finally, the mixture is heated on a water bath to almost dryness. This is followed by addition of about 5 mL conc. HCl. This mixture is gently heated (4-5 min.), diluted with water (10 mL), stirred, boiled, cooled and filtered. The residue may contain AgCl, $PbCl_2$ and SiO_2. The filrate is examined for metallic radicals of groups I to VI.

(*v*) In case, the alloy or the metal resists the action of even aqua-regia, it is fused with NaOH pellets in a silica dish (caution). The mixture is finally cooled, and the contents extracted with water. The solution is acidified (HNO_3), evaporated to almost dryness on a water bath and analysed as given above (step *iii*).

The composition of some of the alloys are given below:

Brass : Cu, Zn, Sn, Pb

German silver : Cu, Ni, Zn

Nichrome : Ni, Fe, Cr

Woods alloy : Bi, Pb, Sn, Cd

1.13 ANALYSIS OF SOME RARE ELEMENTS

So far we have learnt about the separation and identification of common cations and anions in a mixture. Besides the common cations and anions, the mixture may also contain some less common ions. A number of such elements like tungsten, molybdenum, titanium, vanadium and beryllium are known for their industrial applications. Due to this the analysis of such elements has assumed considerable importance.

The less common ions or rare elements, during the precipitation and separation of cations (groups I to VI) get precipitated alongwith them. Following (Table 1.20) are given the less common ions which get precipitated in various groups.

Table 1.20. *Less Common Ions in Different Groups.*

Group	Normal cations	Less common cations
I	Pb^{2+}, Ag^+, Hg_2^{2+}	W, Tl
II A	Hg^{2+}, Pb^{2+}, Bi^{3+}, Cu^{2+}, Cd^{2+}	Pt, Au, Pd
II B	As^{3+}, Sn^{2+}, Sb^{3+}	Pt, Au, Te, Mo
III	Fe^{3+}, Al^{3+}, Cr^{3+}	Ti, Zr, Ce, Th, U
IV	Ni^{2+}, Co^{2+}, Zn^{2+}, Mn^{2+}	Tl

As seen some of the rare elements get precipitated in more than one group.

1.13.1 Separation, Identification of Group I Cations (Pb^{2+}, Ag^+, Hg_2^{2+}), Tl and W

The precipitate of group I cations may conta in $PbCl_2$, AgCl, Hg_2Cl_2, TlCl and tungstic acid ($WO_3 . xH_2O$). The precipitate is washed with 2 M HCl (2×3 mL) and then with water (1×3 mL).

The washings are rejected. The ppt. is boiled with water (10-15 mL) and filtered hot (Scheme 1.16) (see also Scheme 1.2).

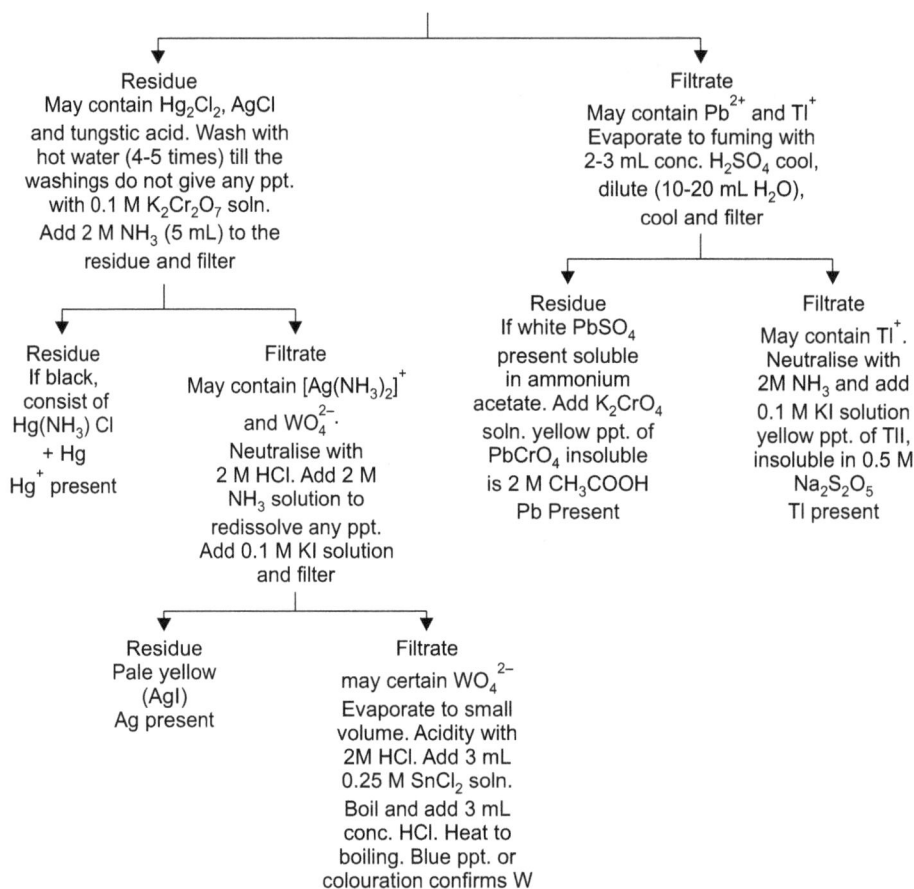

Scheme 1.16. Analysis of group I cations, Tl and W.

The confirmatory tests of only the rare elements is described. Tests for other cations of group I have already been described (Page 45).

Confirmatory Tests of Tl and W

Thallium

(*i*) **Potassium chromate solution:** A yellow ppt. of thallium (I) chromate Tl_2CrO_4 is obtained. The ppt. is insoluble in cold dil. HNO_3 or H_2SO_4.

(*ii*) **Sodium hexanitrocobalt solution [Na$_3$[Co(NO$_2$)$_6$]]:** A light red ppt. of thallium hexanitro-cobaltate, $Tl_3[Co(NO_2)_6]$ is obtained.

(*iii*) **Ammonium thiocyanate solution:** A white ppt. of thallium thiocyanate TlSCN, is obtained. The ppt. is soluble in hot water.

(*iv*) **Using acetylene:** Air flame, thallium can be detected by atomic absorption spectroscopy. A thallium hollow-cathode lamp is used. The wavelength of emitted radiation is 276.79 nm.

(*v*) **Flame test:** Thallium salts on introducing into the colourless Bunsen flame on examining through the spectroscope gives only one sharp line at 535 nm (compared to several lines between 510 and 550 nm in case of barium).

Caution: Thallium salts are highly poisonous and appropriate precaution should be taken in handling these.

Tungsten

(*i*) **Phosphoric acid.** A white ppt. of phospho-tungstic acid $[H_3(PO_4(W_{12}O_{36})]$ or $H_3[P(W_3O_{10})_4]$ (soluble in excess of the reagent) is obtained.

(*ii*) **Zinc-hydrochloric acid.** The solution of tungstate on treatment with HCl-Zn gives a blue ppt. or colouration. This colouration is due to the formation 'tungsten blue' having the composition W_2O_5.

(*iii*) **Tin chloride solution.** A yellow ppt., which becomes blue on warming with conc. HCl is obtained.

This reaction is used for spot test detection of tungsten. It involves mixing 1-2 drops of the test solution with 2-5 drops of $SnCl_2$ reagent on a spot plate. A blue ppt. or colouration of tungsten blue, W_2O_5 is obtained.

Molybdenum also gives a similar reaction. In presence of molybdenum, tungsten is tested as follows. A drop of test solution is placed on a spot plate followed by addition of a drop of conc. HCl. Tungstate produces a yellow stain. Additon of a drop of 10% potassium thiocyanate solution and a drop of Sat'd $SnCl_2$ solution gives a red spot due to formation of $[Mo(SCN)_6]^{3-}$. The red spot disappears on addition of a drop of conc. HCl and a blue colour due to tungsten blue remains.

(*iv*) **Ferrous sulphate.** A brown ppt. is formed, which turns white on addition of dil. HCl and then yellow on heating.

(*v*) In atomic absorption spectrometry, using a tungsten hollow cathode and acetylene-nitrous oxide flame absorption is observed at 255.13 nm.

Molybdate ion

(*i*) **Zinc-tin chloride-hydrochloric acid.** Addition of zinc and $SnCl_2$ and HCl to the test solution gives a blue colour (due to the formation of molybdenum blue, Mo_2O_5) which changes to green and finally brown.

(*ii*) **Potassium thiocyanate solution (spot test).** To a drop of the test solution placed on a drop-reaction paper is added a drop of 10 per cent potassium thiosulphate solution. This is followed by addition of a drop of sat'd. $SnCl_2$ solution. A red spot is produced.

(*iii*) **Disodium hydrogen phosphate solution.** Addition of Na_2HPO_4 to the test solution gives a yellow ppt. of ammonium molybdate in presence of HNO_3.

(*iv*) **Benzoin α-oxime reagent (Cupron reagent).** Addition of Cupron reagent to a test solution acidified with dilute H_2SO_4 gives a white ppt.

(*v*) **Phenyl hydrazine reagent (spot test).** To a drop of the test solution placed on a spot plate is added a drop of the reagent. Red colour is produced.

(*vi*) In atomic absorption spectrometry, using a molybdenum hollow-cathode lamp and acetylene-nitrous oxide or acetylene-air flame, absorption is observed at 313.26 nm.

1.13.2   Separation, Identification of Group II A Cations (Hg^{2+}, Pb^{2+}, Bi^{3+}, Cu^{2+} and Cd^{2+}), Pt, Au and Pd

Group II precipitate (Section 1.4.1.3) is washed with M NH_4Cl solution which was saturated with H_2S. This is followed by addition of 5-10 mL of ammonium polysulphide solution, mixture heated (50-60°, 3-4 minutes) and filtered. (Scheme 1.17)

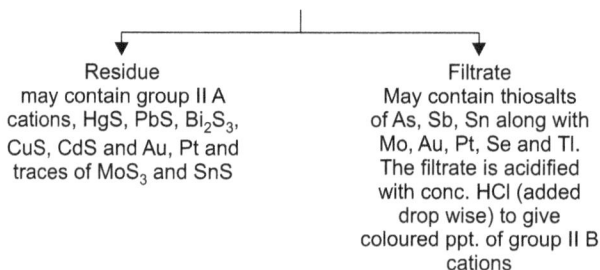

Residue	Filtrate
may contain group II A cations, HgS, PbS, Bi_2S_3, CuS, CdS and Au, Pt and traces of MoS_3 and SnS	May contain thiosalts of As, Sb, Sn along with Mo, Au, Pt, Se and Tl. The filtrate is acidified with conc. HCl (added drop wise) to give coloured ppt. of group II B cations

Scheme 1.17. Separation of group II cations into group II A and group II B cations.
(see also Scheme 1.2a)

The residue (ppt.) of group II A cations (Scheme 1.17) is boiled with 5-10 mL 2 M HNO_3 and filtered (Scheme 1.18).

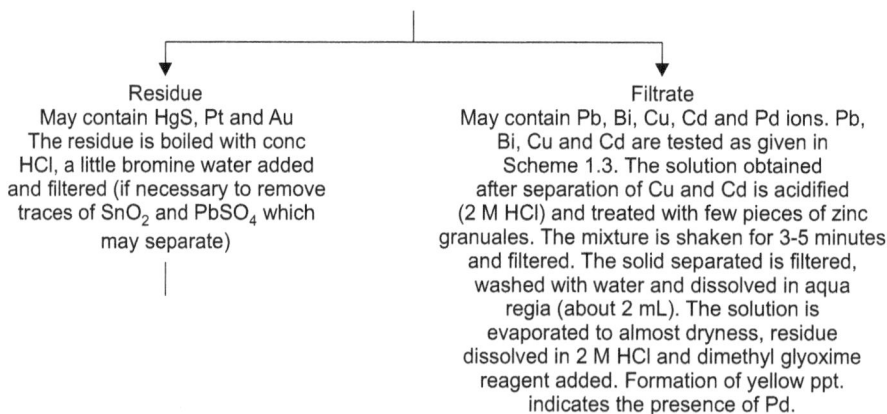

Residue	Filtrate
May contain HgS, Pt and Au The residue is boiled with conc HCl, a little bromine water added and filtered (if necessary to remove traces of SnO_2 and $PbSO_4$ which may separate)	May contain Pb, Bi, Cu, Cd and Pd ions. Pb, Bi, Cu and Cd are tested as given in Scheme 1.3. The solution obtained after separation of Cu and Cd is acidified (2 M HCl) and treated with few pieces of zinc granuales. The mixture is shaken for 3-5 minutes and filtered. The solid separated is filtered, washed with water and dissolved in aqua regia (about 2 mL). The solution is evaporated to almost dryness, residue dissolved in 2 M HCl and dimethyl glyoxime reagent added. Formation of yellow ppt. indicates the presence of Pd.

Residue
yellow crystalline
$K_2[PtCl_6]$
Pt present

Filtrate
May contain $AuCl_4$ and $HgCl_2$.
The filtrate is boiled and rendered
alkaline (with 2 M NaOH solution)
It is boiled with excess 0.5 M oxalic
acid and filtered

Residue
Brownish-black
or purplish-black.
Au present

Filtrate may
contain $HgCl_2$.
Add few drops of
0.25 M $SnCl_2$
solution. White
or grey ppt. shows
the presence of Hg

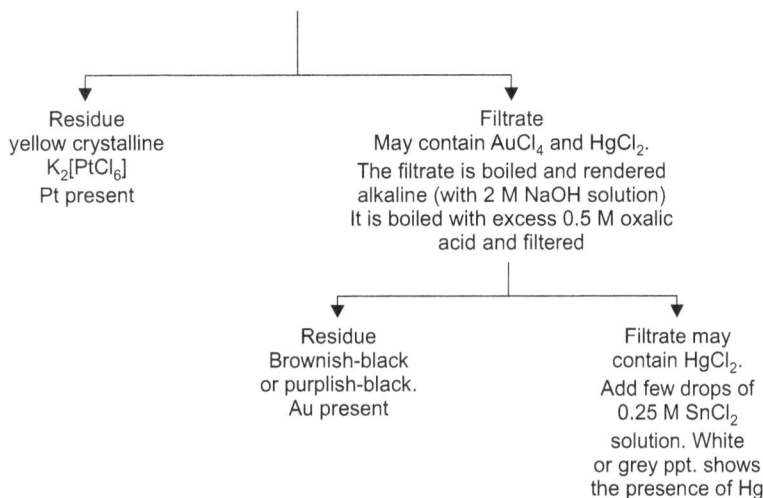

Scheme 1.18. Separation, analysis of group II A cations alongwith
Au, Pt and Pd (see also Scheme 1.3).

Confirmatory Tests of Pt, Au and Pd

The confirmatory tests of Hg, Pb, Bi, Cu and Cd have already been described (page 49-51). Here, the conformatory test of Pt, Au and Pd are described.

Platinum

(*i*) **Potassium chloride solution.** A yellow ppt. of potassium hexachloroplatinate, $K_2[PtCl_6]$ is obtained.

(*ii*) **Sodium formate.** Boiling the neutral test solution with sodium formate produces black metallic platinum

$$(PtCl_6)^{2-} + 2HCOO^- \longrightarrow Pt\downarrow + 2CO_2\uparrow + 2HCl + 4Cl^-$$

(*iii*) **Silver nitrate solution.** A yellow ppt. of silver hexachloroplatinate, $Ag_2(PtCl_6)$ is produced. The ppt. is sparingly soluble in excess NH_4OH but is soluble in alkali cyanide solution.

(*iv*) **$SnCl_2$ solution (spot test).** To a drop of the test solution placed on a drop reaction paper is added a drop of thallium nitrate solution. The formed ppt. is washed with NH_4OH and a drop of $SnCl_2$ solution added to the ppt. A yellow or orange spot remains.

(*v*) **Rubeanic acid.** To a drop of the test solution (acidified with HCl) placed on a spot plate is added a drop of the reagent. A purple-red ppt. is obtained.

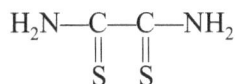

$$H_2N-\underset{\underset{S}{\|}}{C}-\underset{\underset{S}{\|}}{C}-NH_2$$

$$Pt\left(\begin{array}{c} H_2N-C=S \\ | \\ S-\!\!\!-C=NH \end{array}\right)_2$$

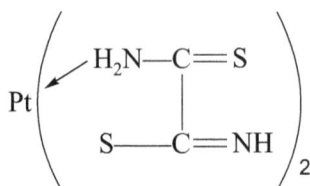

(*vi*) In atomic absorption spectroscopy, using a platinum hollow-cathode lamp and acetylene-air flame an absorption in obtained at 265.95 nm.

Gold

(*i*) **Oxalic acid solution.** To the test solution (obtained by dissolving gold in aqua-regia, heating the solution to almost dryness and dissolving the residue in HCl) is added oxalic acid solution after the test solution is rendered neutral. Gold is obtained in colloidal state as a red, violet or blue solution.

$$2[AuCl_4]^- + 3(COO)_2^{2-} \longrightarrow 2Au\downarrow + 6CO_2\uparrow + 8Cl^-$$

(*ii*) **Rhodamine reagent (spot test).** To the dried drop reaction paper impregnated with the reagent (4-dimethyl aminobenzaldehyde-rhodamine) in placed a drop of neutral or weakly acid test solution [obtained as in (*i*) above]. A violet spot or ring is obtained.

(*iii*) In atomic absorption spectrometry, using a gold hollow-cathode lamp and acetylene-air flame an absorption is obtained at 242.80 nm.

Palladium

(*i*) **Sodium hydroxide solution.** A reddish-brown ppt. of hydrated oxide, $PdO.nH_2O$ is obtained. The ppt. is soluble in excess alkali.

(*ii*) **Ammonium hydroxide solution.** A red ppt. of $[Pd(NH_3)_4][PdCl_4]$ is obtained. The ppt. is soluble in excess NH_4OH to give a colourless solution of $[Pd(NH_3)_4]^{2+}$ ions. The solution on acidification (HCl) gives a ppt. of $[Pd(NH_3)_2Cl_2]$.

(*iii*) **1-Nitroso-2-naphthol solution.** A brown ppt. of $Pd(C_{10}H_6O_2N_2)$ is obtained.

(*iv*) **Atomic absorption spectroscopy.** Using a palladium hollow-cathode lamp and acetylene-air flame absorption at 247.64 nm is observed.

1.13.3   Separation, Identification of Group II B Cations (As, Sb, Sn), Pt, Au, Se, Te and Mo

To the precipitate of group II B cations (scheme 1.17) is added conc. HCl (5 mL), mixture boiled gently (5 min) diluted with water (2-3 mL) and filtered (Scheme 1.19).

Conformatory Tests of Se and Te

The conformatory tests of As, Sb and Sn have already been described (page 52-54).

Conformatory tests of Pt and Mo have also been described (see page 111 and 112 respectively).

Following are given conformatory tests of selenium and tellurium.

Residue
May contain sulphides of As, Au, Pt, Mo, Se and Te. The residue is dissolved in conc. HCl in presence of little $KClO_3$. The solution is concentrated (water bath to reduce loss of Se) to crystalline point and filtered

Filtrate
May contain Sb and Sn as chlorides or complexes of chloro-acids. There are confirmed as described in Scheme 1.4.

Residue
Yellow $K_2[PtCl_6]$. Pt present. Confirmed by dissolving in warm water and adding 0.1 M KI solution. Red or brown ppt. is obtained

Filtrate
May contain chlorides of As, Au, Mo, Se, and Te. The filtrate is rendered alkaline (2 M NH_4OH solution) and $Mg(NO_3)_2$ reagent added. The mixture is allowed to stand (5 min) and filtered

Residue
White crystalline $Mg(NH_4) AsO_4 . 6H_2O$ As present (confirmed as given in Scheme 1.4)

Filtrate
May contain chlorides of Au, Mo, Se and Te. The filtrate is concentrated (to remove NH_3), boiled with sat'd oxalic acid solution (5-6 mL), solution diluted, boiled and filtered. The ppt. obtained is extracted with 2 M HCl (to remove coprecipitated tellurous acid)

Residue
Brownish or purplish-black Au confirmed

Filtrate
concentrated with 6 M HCl (water bath) (to remove ppted. KCl) and treated with excess of solid Na_2SO_4 and filtered

Residue
Se present

Filtrate
Diluted with H_2O followed by addition of a little 0.1 M KI solution and excess solid Na_2SO_3 [(TeI_6^{2-}) ions are reduced to Te]. Filter

Residue
Black
Te present

Filtrate
Boiled with 2 M HCl (to remove dissolved SO_2) and treated with 10% KCNS solution and a little 0.5 M $SnCl_2$ solution. Red colouration is obtained. Mo present

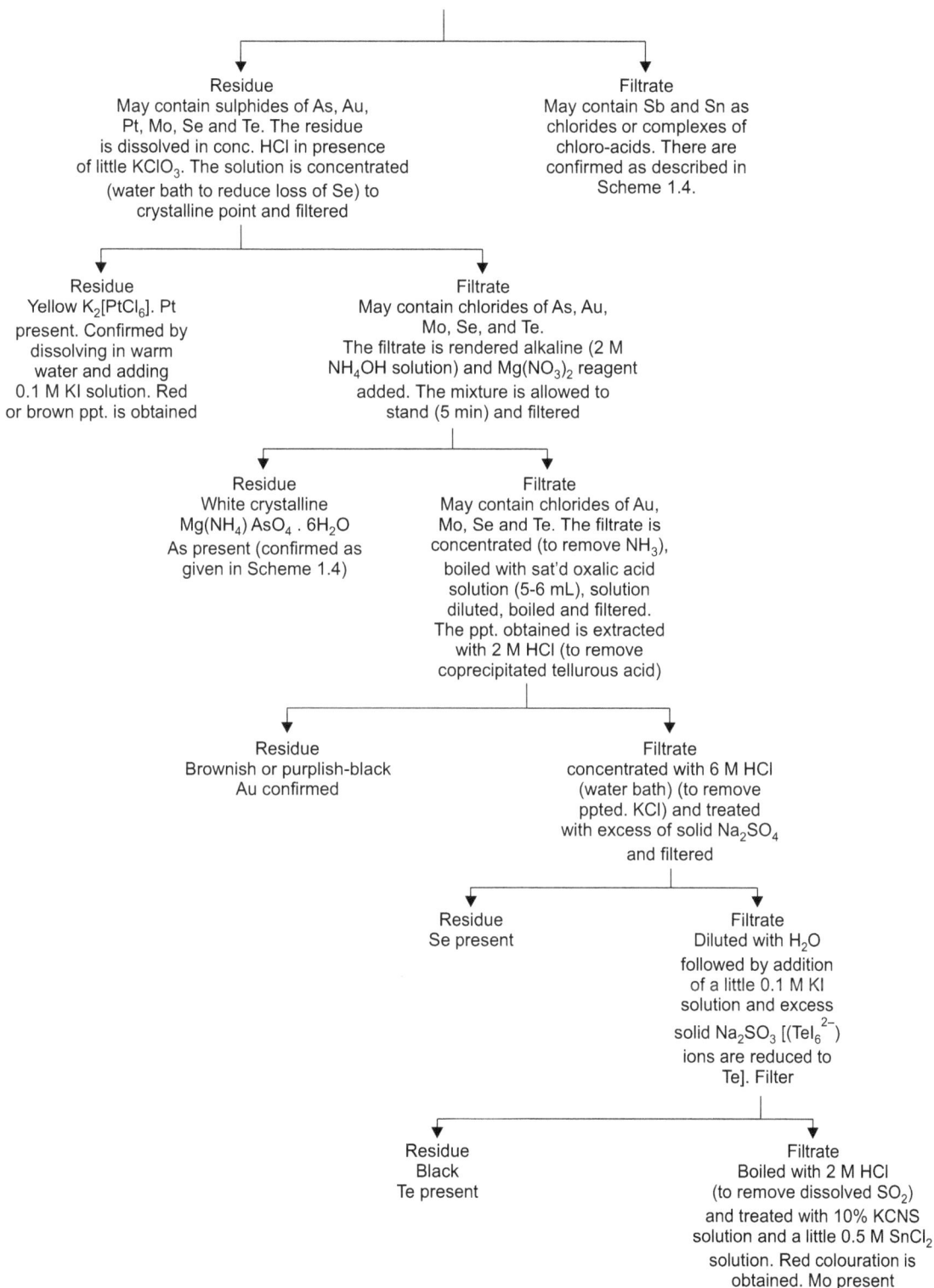

Scheme 1.19. Separation, identification of group II B cations, Pt, Au, Se, Te and Mo.

Selenium

Selenium is obtained as residue (see scheme 1.19). It is dissolved in conc. HNO_3, solution rendered neutral and used as test solution.

(*i*) **Copper sulphate solution.** A bluish-green crystalline ppt. of copper selenite, $CuSeO_3$ is obtained. The ppt. dissolves in dil. CH_3COOH.

(*ii*) **Barium chloride solution.** A white ppt. of barium selenite, $BaSeO_3$ is obtained. The ppt. is soluble in dilute mineral acid.

(*iii*) **Potassium iodide solution (spot test).** To a drop of conc. HCl on a drop reaction paper is added a drop of 6 M KI solution followed by addition of a drop of test solution. A brownish-black spot develops. Subsequent addition of a drop of 0.5 M sodium thiosulphate solution reddish brown stain of elemental selenium remains.

(*iv*) **Pyrrole reagent (spot test).** To a drop of 5 M $FeCl_3$ solution placed on a spot plate is added 5-7 drops of conc. phosphoric acid (d 1.75 cm^{-1}) and a drop of test solution. The mixture on stirring and addition of the pyrrole reagent produces greenish-blue colour.

Tellurium

The reactions of tellurium are studied with potassium tellurite K_2TeO_3, which is obtained by fusing tellurium residue (scheme 1.19) with KCN. Potassium telluride is water soluble and the aqueous solution is used for testing tellurium.

(*i*) **Barium chloride solution.** A white ppt. of barium tellurate, $BaTeO_4$ is obtained. The ppt. is soluble in dilute HCl.

(*ii*) **Potassium iodide solution.** A yellow to red colour due to hexaiodotellurate ion, $(TeI_6)^{2-}$ is obtained.

1.13.4   Separation, Identification of Group III Cations (Fe, Al, Cr), Ti, Zr, Ce, V, Th and U

Before proceeding for the testing of group III cation, the solution (filtrate after separation of group II cations) is tested for phosphate. In case PO_4^{3-} is present (See page 32) it is removed as described in Section 1.4.1.3, page 55. To the solution obtained after removal of PO_4^{3-} is added. Solid NH_4Cl and NH_4OH. The cations of group III are precipitated as hydroxide. The filtrate is used for subsequent groups (Scheme 1.20).

Precipitate of group III cations is dissolved in minimum volume of 2 M HCl. The solution (which is weakly acidic) is poured into equal volume of a solution containing 2 mL 30% H_2O_2 and 0.25 M with respect to NaOH. The solution is boiled for 5 min only. Filter and wash the ppt. with 2% NH_4NO_3 solution.

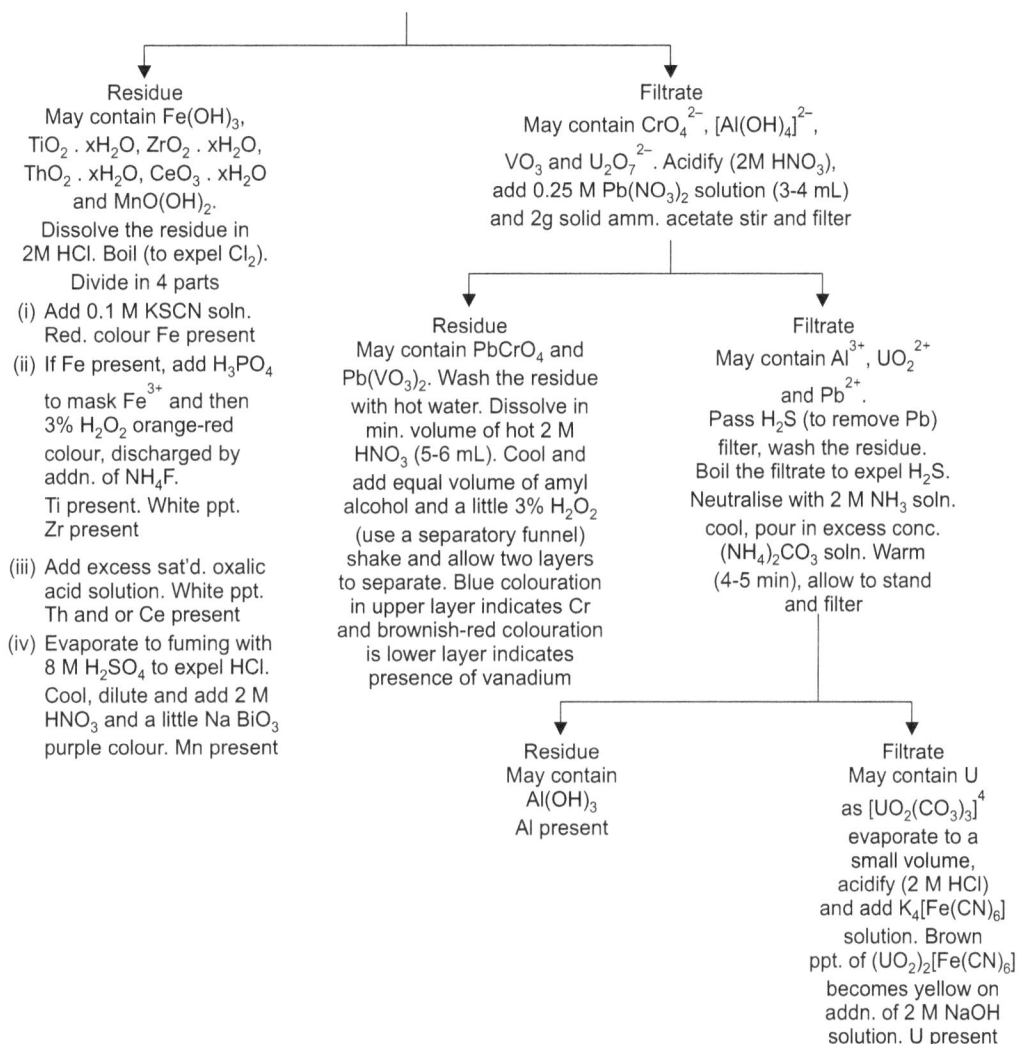

Scheme 1.20. Separation, identification of group III cations, Ti, Zr, Ce, Tn and U.

Confirmatory Tests of Ti, Zr, Ce, Th and U

The confirmatory tests of Al, Fe and Cr have already been described (page 57–58) confirmatory test sof Ti, Zr, Ce, Th and U are described.

Titanium

(*i*) **Sodium hydroxide solution.** White gelatinuous ppt. of orthotitanic acid, H_4TiO_4 is obtained. The ppt. is insoluble in excess NaOH but soluble in mineral acids.

(*ii*) **Zinc or cadmium.** Addition of zinc or cadmium to titanium salt solution containing excess HCl gives violet colour due to reduction to Ti(III) ions.

(*iii*) **Cupferron reagent.** Yellow ppt. of $Ti(C_6H_5O_2N_2)_4$ in acid solution is obtained. If Fe is present, it is removed by precipitation with NH_4OH and $(NH_4)_2S$ solution in presence of a tartarate; subsequently. Ti is ppted from acidified solution by cupferron.

(*iv*) **Chromotropic acid.** Adding chromotropic acid (1, 8-dihydroxynaphthalene-3, 6- disulphonic acid) to the test solution gives reddish-brown colouration in presence of HCl or H_2SO_4.

(*v*) **Catechol (spot test).** A drop of acidified test solution is placed on a drop-reaction paper impregnated with catechol solution. A yellow or yellowish-red spot results.

(*vi*) **Atomic absorption spectroscopy.** Using a titanium hollow-cathode lamp and acetylene-nitrous oxide flame, absorption is observed at 365.35 nm.

Zirconium

(*i*) **Sodium hydroxide solution.** A white gelatinous ppt. of hyroxide, $Zr(OH)_4$ (or $ZrO_2.xH_2O$) in cold, (insoluble in excess NaOH solution) is obtained.

(*ii*) **Disodium hydrogen phosphate solution.** A white ppt. of zirconium hydrogen phosphate, $Zr(HPO_4)_2$ or $ZrO(H_2PO_4)_2$ is obtained.

(*iii*) **Hydrogen peroxide.** A white ppt. of peroxyozirconic acid, $HOO-Zr(OH)_3$ from slightly acidic solution is obtained.

(*iv*) **Oxalic acid solution.** A white ppt. of zirconium oxalate (soluble in excess oxalic acid) is obtained.

(*v*) **Atomic absorption spectroscopy.** Using a zirconium hollow-cathode lamp and acetylene-nitrous oxide flame, absorption at 360.12 nm is observed.

Cerium

(*i*) **Sodium hydroxide solution.** A yellow ppt. of $Ce(OH)_4$ is obtained, which on warming with HCl evolves chlorine.

(*ii*) **Oxalic acid.** On warming with oxalic acid solution a ppt. of cerium oxalate is obtained.

(*iii*) **Ammonical silver nitrate reagent (spot test).** A drop of the neutral test solution on warming (on a watch glass) with one drop of the reagent gives a black or brown colouration.

(*iv*) **Anthranilic acid reagent (spot test).** A drop of the slightly acidic test solution (HNO_3) placed on a spot plate and addition of a drop of the reagent gives a blackish-blue ppt.

Thorium

(*i*) **Oxalic acid solution.** A white crystalline ppt. of thorium oxalate, $Th(C_2O_4)_2$ (insoluble in the reagent and 0.5 M HCl) is obtained.

(*ii*) **Hydrogen peroxide.** White ppt. of hydrated thorium heptoxide, $Th_2O_7.4H_2O$ (in neutral or faintly acidic solution) is obtained.

(*iii*) **Saturated sebacic acid.** A white ppt. of thorium sebacate, $Th(C_{10}H_{16}O_4)_2$ is obtained.

(*iv*) **3-Nitrobenzoic acid reagent.** A white ppt. of the salt $Th(NO_2.C_6H_4.COO)_4$ is obtained by addn. to neutral tests solution and heating to 70–80°C.

Uranium

(*i*) **Ammonium hydroxide solution.** A yellow ppt. of ammonium diuranate (insoluble in excess reagent) is obtained. The ppt, however is soluble in NH_4CO_3 solution.

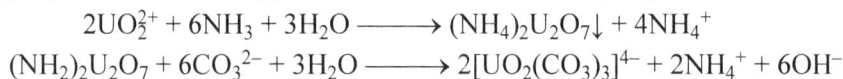

$$2UO_2^{2+} + 6NH_3 + 3H_2O \longrightarrow (NH_4)_2U_2O_7\downarrow + 4NH_4^+$$
$$(NH_2)_2U_2O_7 + 6CO_3^{2-} + 3H_2O \longrightarrow 2[UO_2(CO_3)_3]^{4-} + 2NH_4^+ + 6OH^-$$

(*ii*) **Hydrogen peroxide.** A pale yellow ppt. of uranium tetroxide, $UO_4.2H_2O$ (soluble in ammonium carbonate giving deep yelow solution) is obtained.

(*iii*) **Atomic absorption spectroscopy.** Using a uranium hollow-cathode lamp and acetylene nitrous oxide flame, absorption occurs at 358.49 nm.

1.13.5 Separation, Identification of Group IV Cation (Ni, Co, Zn, Mn) and Thorium

The filtrate obtained after separation of group III cation (Section 1.13.4) contains cations of group IV. H_2S gas is passed into the solution. The precipitate obtained is dissolved in 2M HNO_3. The solution is boiled (to expell H_2S) and saturated solution of H_2SO_3 added (SO_2 is expelled by boiling the solution). The remaining solution is poured into excess of saturated Na_2CO_3 solution and filtered (Scheme 1.21).

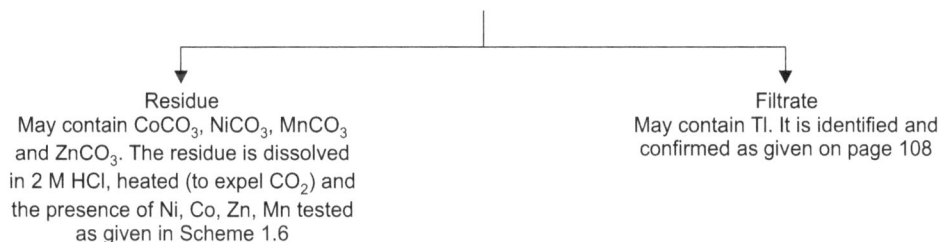

Residue
May contain $CoCO_3$, $NiCO_3$, $MnCO_3$ and $ZnCO_3$. The residue is dissolved in 2 M HCl, heated (to expel CO_2) and the presence of Ni, Co, Zn, Mn tested as given in Scheme 1.6

Filtrate
May contain Tl. It is identified and confirmed as given on page 108

Scheme 1.21. Separation, identification of group IV cations and thorium.

EXERCISES

1. For selecting a method for the identification of cations what should be kept in mind?
2. How different cations are classified into various groups in the usual analysis of cations using H_2S?
3. How the solution of the mixture is obtained for the analysis of cations?
4. Before preceding for the analysis of group III cations, which radicals interfere and how these are removed?
5. Comment on the use of spot test for confirming the cations.
6. Why semi-micro scale analysis is preferred?
7. How are dry tests useful for inferring the presence of cations and anions?

8. Write a note on: (*i*) Flame test, (*ii*) Boraxbead test, and (*iii*) Charcoal cavity tests.

9. For testing of anions, why sodium carbonate extract is prepared.

10. How will you test:

    (*i*) Carbonate in presence of bicarbonate

    (*ii*) Carbonate in presence of sulphite and thiosulphate

    (*iii*) Sulphide, sulphite, sulphate and thiosulphate in presence of each other

    (*iv*) Chloride, bromide and iodide in presence of each other

    (*v*) Nitrate in presence of bromide and iodide

    (*vi*) Sulphide in presence of fluoride

    (*vii*) Oxalate in presence of carbonate.

11. How flame photometry and atomic absorption spectroscopy used for conforming various cations.

12. Write a note on chromatographic separation and identification of cations.

13. How will you proceed to identify the insolubles present in a mixture.

14. How is analysis of metal or an alloy carried out.

OOO

PART II

Volumetric Analysis
Instrumental Methods of Volumetric Analysis

2. Volumetric Analysis

Contents

Volumetric Analysis

2.1 INTRODUCTION

Volumetric analysis, also known as titrimetric analysis is a type of quantitative analysis involving measurement of volumes. The method is less laborious, fast and accurate compared to other methods of quantitative analysis.

This method of quantitative analysis involves determination of an unknown amount of a substance (chemical) by reaction with another reagent to give a known product and determination of the amount of reagent needed to complete the reaction. As an example in the following reaction

$$NaOH + HCl \longrightarrow NaCl + H_2O$$

One formula weight of the base (NaOH) neutralises one formula weight of acid (HCl). It can also be said that one formula weight of acid is equivalent to one formula weight of the base. Thus, the unknown weight of sodium hydroxide can be found out by finding the amount standard of acid required to react completely with it.

Two requirements for volumetric estimations are (*i*) the reaction should be fast and (*ii*) the reaction should go to completion, which is observed by using an indicator. The point at which the completion of the reaction is observed by using an indicator is called the end point. The end point, also called the equivalence point must be sharp.

In volumetric analysis a standard solution is used. A standard solution is a solution whose strength is known. It is prepared by dissolving an accurately weighed quantity of a pure substance (called the primary standard) in water or some other solvent in a volumetric flask so as to get for example 100 mL of the solution. In case, the substance to be dissolved is not a primary standard, then the solution is prepared approximately of the desired concentration. The solution thus obtained is standardised by titrating with another standard solution. Such solutions which are not directly obtained as primary standard are called secondary standards.

The requirement for a substance to be a primary standard are:
- It must be available in pure state
- It must be easily soluble in the desired solvent

- It must be stable and should not undergo decomposition in presence of the solvent
- It must be unaffected by the atmosphere
- It should not deteriorate on keeping.

To sum up, volumetric analysis is the determination of the quantities of the constituent present in a given unknown solution by measuring the volumes of solutions taking part in a chemical reaction. The process is called **titration**, which means determination of the volume of a reagent to bring about a reaction to go to completion.

Law of Equivalence

Volumetric analysis is governed by the law of equivalence, according to which, the number of equivalents of a substance to be titrated is equal to the number of equivalent of the titrant used.

As an example, in a neutralisation reaction, let V_1 mL of an acid solution of normality N_1 require V_2 mL of base of normality N_2

1000 mL of $1N$ acid solution contains $= 1$ gram equivelant of acid

$\therefore V_1$ mL of N_1 acid solution contains $= \dfrac{1}{1000}$ of $N_1 \times V_1$ gram equivalent

In a similar way, number of gram equivalent of a base in V_2 mL of N_2 normality

$$= \dfrac{N_2 V_2}{1000}$$

By the law of equivalents at the end point

$$\dfrac{N_2 V_2}{1000} = \dfrac{N_1 V_1}{1000} \text{ or } N_1 V_1 = N_2 V_2$$

Terms Used in Volumetric Analysis

Following are given various terms used in volumetric analysis:

(*i*) **Equivalent Mass of Substances:** The equivalent mass of an acid is obtained by dividing the molecular mass of the acid by the number of replaceable hydrogen atoms (the number of replaceable H atoms in called **basically of the acid**).

e.g., $H_2SO_4 \longrightarrow 2H^+ + SO_4^{2-}$; it is a dibasic acid, its equivalent mass

$$= \dfrac{\text{Molecular mass}}{\text{Basicity}} = \dfrac{98}{2} = 49$$

Equivalent mass of alkali is its weight which reacts with the equivalent mass of acid.

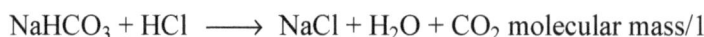

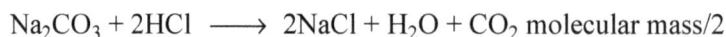

$$NaOH + HCl \longrightarrow NaCl + H_2O \text{ molecular mass/1}$$

$$NaHCO_3 + HCl \longrightarrow NaCl + H_2O + CO_2 \text{ molecular mass/1}$$

$$Na_2CO_3 + 2HCl \longrightarrow 2NaCl + H_2O + CO_2 \text{ molecular mass/2}$$

The following table gives the equivalent weights of some important acids:

Acid	Molecular Weight	No. of Replaceable H Atoms	Equivalent Weight
HCl	36.5	1	36.5
H_2SO_4	98	2	49
HNO_3	63	1	63
$(COOH)_2 \cdot 2H_2O$	126	2	63
CH_3COOH	60	1	60

In a similar way, the equivalent mass of a base is equal to the molecular mass divided by its acidity (*i.e.*, the number of replaceable hydroxyl ions). Thus,

$$\text{Equivalent mass of a base} = \frac{\text{Molecular mass}}{\text{Acidity}}$$

As an example the equivalent mass of NaOH $= \dfrac{40}{1} = 40$

Following table gives the equivalent mass of some commonly used bases:

Base	Molecular Weight	Acidity	Equivalent Weight
NaOH	40	1	40
KOH	56	1	56
$Ba(OH)_2$	171	2	85.5
Na_2CO_3	106	2	53
$NaHCO_3$	84	1	84

In an oxidation-reduction reaction, the equivalent mass is calculated as follows:

It is known that an oxidising agent is a substance that gains electrons and gets reduced. On the other hand, a reducing agent is a substance that loses electron and gets oxidised.

$$\text{Equivalent mass of an oxidising agent} = \frac{\text{Molecular mass or formula mass}}{\text{No. of electrons gained by one molecule}}$$

$$\text{Equivalent mass of a reducing agent} = \frac{\text{Molecular mass or formula mass}}{\text{No. of electrons lost by one molecule}}$$

Equivalent mass of a subtance depends upon oxidation state of the substance undergoing oxidation or reduction according to the specified equation.

Equivalent Mass of Some Substances

Substance	Ionic Equation	Molecular Mass	No. of Electron lost or gained (n)	Eq. mass = Mol. mol. mass / n
1. Ferrous sulphate (Anhydrous) $FeSO_4$	$Fe^{2+} \longrightarrow Fe^{3+} + e^-$	152	1	152
2. Hydrated ferrous sulphate $FeSO_4.7H_2O$	$Fe^{2+} \longrightarrow Fe^{3+} + e^-$	278	1	278
3. Mohr's salt $FeSO_4(NH_4)_2SO_4.6H_2O$	$Fe^{2+} \longrightarrow Fe^{3+} + e^-$	392	1	392
4. Oxalic acid $H_2C_2O_4.2H_2O$ COOH | .2H$_2$O COOH	$C_2O_4^{2-} \longrightarrow 2CO_2 + 2e^-$	126	2	63
5. Potassium permanganate $KMnO_4$	$MnO_4^- + 8H^+ + 5e^- \longrightarrow Mn^{2+} + 4H_2O$	158	5	31.6

(*ii*) **Molarity:** It is defined as the number of gram moles of a solute per litre of solution. Infact, the concentration of a solution is expressed as molarity. The concentration of a solution can also be expressed as normality [see (*iii*) below]

$$\text{Molarity} = \frac{\text{Number of gram moles of solute}}{\text{Volume of solution in litre}}$$

$$= \frac{\text{Weight of substance/Molecular mass}}{\text{Volume of solution in litre}}$$

$$= \frac{\text{Weight of solute in gram per litre of solution}}{\text{Molecular mass}}$$

(*iii*) **Normality:** It is defined as the number of gram equivalents of solute per litre of a solute

$$\text{Normality} = \frac{\text{Number of gram equivalents of solute}}{\text{Volume of solution in litre}}$$

$$= \frac{\text{Weight of solute/Equivalent mass}}{\text{Volume of solution in litre}}$$

$$N = \frac{\text{Weight of solute per litre of solution}}{\text{Equivalent mass}}$$

As an example one normal (IN) solution of oxallic acid is prepared by dissolving 0.63 *g* of oxalic acid in 100 mL water. In this case the normality is calculated as follows:

$$\text{Equivalent mass of oxalic acid} = \frac{\text{Molecular mass}}{\text{Basicity}} = \frac{126}{2} = 63$$

Oxalic acid, $\begin{array}{c} COOH \\ | \\ COOH \end{array}$ $2H_2O$ is a dibasic acid and has two replaceable H atoms.

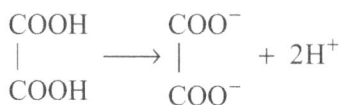

$$\begin{array}{c} COOH \\ | \\ COOH \end{array} \longrightarrow \begin{array}{c} COO^- \\ | \\ COO^- \end{array} + 2H^+$$

Relationship between normality (N) and molarity (M)

$$N = M \times n$$

where n = no. of e^- change.

(*iv*) **Strength:** As in the case of normality, strength is also a concentration term and is defined as the weight of a solute in grams dissolved in a litre of the solution

$$\text{Strength} = \text{Normality} \times \text{Equivalent weight}$$

$$= \frac{1}{10} \times 63 = 6.3 \ g/L$$

Thus, for preparation of one litre of $\frac{N}{10}$ oxalic acid solution, 6.3 g oxalic acid are dissolved in 1 litre water.

$\therefore$ For preparing 250 mL of $\frac{N}{10}$ oxalic acid, the amount of oxalic acid required

$$= \frac{6.3}{1000} \times 250 = 1.575 \ \text{gram}$$

Preparation of 250 mL of $\frac{N}{10}$ oxalic acid involves following steps:

(*i*) A clean and dried weighing bottle is exactly weighed

(*ii*) About 1.575 g of oxalic acid is added to the weighing bottle and the weight again found.

(*iii*) The weighed oxalic acid is transferred to a 250 mL measuring flask using a funnel. The weighing bottle is weighed again.

(*iv*) 75-100 mL distilled water is added and the measuring flask is shaken till whole of oxalic acid has dissolved. Finally, the volume is made up with distilled water until the lower meniscus is at level with the marking on the measuring flask. The solution is shaken to get a uniform solution.

(*v*) The exact normality of the oxalic acid solution is calculated as follows:

The amount of oxalic acid transferred to the measuring flask

$$= W_2 - W_1 = 1.5702 \ g.$$

where W_1 and W_2 are the weight of the weighing bottle + oxalic acid and W_2 is the weight of weighing bottle after transfer of oxalic acid.

$\therefore$ 250 mL of the solution contains 1.5702 g oxalic acid.

$\therefore$ 1000 mL of this solution will contain $\dfrac{1.5702 \times 1000}{250}$

$$= 6.280 \ g \ \text{oxalic acid}$$

$$\text{Strength} = \text{Normality} \times \text{Equivalent Weight}$$

$$\therefore \qquad \text{Normality} = \frac{\text{Strength}}{\text{Equivalent weight}} = \frac{6.280}{63} = 0.0996 \ N$$

In a similar way, 250 mL of $\frac{N}{20}$ Mohr's salt [$FeSO_4(NH_4)_2SO_4.6H_2O$] is prepared as follows:

The ionic equation of the oxidation of Mohr's salt is

$$Fe^{2+} \longrightarrow Fe^{3+} + e^-$$

$\therefore$ Equivalent weight of Mohr's salt $= \dfrac{\text{Molecular weight}}{1}$

$$= \frac{392}{1} = 392$$

Strength = Normality × Equivalent weight

$$= \frac{1}{20} \times 392 \qquad \qquad \text{(if normality is } N/20)$$

$$= 19.6 \ g/l$$

For preparing 250 ml of $\frac{N}{20}$ Mohr's salt solution, Mohr's salt required

$$= \frac{19.6}{1000} \times 250 = 4.9 \ g$$

The procedure involved for the preparation of Mohr's salt solution is similar to that used for the preparation of standard oxalic acid solution (page 127).

Apparatus used in Volumetric Analysis

Different apparatus used in volumetric analysis are given below:

(*i*) **Burette:** It is a long cylindrical tube of uniform bore having a stop cock at the lower end. The stop cock is either fused to the lower end or a separate stop cock is attached with a rubber tube. The burette is graduated in millimeters from 0 to 50 and each division is subdivided into 10 equal parts. (*see Fig. 2.1*)

Before filling the burette with the soluton, it is washed with water and then rinsed with the solution to be filled. This is done by holding the burette in horizontal position any rotating it to remove any water sticking inside. The solution is discarded.

The burette is filled with the solution using a funnel (*see Fig. 2.1*). Adequate precaution is taken to remove air bubbles remaining in the narrow bottom tip of the burette. Finally, the level of the solution in the burette is adjusted to zero mark or any other convenient mark (this is achieved by opening the stop cock slightly and allowing the solution to run down).

The burette reading is noted by reading the lower meniscus in case of a colourless solution and upper meniscus in case of coloured solution. This is achieved by using an anti parallex card to eleminate errors in reading due to parallex. An anti parallex card is prepared by taking a piece of paper and folding it half and giving two cuts as shown in (Fig. 2.2). The fold of the paper is opened and mounted on the burette. The reading in the burette is taken by placing the eye exactly in the level of the meniscus.

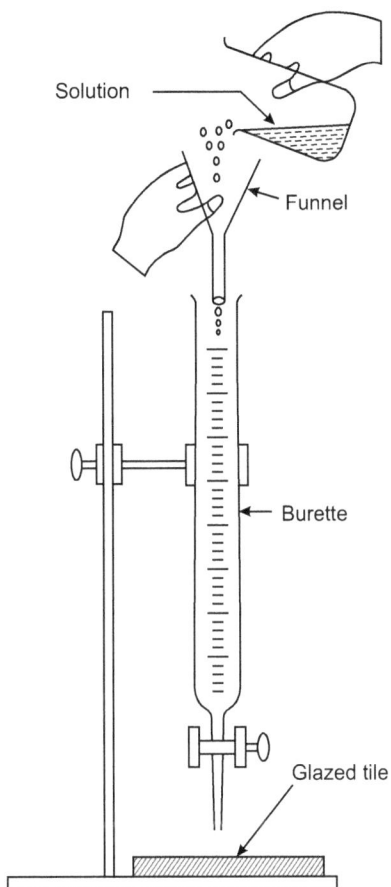

Fig. 2.1. Filling the burette.

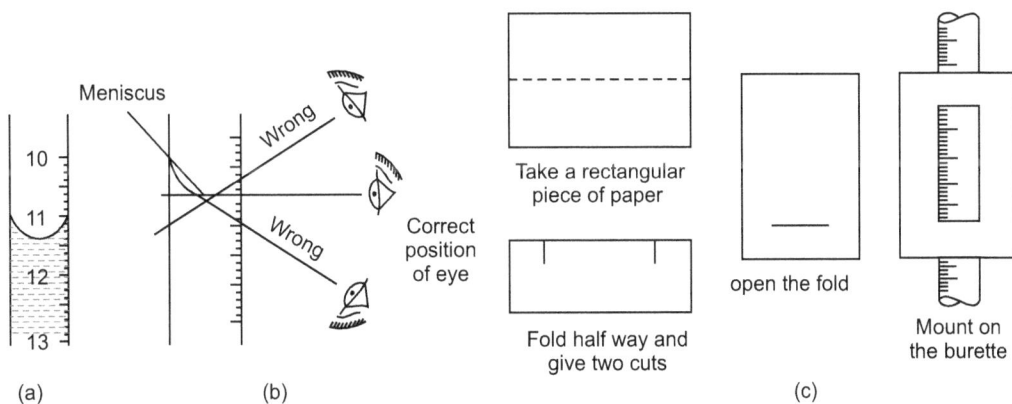

Fig. 2.2. (*a*) Lower meniscus (*b*) correct way of reading burette
(*c*) making and mounting an anti parallex card.

(*ii*) **Pipette:** The pipette is a long narrow glass tube having a cylindrical bulb in the middle and a jet at its lower end. It is used for the measurement of a definite volume of a solution. An etched circular mark is on the upper part of the pipette. The bulb has a mark like 10, 20 etc. indicating the volume of solution (in mL) it can deliver when filled upto the circular mark.

The pipetle is first cleaned with water and then rinsed with the solution. It is then filled with the solvent by holding the stem of the pipette by the thumb and middle finger of the right hand, dipping the lower end in the solution and sucking the solution in the pipette just above the etcheed mark. The open end of the pipette is closed with the index finger, the pipette is raised.

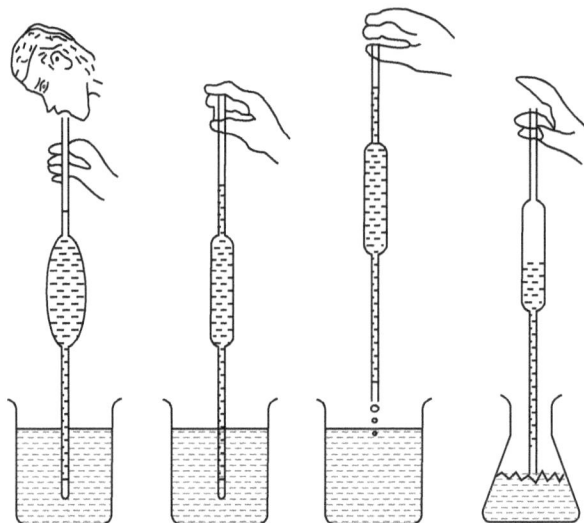

Fig. 2.3. Correct use of a pipette.

So that the mark is at eye level and the liquid is allowed to fall slowly (by controlled release of the finger from the stem of the pipette) till the lower meniscus (in case of colourless solution) is at the mark. Subsequently, the solution in the pipette is allowed to run freely in the titration flask. A very small amount to the liquid is invariably retained in the tip of the pipette and no attempt be made to expel it by blowing.

(*iii*) **Titration Flask:** It is usually a conical flask and is used for titration. The titration flask should be rinsed only with distilled water and **NOT** with the solution in order to avoid volume error.

Titration Process

The process of titration is useful to find the volume of a solution required to react completely with a certain known volume of another solution.

The burette is clamped in a clamp stand and then filled with the solution to a convenient mark. Normally, a solution of unknown concentration called the titre is taken in burette. The solution to be titrated is pipetted into the titration flask and indicater (1-2 drops) added. A glazed tile (or white paper) is placed below the titration flask, which is placed on the tile below the burette nozzle. The initial reading of the burette is noted. The solution from the burette is added to the titration flask by opening the stop cock of the burette (Fig. 2.4). The titration flask is rotated throughout the titration. The addition to the solution from the burette is stopped when the end point is reached. Final reading in the burette is noted. The difference between the initial and final reading gives the volume of the solution required for completion of the reaction.

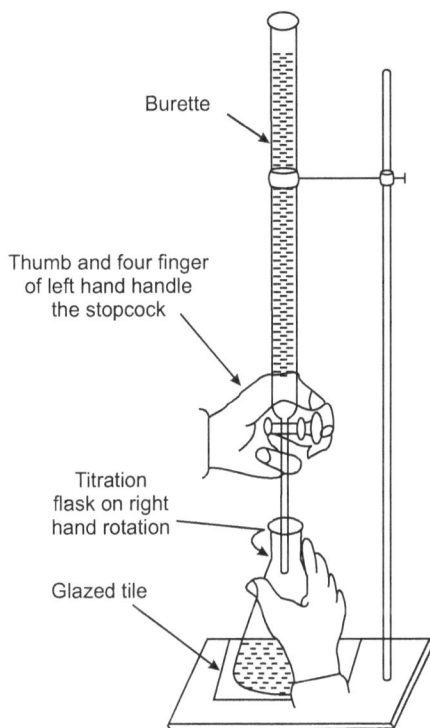

Fig. 2.4. Correct way of handling a burette.

The process of titration is repeated with the cleaned titration flask until two successive concordant reading are obtained.

(*iv*) **Measuring Flask:** It is used for making standard solution and is generally available in 100, 250, 500 and 1000 mL capacity. The measuring flask in flat bottomed having a long neck and has a circular mark etched on it. The measuring flask should never be heated. It can be rinsed with distilled water before using.

(*v*) **Weighing Tube:** It is a flat bottomed cylindrical glass tube with a glass stopper and is used for weighing chemical substances for making standard solutions. The weighing bottle to be used must be clean and dry. In place of weighing bottle a watch glass can also be used.

Note: All glass apparatus to be used for volumetric analysis must be thoroughly cleaned by soap solution or detergent (for removing ordinary contaminats by aquea regic (3:1 mixture of conc. HNO_3 and conc. HCl) (for removing carbonaceous matter. For removing iodine stains, sodium, thiosulphate. For removing $KMnO_4$ stains, ferrous sulphate-sulphuric acid mixture is used. Finally the apparatus is washed with water.

(*vi*) **Analytical Balance:** It is used for accurate weighing and is the most important part of volumetric analysis. All weighings must be done upto 3 or 4 significant figures, which is only possible if the balance is sensitive and accurate. Two types of analytical balances, *viz.*, chemical balance and single pan electrical balance are available.

Chemical Balance

A chemical balance must have a high degree of precision and can be used for weighing objects to a precision of 0.0002 *g*. A usual design of a balance is shown in Fig. 2.5.

A chemical balance has three knife edges made of agate of hard steel. The knife edge rests on a smooth agate plate on the top of a balance column. The pans of the balance are suspended from terminal knife edges by means of stirrups. A pointer is fixed to the center of the beam. As the balance pans swing, the lower end of the pointer moves to the left or right on the scale (fixed at the bottom of the balance. When not is use, the balance must be arrested. A glass case encloses the balance to protect it from dust and air movement etc. The glass case has two side doors. The substance are weighed in a weighing bottle.

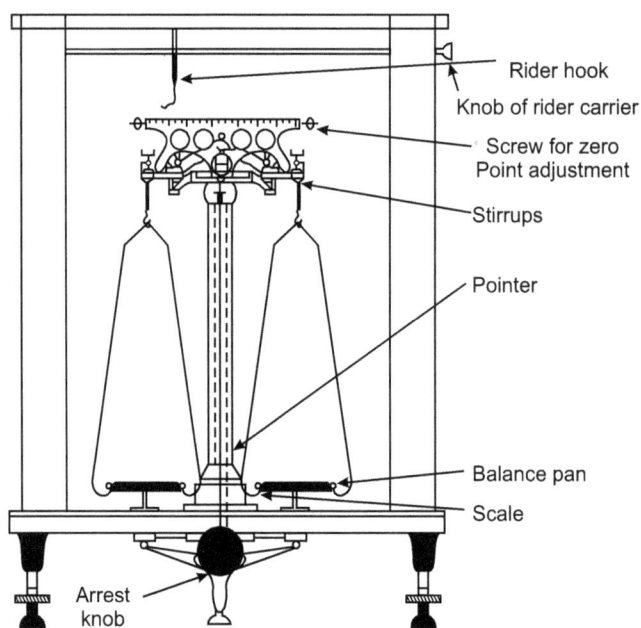

Fig. 2.5. Chemical balance.

The weight box contains a pair of forceps (for lifting the weights of 50, 20, 10, 5, 2 and 1 *g*). It also contains fractional weights of 500, 200, 100, 50, 20, 10 mg and a rider. The rider is a thin bent wire and is used to weigh 0.001 *g* (as shown in Fig. 2.6).

In order to avoid damage to the chemical balance and to ensure accurate weighing, following precautions should be taken:

• The unarrested balance should not be touched,

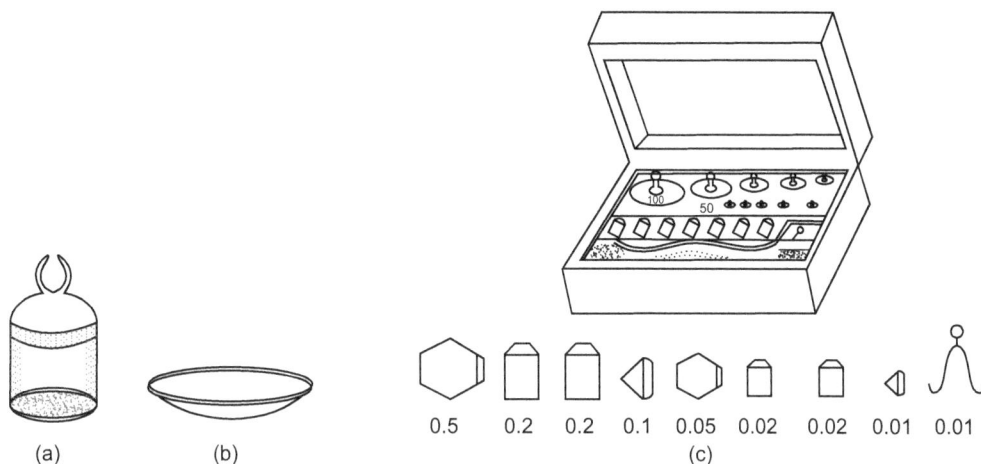

Fig. 2.6. (a) Weighing bottle (b) Watch glass (c) Weight box and Fractional weights.

- The balance should be arrested before the objects and weights are put on the pans or taken off,
- The substance to be weighed should not be placed directly on the balance pan,
- Do not weigh hot objects,
- Only the side doors should be used while weighing,
- Do not touch the weights or rider with fingers, use forcep for lifting the weights and rider.

Single Pan Electrical Balance

A single pan electrical balance has only two knife edges. A simplified representation of a single pan balance is shown in Fig. 2.7. The knife edge is situated at the centre of the beam in such a way that one balance arm is about twice the length of the other. The pan is suspended from the shorter arm of the beam. A set of weights are placed on the side of the knife edge. A constant weight is placed at the opposite end of the beam in order to encounter balance the weights and weight of the pan on placement of an object on the pan, the beam is heavier on the side of the knife edge due to weight

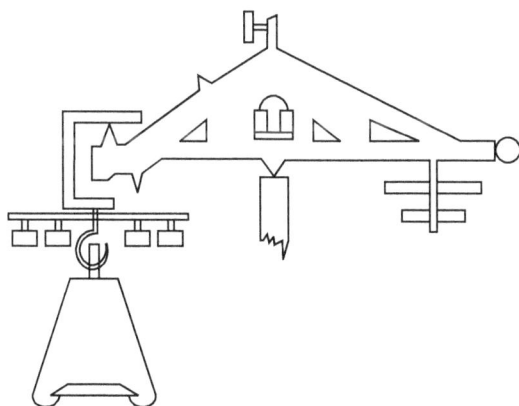

Fig. 2.7. A single pan electrical balance.

of the object. Subsequently, the weights are removed incrementally from the side of the beam till the sum of the weights exactly equals the weight of the object. Thus, the load on the beam when weighing an object is the same as there is an object on the pan and no weights have been removed. Such a types of balance is called a constant load balance.

A single pan electrical balance is useful for accurate weighing in a much shorter time. The sensitivity of the balance remains constant as the load on the beam is maintained constant.

Applications of Volumetric Analysis

The volumetric analysis is useful for many types of applications including determination of solubilities, purities and composition of mixtures. Finding the composition of substances like molecular formula, water of crystallization, composition of ores and alloys, and determination of the stoichiometry of equations, *e.g.,* oxidizing power of substances, number of replaceable hydrogen atoms, etc.

Types of Volumetric Analysis

Four main types of volumetric analysis are generally used. These are:

1. Acidimetry and alkalimetry (Neutralization titrations).
2. Redox titrations.
3. Precipitation titrations.
4. Complexometric titrations.

2.2   ACIDIMETRY AND ALKALIMETRY

Determination of the strength of an acid solution by titration against a standard solution of a base is called **acidimetry**. On the other hand, determination of the strength of an alkali by titration against a standard solution of an acid is called **alkalimetry**.

The reaction between an acid and an alkali (called **neutralization**) proceeds with the formation of a salt and water in accordance with the general equation.

$$\text{Acid} + \text{Base} \longrightarrow \text{Salt} + \text{Water}$$

The completion of the reaction is indicated by the **end point**, which is assisted by addition of an Indicator. The indicator which is used in these titrations is required to indicate the equivalence point rather than the neutral point (pH 7). At the equivalence point, the pH of the solution could be equal to, greater or less than pH 7, depending on the relative strength of the acid and alkali. This is because the salt formed in solution at the end of the reaction undergoes slight hydrolysis and so the pH of the solution will be either less or more than 7. The solution is not exactly neutral. In case of titration of a strong acid and a strong base, the degree of hydrolysis is negligible and the pH of the solution will be approximately 7. However, if one or both of the acid and base form a weak salt, for example, an aqueous solution of sodium acetate (which is alkaline) forms ammonium chloride which is acidic. The selection of a indicator for a partitular titration is decided by these considerations.

Indicators

The indicators used in acidimetry and alkalimetry are either weak organic acids or weak organic bases. The degree of dissociation of these indicators is greatly affected by the alternation of hydrogen ion concentration of the solution. An acid indicator is generally expressed by the formula HIn, as for phenolphthalein and a basic indicator is expressed as InOH (as in the case of methyl orange). Their respective dissociation is expressed in the following way:

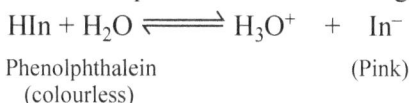

$$HIn + H_2O \rightleftharpoons H_3O^+ + In^-$$

Phenolphthalein   (Pink)
(colourless)

$$K_{In} = \frac{\left[H_3O^+\right]\left[In^-\right]}{\left[HIn\right]}$$

and

$$InOH \rightleftharpoons In^+ + OH^-$$

(Yellow)   (red)

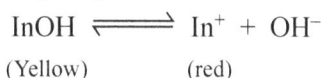

Methylorange

$$K_{In} = \frac{\left[In^+\right]\left[OH^-\right]}{\left[InOH\right]}$$

As seen, the dissociated and the undissociated forms are in dynamic equilibrium mixture of two tautomeric forms, which have different structural formulae and colours.

One of these tautometric form exists in acidic medium and the other form in alkaline medium. Usually, one of the forms in benzenoid and the other form is quinonoid in structure. The latter (*viz.* quinonoid form) is deeper in colour. The change in pH causes the transformation of benzenoid form to the quinonoid form and *vice-versa* and consequently leads to a change in colour. As an example, the tautometric forms of two indicators phenolphthalein and methyl orange are represented as shown below:

(*i*) Phenolphthalein

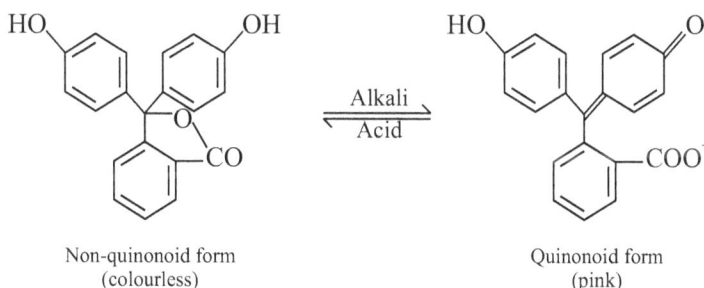

Non-quinonoid form   Quinonoid form
(colourless)   (pink)

(*ii*) Methyl orange

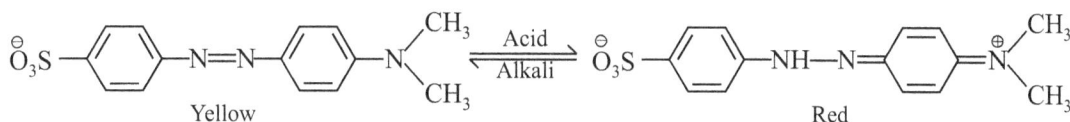

Following are given some examples of acid-base titrations.

2.2.1 Determine the Strength of Hydrochloric Acid using $\frac{N}{20}$ Sodium Hydroxide

Theory

It is a titration between a strong acid and a strong base

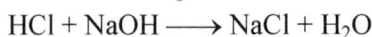

$$HCl + NaOH \longrightarrow NaCl + H_2O$$

At the end point, with a slight excess of NaOH added from the burette, the pH of the solution reaches between 8.5 – 10 and hence, phenolphthalein is the indicator of the choice. The colourless solution acquires a pink colour at the end point.

Procedure

Following steps are involved:

(i) Prepare a standard solution of oxalic acid (dissolve ~ 0.3 g, weigh accurately) in 100 mL distilled H_2O to get an $N/20$ solution).

(ii) Pipette out 10 mL of the standard oxalic acid solution in a conical flask, add a drop of phenolphthalein and titrate with sodium hydroxide (taken in the burette) to obtain a permanent light pink colour. Repeat to get three concordant values.

(iii) Pipette out 10 mL of given hydrochloric acid in the conical flask; add a drop of phenolphthalein and titrate against standardized sodium hydroxide solution as described above.

Calculation

Amount of oxalic acid in the standard solution $= y \times 10$ g/litre

(y g is the amount of oxalic acid in 100 ml of standard solution)

$$\text{Normality } (N) \text{ of oxalic acid} = \frac{\text{Amount per litre}}{\text{Equivalent}} = \frac{y \times 10}{63} = N$$

Normality (N_1) of NaOH solution

(Oxalic acid) $V \times N = V_1 \times N_1$ (NaOH)

$$N_1 = \frac{N \times 10}{V_1}$$

(V_1 = volume of NaOH used for titration with 10 ml of oxalic acid)

Normality (N_2) of given hydrochloric acid

(NaOH) $V_1 N_1 = V_2 N_2$ (HCl)

$$N_2 = \frac{N_1 \times V_1}{V_2}$$

Amount of HCl $= N_2 \times 36.5$ g litre,

where 36.5 g is the equivalent weight of HCl.

2.2.2 Determine the Strength of Oxalic Acid Solution Using $\frac{N}{20}$ Sodium Hydroxide

Theory

This estimation involving the titration of a weak acid (oxalic acid) against strong base (sodium hydroxide) and phenolphthalein is used as an indicator

$$(COOH)_2 + 2NaOH \longrightarrow (COONa)_2 + 2H_2O$$

As sodium hydroxide is not a primary standard, a standard solution of oxalic acid is prepared and used for standardization of sodium hydroxide.

$$\text{The equivalent weight of NaOH} = \frac{Molecular\ weight}{1} = \frac{40}{1} = 40$$

and equivalent weight of oxalic acid dihydrate

$$(COOH)_2.2H_2O = \frac{Mol.\ weight}{2} = \frac{126}{2} = 63.$$

Procedure

Following steps are involved:

1. Prepare a standard solution of oxalic acid, $N/20$ (dissolve 0.3 g of oxalic acid dihydrate in 100 mL distilled water).

2. Rinse and fill the burette with the given solution of sodium hydroxide.

3. Pipette out 10 mL of the standard oxalic acid solution in a 100 mL conical flask and add a drop of phenolphthalein.

4. Titrate against sodium hydroxide till a light permanent pink colour is obtained.

5. Repeat the titration to obtain three concordant readings.

6. Now pipette out 10 mL of the given oxalic acid solution, add a drop of phenolphthalein and titrate it against sodium hydroxide to obtain three concordant readings.

Calculations

Mass of oxalic acid dissolved in 100 mL of standard solution = y g

$$\text{Amount of oxalic acid} = y \times 10 \text{ g/litre}$$

$$\text{Normality } (N) \text{ of standard oxalic acid} = \frac{\text{Amount/litre}}{\text{Eq. wt.}} = \frac{y \times 10}{63} = N$$

Normality (N_1) of sodium hydroxide solution

$$N_1 \times V = N \times 10$$

$$\text{(NaOH)} \quad \text{(Oxalic acid) (std)}$$

$$\therefore \qquad N_1 = \frac{N \times 10}{V}$$

Normality (N_2) of given oxalic acid solution

$$N_2 \times 10 = N_1 \times V_1$$

(Oxalic acid)   (NaOH)

$$N_2 = \frac{N_1 \times V_1}{10} = \frac{N \times 10 \times V_1}{10} = N_2$$

Amount of given oxalic acid = $N_2 \times 63$ g / litre.

2.2.3 Determine the Amount of Sodium Carbonate and Sodium Hydroxide in the given Solution Using $\frac{N}{20}$ Hydrochloric Acid

Theory

The reactions taking place in this titration are:

$$NaOH + HCl \longrightarrow NaCl + H_2O \qquad ...(i)$$
$$Na_2CO_3 + HCl \longrightarrow NaHCO_3 + NaCl \qquad ...(ii)$$
$$NaHCO_3 + HCl \longrightarrow NaCl + H_2O + CO_2 \qquad ...(iii)$$

Two sets of titrations are carried out between the solution of Na_2CO_3 + NaOH with standardized HCl. In the first titration methyl orange is added as indicator to a known volume of the given solution and titration is carried out till the solution acquires a orange colour (yellow $\longrightarrow$ orange). Methyl orange changes colour at pH, ~ 3.6 i.e., when whole of NaOH and Na_2CO_3 [e.g. (i), (ii), (iii)] have been neutralized (let the volume of HCl required in this titration be V_2).

In the second titration, phenolphthalein is added as the indicator and titration is carried out till the disappearance of the pink colour. This colour change occurs at pH (~ 9.5) and corresponds to the neutralization of NaOH [equation (i) above] and conversion of Na_2CO_3 to $NaHCO_3$ [equation (ii) above] (let the volume of HCl required is V_3).

So the volume of HCl required for complete neutralization of Na_2CO_3 = $2(V_2 - V_3)$ and the volume of HCl required for neutralization of NaOH = $V_2 - 2(V_2 - V_3)$.

Procedure

Following steps are involved:

(i) **Standardization of HCl using Std. Na_2CO_3:** Prepare a standard solution of sodium carbonate ($N/20$) by accurately weighing ~ 0.26 g of Na_2CO_3 and dissolving in 100 ml water. Pipette out 10 ml of standard sodium carbonate solution in a conical flask, add a drop of methyl orange. (The solution acquires a yellow colour). Titrate against hydrochloric acid solution till the solution becomes orange—red in colour. Repeat to obtain three concordant readings.

(*ii*) Titration of give solution of Na_2CO_3 and NaOH against standardised HCl using methyl orange as indicator.

Pipette out 20 mL of the given solution (containing NaOH + Na_2CO_3) in a conical flask and add few drops of methyl orange indicator. Titrate against standardised HCl till a permanent colour is obtained. Repeat the titration to get three concordant readings.

(*iii*) Titration of given solution of Na_2CO_3 and NaOH against standardised HCl using phenol-phthalein as indicator.

Pipette out 20 mL of the given solution (containing NaOH + Na_2CO_3) in a conical flask and add few drops of phenolphthalein indicator. Titrate against standardised HCl till the pink colour just disappears. Repeat the titration to get three concordant reading.

In an alternative procedure, instead of two sets of titration, only one set may be carried out. Various steps involved are:

1. Pipette out 20 mL of the given solution, add a drop of phenolphthalein and titrate till the pink colour disappears (this will give the volume of HCl required to neutralized NaOH + conversion Na_2CO_3 to $NaHCO_3$). Now add two drops of methyl orange to the above colourless solution and continue the titration with HCl till a red colour is obtained. This will give the total volume of HCl required to neutralize NaOH + total Na_2CO_3.

2. Methyl red can also be used instead of methyl orange as the indicator.

3. A mixture of thymol blue and cresol red (6:1) may be used instead of phenolphthalein. The colour change at the end point is violet to blue (pH 8.4 – 8.3).

Calculations

Mass of sodium carbonate = y g/100 mL

Normality (N) of standard sodium carbonate solution = $\dfrac{y \times 10}{53} = N$

Normality (N_1) of hydrochloric acid

$$N_1 V_1 = NV$$

$$\text{HCl} \quad \text{Standard}$$
$$\text{Na}_2\text{CO}_3$$

$$N_1 = \dfrac{NV}{V_1} = N_1$$

Total volume of HCl used for neutralization of 20 mL of given solution when methyl orange is used as the indicator = V_2.

Volume of HCl used when phenolphthalein is used as the indicator = V_3.

∴ Volume of HCl used to neutralic sodium carbonate to sodium bicarbonate = $V_2 - V_3$.

Since each equivalent of Na_2CO_3 gives two equivalent of $NaHCO_3$.

∴   Volume of HCl required to neutralize sodium carbonate in the given solution = $2(V_2 - V_3)$ mL

Hence, the normality (N_2) of Na_2CO_3 in the given solution

$$N_2 \times 20 = N_1 \times 2 \, (V_2 - V_3)$$

$$\text{(Na}_2\text{CO}_3) \qquad \text{(HCl)}$$

$$N_2 = \frac{2(V_2 - V_3) \times N_1}{20}$$

$$\text{Amount of Na}_2\text{CO}_3 = \frac{N_1 \times 2(V_2 - V_3)}{20} \times 53 \; g\,/\,\text{litre}$$

Volume of HCl used for neutralization of NaOH $= V_2 - 2(V_2 - V_3)$

$\therefore$ Normality (N_3) of NaOH is

$$N_3 \times 20 = [V_2 - 2(V_2 - V_3)] \times N_1$$

$$N_3 = \frac{[V_2 - 2(V_2 - V_3)] \times N_1}{20}$$

Amount of NaOH $= N_3 \times 40 \; g/\text{litre}$

2.2.4   Determine the Amount of Sodium Carbonate and Sodium Bicarbonate Present in the given Solution Using $\dfrac{N}{20}$ Hydrochloric Acid

Theory

The reaction between Na_2CO_3 and HCl takes place in two steps. In the first step Na_2CO_3 is converted to $NaHCO_3$ and in the second step $NaHCO_3$ reacts with HCl.

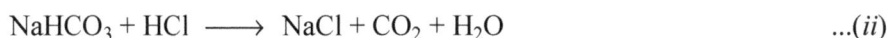

$$Na_2CO_3 + HCl \longrightarrow NaHCO_3 + NaCl \qquad \qquad ...(i)$$

$$NaHCO_3 + HCl \longrightarrow NaCl + CO_2 + H_2O \qquad \qquad ...(ii)$$

The sodium bicarbonate present in the mixture also reacts [equation (ii)] with HCl to form NaCl and CO_2.

The strength of Na_2CO_3 in the given solution is determined by titrating with standardized HCl in presence of phenolphthalein as the indicator. The solution (pink) becomes colourless when Na_2CO_3 is converted to $NaHCO_3$ [equation (i)]. Thus, the volume of HCl used in this titration when multiplied by two will be equivalent to the total volume of HCl required to neutralize the total amount of sodium carbonate in the solution.

The second set of titration is carried out using methyl orange as the indicator; methyl orange changes colour from yellow to red when whole Na_2CO_3 and $NaHCO_3$ have been neutralized.

Procedure

Following steps are involved:

(i) **Standardization of HCl:** Weight accurately ~ 0.26 g of anhydrous sodium carbonate and dissolve in water to make 100 mL solution in a standard flask.

Pipette out 20 mL of standard sodium carbonate in a conical flask and add 1-2 drops of methyl orange. The solution acquires a yellow colour. Titrate against hydrochloric acid slowly till the yellow solution becomes orange—red. Repeat the titration to obtain three concordant readings. (Let the volume of HCl be V_1).

(*ii*) **Titration of Na$_2$CO$_3$ + NaHCO$_3$ using Std. HCl:** Pipette out 20 mL of the given solution of Na$_2$CO$_3$ and NaHCO$_3$ in a conical flask, after adding a drop of phenolphthalein (the solution requires a pink colour) and titrate against HCl till a colourless solution is obtained (let the volume of HCl be V_2). Repeat the titration to obtain three concordant readings.

(*iii*) Finally pipette out 20 mL of the given solution of sodium carbonate + sodium bicarbonate, adding 1–2 drops of methyl orange and titration with HCl till a red colour is obtained (let the volume of HCl be V_3). Repeat to obtain three concordant readings.

Calculations

Mass of Na$_2$CO$_3$ in 100 ml solution = y g

Normality (N) of standard sodium carbonate = $\dfrac{y \times 10}{53} = N$

Normality (N_1) of HCl is $\qquad N_1 V_1 = NV$

$\qquad\qquad\qquad$ (HCl) $\qquad$ (Standard Na$_2$CO$_3$)

$$N_1 = \frac{NV}{V_1} = N_1$$

Volume of HCl used for neutralization of 20 mL of given solution of (Na$_2$CO$_3$ + NaHCO$_3$) when phenolphthalein is used as an indicator = V_2

$\therefore$ Volume of HCl required to neutralize the total amount of Na$_2$CO$_3$ present in the given solution of (Na$_2$CO$_3$ + NaHCO$_3$) = $2V_2$

$\therefore$ Normality N_2 of Na$_2$CO$_3$ in the given solution (Na$_2$CO$_3$ + NaHCO$_3$)

$$N_2 \times 20 = 2V_2 \times N_1$$

$\qquad$ (Na$_2$CO$_3$ in given solution) $\quad$ (Standardized HCl)

$$\text{Amount of Na}_2\text{CO}_3 = \frac{2V_2 N_1 \times 53}{20} \ g/\text{liter}$$

Volume of HCl required to neutralize total amount of Na$_2$CO$_3$ + NaHCO$_3$ present in the given solution = V_3 (*i.e.*, when methyl orange is used as indicator).

So the volume of HCl required to neutralize NaHCO$_3$ only in the solution = $V_3 - 2V_2$

Normally (N_3) of NaHCO$_3$

$$N_3 \times 20 = N_1(V_3 - 2V_2)$$

$$N_3 = \frac{N_1 (V_3 - 2V_2)}{20} ;$$

where N_3 is the Normality of NaHCO$_3$ in the given solution.

Amount of NaHCO$_3$ in the given solution = $N_3 \times 84$ g/litre

$$= \frac{N_1(V_3 - 2V_2) \times 84}{20} \ g/\text{liter}.$$

2.3 REDOX TITRATIONS

Introduction

Redox titrations include oxidation-reduction titrations and also iodine titrations. These involve transfer of electron from a substance which releases electrons to another substance which accepts the electrons.

The process of releasing electrons is oxidation and the process of accepting electrons is called reduction. Both oxidation and reduction are complementary to one another and take place simultaneously.

The reagent undergoing reduction is called **oxidizing agent** (oxidant) *i.e.,* oxidizing agent gains electrons and is reduced to lower valency state. The reagent which undergoes oxidation is called **reducing agent** (reductant), *i.e.,* reducing agent loses electrons and is oxidised to higher valency state. Examples of reducing agents include sodium oxalate, oxalic acid, ferrous sulphate, ferrous ammonium sulphate (Mohr's salt) etc., and examples of oxidizing agents include $KMnO_4$, $K_2Cr_2O_7$, potassium iodate, potassium bromate, ceric salts and I_2 solution.

Indicators

The indicators which are used in oxidation-reduction titrations are of following types:

(*i*) **Internal Oxidation-Reduction Indicators:** They exist in two different colours in their oxidised and reduced forms, and undergo the redox reaction only after the titration reaction is completed. This can happen only if their reduction potential is close to the reduction potential of the reacting reagents at the equivalence point. This would result in a sharp change in colour and would minimize the indicator error.

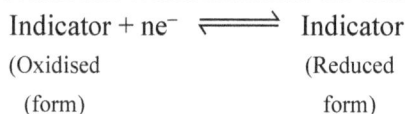

$$\text{Indicator} + ne^- \rightleftharpoons \text{Indicator}$$

(Oxidised   (Reduced

(form)   form)

$$E = E_0 - \frac{0.059}{N} \log \frac{[\text{In}]\,(\text{reduced form})}{[\text{In}^{n+}]\,(\text{oxidised form})}$$

Some examples of these indicators include diphenylamine, ferroin, N-phenyl anthranilic acid etc. Diphenylamine is a colourless molecule in the reduced form. On oxidation by an oxidizing agents like $KMnO_4$, $K_2Cr_2O_7$, $KClO_3$ etc. it gives a deep violet oxidised form.

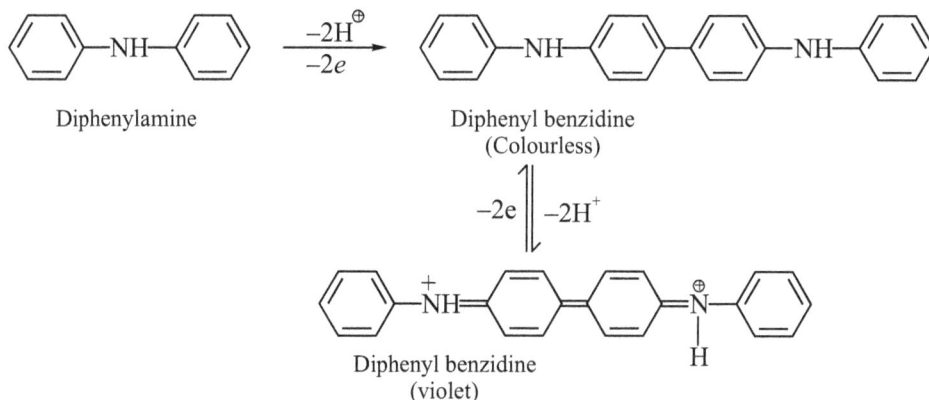

Diphenylamine   Diphenyl benzidine
                                         (Colourless)

Diphenyl benzidine
(violet)

(Since diphenylamine is sparingly soluble in water so usually sodium salt of diphenyl amine sulphonic acid is used).

The reduced form of the indicator is predominant at potentials below 0.73 V and the solution remains colourless. The oxidised form is predominant at $E = 0.79$ V and at higher potential the solution is violet. So the colour changes gradually from 0.73 V to 0.79 V. Hence, this indicator is very suitable for the titration of Fe^{++} with dichromate in which the reduction potential for the ferrous-ferric transition is 0.77 V (that for dichromate system being 1.33 V). However, this indicator is not suitable for the titration of ferrous ions with potassium permanganate where the change of potential extends from 0.94 V to 1.47 V.

(ii) **Self-indicator:** In the titrations using $KMnO_4$ as a titrating agent the reduced form Mn^{2+} forms an almost colourless solution. So as long as there is some reducing agent in the reaction mixture the solution remains colourless. But at the end of the titration all the reducing agent has been consumed. The next drop of $KMnO_4$ is in excess and remains unreduced. Thus, the solution turns pink and indicates the end of the titration by changing the colour of the solution from colourless to pink.

(iii) **Starch Indicator:** Iodine complexes with starch to form a deep blue coloured complex. The solution remains colourless till iodine is getting consumed. The first drop of excess iodine will change the solution to deep blue. If the iodine is in the reaction mixture then starch is added towards the end of the titration (when colour has became pale yellow) and the end point is indicated by a colour change from blue to colourless.

(iv) **External Indicator:** These are used when no internal indicator is available. These types of indicators are not added to the reaction medium because of two difficulties:

(a) In case of dark coloured liquids, a sharp change in colour at the end point is not visible;

(b) When the indicator forms an insoluble precipitate with an ion present in the reaction mixture solution to which it is added.

An example of external indicator is potassium ferricyanide $K_3Fe(CN)_6$ in the titration of $K_2Cr_2O_7$ and $FeSO_4(NH_4)_2SO_4.6H_2O$ in acidic medium. The reaction is

$$2K_3Fe(CN)_6 + 3FeSO_4 \longrightarrow Fe_3[Fe(CN)_6]_2 + 3K_2SO_4$$
Ferro-ferricyanide
(Deep blue colour)

If in case this indicator is added directly to the titration flask, some of the ferrous ions would be removed from the solution and a permanent deep blue colour would result.

To avoid these difficulties a freshly prepared 0.1% solution of potassium ferricyanide is placed in grooves of a white tile. A drop of the reaction mixture would not give a blue colour once the titration is complete, *i.e.,* all the ferrous ions have been oxidised to ferric ions.

Determination of Equivalent Weight

The equivalent weight of an oxidizing or reducing agent is equal to its molecular weight divided by the number of electrons gained or lost by it per molecule (or ion) during the course of a particular reaction.

In order to determine the equivalent weight of an oxidizing or reducing agent, a completely balanced ionic equation is written for the oxidation or reduction process and the number of electrons gained or lost per molecule is ascertained. Thus,

$$\text{Gram equivalent weight} = \frac{\text{Molecular weight of oxidising or reducing agent}}{\text{No. of electrons gained or lost per molecule of the substance}}$$

For example for the reactions:

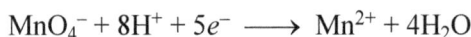

$$MnO_4^- + 8H^+ + 5e^- \longrightarrow Mn^{2+} + 4H_2O$$

Since, the reaction involves five electrons per ion of permanganate, the gram equivalent weight of $KMnO_4$ will be $1/5^{th}$ of its molecular weight.

$$\therefore \quad \text{Gram equivalent weight of } KMnO_4 = \frac{\text{Molecular weight of } KMnO_4}{5} = \frac{158.03}{5} = 31.606$$

2.3.1 Determine the Strength of given Oxalic Acid Solution by Titrating Against about $\dfrac{N}{20}$ Solution of Potassium Permanganate

Theory

The titration is a Redox Titration involving oxidation of oxalic acid by $KMnO_4$. In this reaction oxalic acid is oxidised to CO_2 and $KMnO_4$ gets reduced to $MnSO_4$. The reaction takes place in acidic medium (H_2SO_4) at $60 - 70°C$.

$$2KMnO_4 + 5C_2H_2O_4 + 3H_2SO_4 \longrightarrow K_2SO_4 + 2MnSO_4 + 10CO_2 + 8H_2O$$

The ionic form may be written as

$$2MnO_4^- + 16H^+ + 5C_2O_4^{2-} \longrightarrow 2Mn^{2+} + 10CO_2 + 8H_2O$$

Oxalic acid oxidises Mn^{+7} to Mn^{2+} (as $MgSO_4$)

$$MnO_4^- + 8H^+ + 5e^- \longrightarrow Mn^{2+} + 4H_2O$$

So, the equivalent weight of $KMnO_4 = \dfrac{\text{Mol. weight}}{5} = \dfrac{158.03}{5} = 31.606$

The oxalate ion loses two electrons to form CO_2

$$C_2O_4^{-2} \longrightarrow 2CO_2 + 2e^-$$

So, the equivalent weight of oxalic acid $= \dfrac{\text{Mol. wt.}}{2} = \dfrac{126}{2} = 63$

The reaction between potassium permanganate and oxalic acid takes place at 60–70°C. Once a small amount of Mn^{2+} is formed, it acts as a catalyst and the reaction goes to completion.

Potassium permanganate acts as a self-indicator and an extra drop of $KMnO_4$ produces a permanent pink colour at the end point.

Procedure

Following steps are involved:

(i) Prepare a standard solution $\left(\dfrac{N}{20}\right)$ of oxalic acid by dissolving $\sim 0.30\,g$ oxalic acid dihydrate in water in a 100 mL volumetric flask.

(*ii*) Pipette out the standard oxalic solution [step (*i*)] (10 mL) in a conical flask and add equal volume of dil. H_2SO_4.

(*iii*) Heat the solution to 60 – 70° and titrate against $KMnO_4$ (for coloured solution like $KMnO_4$, the upper miniscus is read) till a permanent pink colour is obtained. Repeat the titration till three concordant readings are obtained.

(*iv*) Finally pipete out the given oxalic acid solution (10 mL), add an equal volume of dil. H_2SO_4, heat (60–70°) and titrate against $KMnO_4$ solution as in (*iii*) above. Repeat the titration to obtain three concordant readings.

Notes:

1. In place of sulphuric acid (used for acidifying $KMnO_4$), nitric acid cannot be used as it is a powerful oxidizing agent. Hydrochloric acid is not used because it is oxidised by $KMnO_4$ to Cl_2.

$$2MnO_4^- + 10\ Cl^- + 16H^+ \longrightarrow 2Mn^{+2} + 5Cl_2 + 8H_2O$$

2. Though potassium permanganate is a self-indicator, other redox indicators like ferroin or N-phenyl anthranilic acid may be used in case a very dilute solution of $KMnO_4$ has been used for the titration.

3. Arsenious oxide or sodium oxalate solution are the best reagents for standardization of $KMnO_4$. However, for elementary work, oxalic acid dihydrate ($H_2C_2O_4.2H_2O$) or Mohr's salt ($FeSO_4(NH_4)_2SO_4.6H_2O$) can also be used for standardization of $KMnO_4$. These reagents are generally not used for high precision estimations as their water contents may vary.

Calculations

Mass of oxalic acid dissolved in 100 ml solution $= y\ g$

Mass of oxalic acid dissolved in 1000 ml solution $= y \times 10\ g/liter$

Normally (N) of standard oxalic acid $= \dfrac{\text{Mass}}{\text{Equivalent weight}} = \dfrac{y \times 10}{63} = N$

Normality (N_1) of Oxalic Acid

$$N_1 V_1 = NV$$

$$\text{(KMnO}_4\text{)} \quad \text{(Standard oxalic acid)}$$

$$N_1 = \frac{N \times V}{V_1}$$

Normality (N_2) of Given Oxalic Acid

$$N_2 V_2 = N_1 V_3$$

$$\text{(given oxalic acid)} \quad \text{(KMnO}_4\text{)}$$

$$N_2 = \frac{N_1 \times V_3}{V_2} = \frac{N_1 V_3}{10} \quad \text{where } V_3 = \text{Volume of KMnO}_4$$

Amount of the given oxalic acid $= N_2 \times 63\ g/litre$.

2.3.2   Dtermine the Strength of a given Mohr's Salt Solution (Ferrous Ammonium Sulphate) by Titrating Against Approximately $\dfrac{N}{20}$ Solution of Potassium Permanganate

Theory

This involves Redox Titration involving oxidation reduction reaction. In this titration Mohr's salt is the reducing agent and $KMnO_4$ is the oxidizing agent.

The ionic equation is

$$MnO_4^- + 8H^+ + 5Fe^{2+} \longrightarrow 5Fe^{3+} + Mn^{2+} + 4H_2O$$

The equivalent weight of Mohr's salt in equal to its molecular weight (392) because the oxidation of Fe^{+2} to Fe^{+3} is one electron change, *i.e.*,

$$Fe^{2+} \longrightarrow Fe^{3+} + e^-$$

The equivalent weight of $KMnO_4$ $= \dfrac{\text{Mol. wt.}}{5} = \dfrac{158.03}{5} = 31.606$

because the Mn^{+7} (in $KMnO_4$) is reduced to Mn^{2+} ($MnSO_4$), *i.e.*,

$$Mn^{7+} + 5e^- \longrightarrow Mn^{2+}$$

$KMnO_4$ acts as a self-indicator.

Procedure

Following steps are involved:

(*i*) Prepare a standard solution of Mohr's salt $\left(\dfrac{N}{20}\right)$ by dissolving 1.95 g Mohr's salt in dilute H_2SO_4 (N, 10 mL) and making upto 100 mL with distilled water in a 100 mL volumetric flask.

(*ii*) Pipette out 10 mL of the standard Mohr's salt [step (*i*)] in a conical flask, add dil H_2SO_4 (1N, 10 mL) and titrate against $KMnO_4$ solution till a permanent light pink colour is obtained. Repeat the titration till three concordant readings and obtained.

(*iii*) Finally repeat the titration with 10 mL in the given Mohr's salt solution [as in (*ii*) above].

Calculations

Mass of Mohr's salt in 100 ml solution $= y\ g$

Mass of Mohr's salt in 1000 ml solution $= y \times 10\ g$

Normality (N) of standard Mohr's salt solution $= \dfrac{y \times 10}{392} = N$

Normality (N_1) of $KMnO_4$

$$N \times 10 = N_1 V_1$$

(Standard Mohr's salt solution)   ($KMnO_4$)

$\therefore$ $$N_1 = \dfrac{N \times 10}{V_1}$$

Normality (N_2) of given Mohr's salt soluton $= N_2V_2 = N_1V_3$

$$N_2 = \frac{N_1V_3}{V_2}$$

Amount $= N_2 \times 392$ g/litre

2.3.3 Determine the Strength of Mohr's Salt Solution Using About $\frac{N}{20}$ Potassium Dichromate Solution and using Diphenyl Amine as Internal Indicator

Theory

It is a redox titration involving oxidation-reduction reaction. In this titration $K_2Cr_2O_7$ is the oxidising agent, which oxidises ferrous ions in Mohr's salt to ferric ions. The reaction involved is

$$Cr_2O_7^{-2} + 6Fe^{+2} + 14H^+ \longrightarrow 2Cr^{+3} + 6Fe^{+3} + 7H_2O$$

The Cr^{+6} in dichromate is reduced to Cr^{+3} as chromium sulphate, *i.e.,*

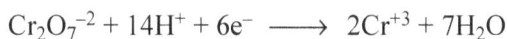

$$Cr_2O_7^{-2} + 14H^+ + 6e^- \longrightarrow 2Cr^{+3} + 7H_2O$$

Hence, the equivalent weight of $K_2Cr_2O_7$ is one-sixth of its molecular weight.

Molecular weight of potassium dichromate $= 294$

∴ Equivalent weight $= \dfrac{294}{6} = 49.0$

Equivalent weight of Mohr's salt is calculated from the reaction

$$Fe^{+2} \longrightarrow Fe^{+3} + e^-$$

Molecular weight of Mohr's salt is $= 392$

Equivalent weight $= \dfrac{392}{1} = 392$

Indicator

Diphenyl amine which is used as the internal indicator undergoes the following oxidation reaction with the slight excess of the dichromate solution at the end point.

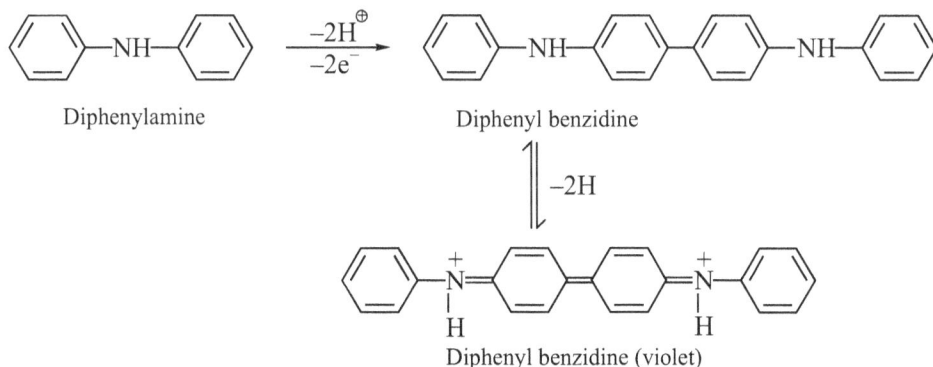

Diphenylamine   Diphenyl benzidine

Diphenyl benzidine (violet)

Procedure

Following steps are involved:

(*i*) Prepare a standard solution of ferrous ammonium sulphate $\left(\dfrac{N}{20}\right)$ by dissolving about 1.9g of ferrous ammonium sulphate in 10 mL dilute H_2SO_4 (2N) and diluting with distilled water to make 100 mL in a 100 mL volumetric flask.

(*ii*) Pipette and 10 mL to the above solution in a conical flask. Add dilute H_2SO_4 (10 mL), phosphoric acid (2 mL) and 2-3 drops of diphenyl amine indicator.

(*iii*) Titrate the above solution with $K_2Cr_2O_7$ solution till a first permanent blue colour is obtained. Repeat the titration till two concordant readings are obtained.

(*iv*) Finally pipette out 10 mL of given Mohr's salt solution and titrate against $K_2Cr_2O_7$ solution as described in (*ii*) and (*iii*) above.

Notes:

1. Phosphoric acid combines with the yellow Fe^{3+} ions (formed by oxidation of Fe^{2+}) to form the complex $[Fe(HPO_4)]^+$ thus rendering the end point more clear.

2. Phosphoric acid also prevents the premature oxidation of the indicator. Diphenyl amine changes colour between the potential $0.73 - 0.79$ V. This potential is reached when only $50 - 60\%$ of Fe^{2+} has been oxidised, pholphoric acid removes Fe^{3+} ions from solution by complexation and thus the potential ($\sim$0.75) is reached only when all Fe^{2+} is oxidised and thus diphenyl changes colour only at the end point.

3. F^- ions may also be used instead of phosphoric acid.

4. Sometimes the blue colour obtained during the titration tends to disappear. This premature oxidation of indicator is due to excess of dichromate. Thus, the solution should be stirred throughout the titration to prevent this. A permanent blue colouration is the correct end point.

Calculations

Normality of Standard Mohr's Salt Solution

Mass of Mohr's salt in 100 mL of the standard solution $= y$ g

Amount of Mohr's salt in 1000 mL $= y \times 10$ g/litre

Normality (N) of standard Mohr's salt solution $= \dfrac{\text{Amount/litre}}{\text{Eq. wt}} = \dfrac{y \times 10}{392} = N$

Normality (N_1) of $K_2Cr_2O_7$ solution (intermediate) $N_1 \times V_1 = N \times V$

$$N_1 \times V_1 = N \times V$$

    ($K_2Cr_2O_7$)   Standard Mohr's salt solution)

$$N_1 = \dfrac{N \times 10}{V_1}$$

Normality (N_2) of given Mohr's salt solution

$$N_1 \times V_2 = N_2 V_3$$

    ($K_2Cr_2O_7$)   (Mohr's salt)

$$N_2 = \frac{N_1 \times V_2}{V_3}$$

Amount of Mohr's salt = $N_2 \times 392$ g/liter.

2.3.4 Determine the amount of Mohr's Salt Solution using Approximately $\frac{N}{20}$ Potassium Dichromates Solution and Potassium Ferricyanide as the External Indicator

Theory see experiment 2.3.3

A solution of potassium ferricyanide is used as the external indicator. It is not added to the titration flask unlike other indicators but is kept outside in a groove tile to monitor the reaction and detect the end point. As long as the solution in the titration flask contains Fe^{2+}, a drop of this solution gives a blue colour with the indicator due to the formation of ferro-ferricyanide.

$$2K_3[Fe(CN)_6] + 3Fe^{2+} \longrightarrow Fe_3[Fe(CN)_6]_2 + 6K^+$$

when the oxidation of Fe^{2+} to Fe^{3+} is complete (end point), no blue colour is obtained with the indicator.

Procedure

Following steps are involved:

(i) Prepare a standard solution of ferrous ammonium sulphate $\left(\frac{N}{20}\right)$ as in experiment 2.3.3, step (i).

(ii) Pipette out 10 mL of the standard solution of ferrous ammonium sulphate in a conical flask, add equal volume of dil H_2SO_4.

(iii) To the above solution add 1 mL of $K_2Cr_2O_7$ solution from the burette. Take out a drop of the solution (use a dropper) and add to a drop of the indicator kept in a groove tile. A blue colour will be obtained. Keep adding 1 mL $K_2Cr_2O_7$ solution at a time to the titration flask and testing a drop of the solution each time with a fresh drop of indicator in a groove tile. Note the burette reading when no blue colour is obtained. This given an approximate volume of $K_2Cr_2O_7$ solution required to titrate (oxidise) 10 mL of Mohr's salt solution.

(iv) Repeat the above titration and test with indicator only near the end point in order to get the exact volume of $K_2Cr_2O_7$ required. Repeat to obtain three concordant values.

(v) Finally titrate the given Mohr's salt solution (10 mL) (as in steps (iii) and (iv) above). Repeat the titration to obtain three concordant values.

Calculation

As in experiment 2.3.3.

2.3.5 Determine the Strength of a given Copper Sulphate Solution Iodometrically using Approximately $\frac{N}{20}$ Sodium Thiosulphate Solution

Theory

In the reaction between copper sulphate and KI, the copper (II) is reduced to Cu(I) and I⁻ is oxidised to iodine and an equivalent amount of iodine thus produced is volumetrically estimated by titrating with sodium thiosulphate solution.

$$2CuSO_4 + 4KI \longrightarrow 2K_2SO_4 + Cu_2I_2 + I_2$$

The ionic equation is

$$2Cu^{2+} + 4I^- \longrightarrow Cu_2I_2 + I_2$$

The reaction between Cu^{2+} and I⁻ takes place in absence of any mineral acid.

The iodine is then titrated with $Na_2S_2O_3$

$$2Na_2S_2O_3 + I_2 \longrightarrow Na_2S_4O_6 + 2NaI$$

or $$2S_2O_3^{2-} + I_2 \longrightarrow S_4O_6^{2-} + 2I^-$$

Equivalents weights of the

(*a*) Copper sulphate = Molecular weight = 249.5
   ($CuSO_4.5H_2O$)

(*b*) Sodium thiosulphate = Molecular weight = 248.2

The copper sulphate solution undergoes hydrolysis. This is prevented by addition of mineral acid like sulphuric acid. However, since the reaction between KI and $CuSO_4$ to liberate iodine does not occur in presence of any mineral acids the acid added to copper sulphate solution is first neutralized by adding a slight excess of ammonia. The excess ammonia is then neutralized with acetic acid.

As seen Cu_2I_2 is the product of reaction between $CuSO_4$ and KI. Cu_2I_2 is a white precipitate insoluble in the medium. The free iodine present in the solution gets adsorbed on the surface of Cu_2I_2 and hence, does not react with $Na_2S_2O_3$. This is prevented by addition of small amount of KCNS or NH_4CNS.

Procedure

Following steps are followed:

(*i*) Prepare $\frac{N}{20}$ copper sulphate solution by dissolving accurately weighed 1.25 g of copper sulphate penta hydrate in water (about 20 mL), adding 1-2 mL dil H_2SO_4 and making up the volume to 100 mL with distilled water using a 100 mL volumetric flask.

(*ii*) Pipette out standard copper sulphate solution (20 mL) in a conical flask and add dilute ammonia solution (1:1) dropwise with shaking till a permanent turbidity in obtained. Then add dilute CH_3COOH (2N) dropwise with shaking to remove the turbidity. Add potassium iodide solution (10 mL, 10%) to the above solution and keep in dark for 2 minutes. The solution acquires a dark brown colour due to liberation of iodine.

(*iii*) Titrate the solution with sodium thiosulphate solution till a light yellow colour is obtained. At this stage add starch solution (~1 mL) and a pinch of solid KCNS or NH₄CNS and continue titration till the blue colour first disappears. Repeat to obtain three concordant reading.

(*iv*) Finally titrate the given copper sulphate solution (20 mL) against sodium thiosulphate solution [as in (*ii*) and (*iii*)].

Notes:

1. The reaction between Cu^{+2} and I^- takes place in absence of any mineral acid. Therefore, it is necessary to neutralize sulphuric acid present with ammonia. Formation of a precipitate (or turbidity) is due to slight excess of NH_4OH, which gives a precipitate of $Cu(OH)_2$, indicating complete neurtalization of sulphuric acid.

2. As iodine volatilizes, the titration must be carried out at room temperature without delay.

3. Excess of KI should be used (*a*) to ensure a stoichiometric reaction between copper sulphate and KI, (*b*) the liberated iodine dissolves in $KI(KI + I_2 \Leftrightarrow KI_3)$ and hence will not be lost due to volatility.

4. The reaction between KI and $CuSO_4$ is slow, the solution is kept in dark for ~2 minutes to ensure the completion of reaction.

5. As the titration proceeds, a precipitate of Cu_2I_2 is formed. Iodine gets adsorbed on the surface of the precipitate. In order to prevent this adsorption a pinch of KCNS or NH₄CNS is added which liberates the adsorbed iodine into the solution.

6. Starch is added only towards the end of titration (*i.e.,* to pale yellow solution, when the reaction is almost complete or when there is a very small amount of iodine present). This is done to prevent permanent adsorption of iodine on the indicator.

Calculations

Mass of $CuSO_4.5H_2O = y$ g/100 mL

Normality (N) of standard copper sulphate solution $= \dfrac{y \times 10}{249.5} = N$

Normality (N_1) of the intermediate sodium thiosulphate solution

$$N_1 V_1 = NV$$

(Thiosulphate)   (Copper sulphate)

$$N_1 = \frac{NV}{V_1}$$

Normality (N_2) of given copper sulphate solution

$$N_2 V_2 = N_1 \times V_3$$

$\quad CuSO_4 \quad Na_2S_2O_3$

$\quad$ Solution $\quad$ Solution

$$N_2 = \frac{N_1 \times V_3}{V_2}$$

Amount of copper sulphate in the given solution $= N_2 \times 249.5$ g litre.

2.3.6 Determine the Strength of Potassium Dichromate Solution using Approximately $\dfrac{N}{20}$ Solution of Sodium Thiosulphate

Theory

An acidified solution of potassium dichromate on treatment with a slight excess of potassium iodide produces an equivalent amount of iodine. The liberated iodine is estimated by titrating with sodium thiosulphate.

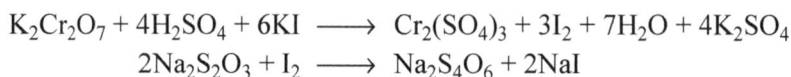

$$K_2Cr_2O_7 + 4H_2SO_4 + 6KI \longrightarrow Cr_2(SO_4)_3 + 3I_2 + 7H_2O + 4K_2SO_4$$
$$2Na_2S_2O_3 + I_2 \longrightarrow Na_2S_4O_6 + 2NaI$$

The ionic equations can be represented as follows:

$$Cr_2O_7^{2-} + 14H^+ + 6I^- \longrightarrow 2Cr^{+3} + 3I_2 + 7H_2O$$
$$2S_2O_3^{2-} + I_2 \longrightarrow {}^\bullet S_4O_6^{2-} + 2I^-$$

Cr^{6+} in dichromate is reduced to Cr^{3+} in $Cr_2(SO_4)_3$. The equivalent weight of potassium dichromate is one sixth its molecular weight.

$$\text{Equivalent wt. of } K_2Cr_2O_7 = \frac{\text{Mol. wt.}}{6} = \frac{294}{6} = 49$$

$$\text{Equivalent wt. of sodium thiosulphate} = \frac{\text{Mol. wt.}}{1} = 248.2$$

Starch is used as the indicator, it forms a blue coloured complex with iodine and the equivalence point is indicated by disappearance of the blue colour.

Procedure

Following steps are followed:

(*i*) Prepare a standard solution of potassium dichromate ($N/20$) by weighing accurately ~0.125 g of $K_2Cr_2O_7$ and prepare 100 mL solution in a standard volumetric flask.

(*ii*) Pipette out the standard potassium dichromate solution (20 mL) in a conical flask, add dilute sulphuric acid (10 mL) and potassium iodide (10 mL, 5%). Cork the flask and keep in dark for 2-3 min, an equivalent amount of iodine is produced and the solution acquires a dark brown colour.

(*iii*) Titrate with sodium thiosulphate solution till the solution becomes light yellow in colour. Add starch (~1 mL) the solution acquires a blue colour due to the formation of starch-iodine complex) and titrate till the blue colour first disappears (the solution at this stage is green in colour due to the presence of Cr^{3+} ions). Repeat the titration to obtain three concordant readings.

(*iv*) Carry out the titration with the given potassium dichromate solution.

Notes:

1. Excess of potassium iodide should be used (*a*) to ensure complete reaction between KI and $K_2Cr_2O_7$, (*b*) the liberated iodine, which is otherwise insoluble in water dissolves in the solution due to the formation of $KI_3(KI + I_2 = KI_3)$ and loss of iodine due to its volatiity is reduced.

2. Reaction between KI and acidified potassium dichromate is slow and hence the solution is allowed to stand for 2 – 3 minutes before titration.
3. Starch should be added only near the equivalence or the end point to reduce permanent adsorption of iodine on starch.

Calculations

Mass of potassium dichromate $= y$ g/100 mL

Normality (N) of standard potassium dichromate solution $= \dfrac{y \times 10}{\text{Eq. wt.}} = \dfrac{10y}{49} N$

Normality (N_1) of sodium thiosulphate solution

$$N_1 V_1 = NV$$

(Thiosulphate)   (Standard dichromate)

$$N_1 = \frac{NV}{V_1}$$

Normality (N_2) of given potassium dichromate solution $N_2 \times V_2 = N_1 \times V_3$

$$N_2 V_2 = N_1 V_3$$

[$K_2Cr_2O_7$ given]   [standard thio soln.]

$$N_2 = \frac{N_1 V_3}{V_2}$$

Amount of the given potassium dichromate

$$= N_2 \times \text{Eq. wt.}$$
$$= N_2 \times 49 \text{ g / litre.}$$

2.3.7 Determine the Strength of the given Tartar Emetic Solution Iodometrically using Approximately $\dfrac{N}{20}$ Iodine Solution

Theory

Tartar emetic is potassium antimonyl salt of tartaric acid, in which antimony is present in +3 oxidation state and on reaction with I_2 is oxidised to +5 oxidation state.

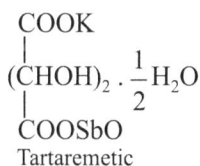

COOK
|
(CHOH)$_2$. $\frac{1}{2}$H$_2$O
|
COOSbO

Tartaremetic

Iodine reacts with tartar emetic in presence of sodium bicarbonate.

$$K(SbO)C_4H_4O_6 + I_2 + 4NaHCO_3 \longrightarrow NaH_2SbO_4 + KNaC_4H_4O_6 + 2NaI + 4CO_2 + H_2O$$

Tartar Emetic

The reaction is reversible. The formed HI is neutralised by the addition of $NaHCO_3$ and so the reaction goes in forward direction

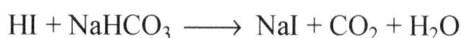

$$HI + NaHCO_3 \longrightarrow NaI + CO_2 + H_2O$$

Procedure

Following steps are involved:

(i) A standard solution of tartar emetic $\left(\dfrac{N}{20}\right)$ is prepared by accurately weighing about 0.83 g and dissolving in water and making the solution to 100 mL using a 100 mL volumetric flask.

(ii) Iodine solution (20 mL) is taken in a conical flask and sodium bicarbonate (2–3g) added.

(iii) The solution obtained in step (ii) is titrated against standard tartar emetic solution (taken in a burette) till a light yellow colour is obtained. Starch solution (10 drops) added and titration continued till the blue colour formed on addition of starch first disappears. Repeat the titration to obtain three concordent reading.

(iv) The given tartar emetic solution (taken in a burette) is titrated against iodine solution (taken in the conical flask) as described in steps (ii), (iii).

Notes:

1. Iodine solution can also be standardised by using standard $K_2Cr_2O_7$ or arsenious oxide (As_2O_3).

2. The iodine or tartar emetic solution should not be pipetted as these are poisonous.

3. Addition of $NaHCO_3$ neutralises the liberated HI in order to make the reaction to go to completion.

4. The titration should be carried out immediately after the addition of $NaHCO_3$. Otherwise, some metal hydroxide may get precipitated, which reacts very slowly with iodine.

Calculations

Mass of tartar emetic = y g/100 mL

Normality (N) of standard tartar emetic solution = $(y \times 10)/166.96$

[Equivalent wt. of tartar emetic is $\dfrac{333.92}{2} = 166.96$]

Normality to the intermediate iodine solution

$$N_1 V_1 = N V$$

$$\underset{\text{Iodine soln.}}{} \qquad \underset{\text{Standard tartar emetic solution}}{}$$

$$N_1 = \frac{NV}{V_1}$$

Normality (N_2) of the given tartar emetic solution.

$$\underset{\substack{\text{Given} \\ \text{tartar emetic}}}{N_2 V_2} = \underset{\text{Iodine}}{N_1 \times V_3}$$

$$N_2 = \frac{N_1 \times V_3}{V_2}$$

Strength = $N_2 \times 166.96$ g/litre.

2.3.8 Determine the Strength of the given Arsenious Oxide Solution Iodometrically using Approximately $\frac{N}{20}$ Iodine Solution

Theory

For preparation of arsenious oxide (As_2O_3) solution, a dilute solution of NaOH is used (since it is insoluble in water).

$$As_2O_3 + 6NaOH \longrightarrow 2Na_3AsO_3 + 3H_2O$$

The excess of NaOH is subsequently neutralised with HCl since titration with iodine cannot be carried out in presence of NaOH. Iodine also reacts with NaOH

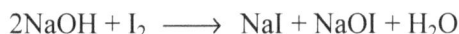

$$2NaOH + I_2 \longrightarrow NaI + NaOI + H_2O$$

The reaction of sodium arsenate with iodine in presnece of HCl is represented as:

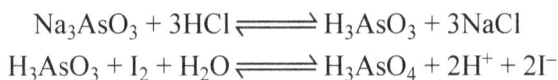

$$Na_3AsO_3 + 3HCl \rightleftharpoons H_3AsO_3 + 3NaCl$$
$$H_3AsO_3 + I_2 + H_2O \rightleftharpoons H_3AsO_4 + 2H^+ + 2I^-$$

As seen the above reaction is reversible. It gets to completion if some $NaHCO_3$ (which neutralises the formed HI) is added to the reaction mixture.

Being poisonous arsenic solution should not be pipetted out. It be measured by using a burette.

Procedure

Following steps are followed:

(*i*) Prepare a standard $\left(\frac{N}{20}\right)$ solution of arsenious oxide by accurately weighing 0.25 g and transferring to 100 mL volumetric flask. Sodium hydroxide solution (10%, 10-15 mL) is added to completely dissolve As_2O_3. To the solution is added a drop of phenophthalein (a dark pink colour develops) and the solution is neutralised by adding HCl (1N) dropwise till the solution just becomes colourless. This is follwed by addition of $NaHCO_3$ (0.2 – 0.3g). When all $NaHCO_3$ has dissolved make up the volume to 100 mL with distilled water.

(*ii*) Standard arsenous oxide solution (20 mL) is taken in a conical flask (do not pipette as it is poisonous; use a burette), water (20 mL) added followed by addition of $NaHCO_3$ (2.5 g) and starch (2 mL). The solution is titrated with iodine solution (taken in a burette) till a blue colour appears. Repeat the titration to obtain three concordant reading.

(*iii*) Finally carry out titration of the given arsenious oxide solution with iodine solution as above.

Calculations

Mass of arsenious oxide = y g/100 mL

Normality (N) of standard arsenious oxide solution = $(y \times 10) / 49.45$

[The equivalent weight of arsenious oxide is one fourth of its molecular weight $= \dfrac{197.82}{4} =$ 49.45].

Normality (N_1) of iodine solution

$$N_1V_1 = NV$$

iodine   Standard

As$_2$O$_3$ solution

$$N_1 = \frac{NV}{V_1}$$

Normality (N_2) of the given As$_2$O$_3$ solution

$$N_2V_2 = N_1V_3$$

Given   Iodine

As$_2$O$_3$ soln.   solution

$$N_2 = \frac{N_1V_3}{V_2}$$

Strength of the given arsenious oxide solution = $N_2 \times 49.45$ g/litre.

2.4   PRECIPITATION TITRATIONS

Introduction

Also known as **argentometric titrations**, the precipitation titrations involve the formation of an insoluble precipitate when the two reacting solutions are mixed together. As an example, addition of AgNO$_3$ solution to a solution of NaCl gives a white ppt. of AgCl.

$$AgNO_3 + NaCl \longrightarrow AgCl\downarrow + NaNO_3$$

In a similar way, addition of AgNO$_3$ solution to KCNS solution gives a while ppt. of AgCNS.

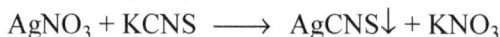

$$AgNO_3 + KCNS \longrightarrow AgCNS\downarrow + KNO_3$$

The precipitation titrations are useful for the volumetric estimation of silver or halides or thiocyanates. For the titration to be quantitative, the formed ppt. should be practically insoluble and the ppt. should be formed rapidly. Also the end point should be easily detectable.

Indicators

Solutions involved in these titrations are mostly colourless and hence, an indicator is required to indicate the end point. The indicator used, depends on the reacting substances.

Following are some of the indicators used:

1. **Estimation of Silver or Thiocyanate Solution:** (Volhard's method). **Ferric alum or ferric nitrate are used as indicators** in the presence of dilute nitric acid.

    In this, when thiocyanate solution is added to a silver nitrate solution in the presence of nitric acid and ferric ions, a white precipitate of silver thiocyanate results. This continues till all silver ions have been removed as precipitate. The next drop of excess thiocyanate added reacts with the ferric ions to form a red coloured ferric thiocyanate complex.

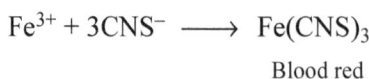

$$Fe^{3+} + 3CNS^- \longrightarrow Fe(CNS)_3$$

Blood red

The appearance of a blood red colouration indicates the end point. A vigorous stirring of the solution is recommended at the end point to dislodge Ag^+ ions that have got adsorbed on the surface of the silver thiocyanate precipitate. Nitric acid used must be free of nitrite ions and other nitrogen oxides, since they are known to interfere in the titration.

2. **Estimation of Silver Ions and Halides using Potassium Chromate as Indicator (Mohr's Method):** This method makes use of the fact that the solubility product of Ag_2CrO_4 is higher than that of the silver halides, *i.e.,* silver halides are more insoluble than silver chromate. So as long as there is any chloride or bromide left in the solution, no silver chromate is formed. When all the halide ions are completely precipitated, the end point is indicated by the formation of brick red colouration or precipitate of silver chromate.

The above method has the following limitations:

(*a*)   A neutral solution is required, since silver chromate is soluble in an acidic solution. The acid in the solution is neutralized by adding powdered $CaCO_3$ before adding the indicator.

(*b*)   In an alkaline solution silver gets precipirated as silver oxide.

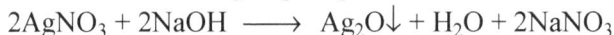

$$2AgNO_3 + 2NaOH \longrightarrow Ag_2O\downarrow + H_2O + 2NaNO_3$$

3. **Adsorption indicators: (Fajan's method)** Commonly used indicators for titration of chloride with silver nitrate are:

(*a*)   **Fluorescein:** Suitable for dilute chloride (acidic) solutions. At the end point, the colour change is from white precipitate in a greenish yellow medium to red.

(*b*)   **Eosin:** Suitable for titration of bromide and iodide with silver nitrate in the presence of acetic acid. Colour change is from pink to reddish violet.

The silver chloride precipitate adsorbs the excess chloride ions present in solution. This primary layer adsorbs oppositely charged positive ions (*e.g.,* Na^+, K^+, etc.). A secondary adsorption of negatively charged ions, such as NO_3^- takes place. But once the Ag^+ ions are in excess of the Cl^- ions, these get adsorbed in and the negatively charged fluorescein ions form a complex of silver, which is pink in colour. Thus, the end point is indicated by the white ppt. colour changing to pink in a greenish yellow medium.

Determination Equivalent Weight

One gm mole of $AgNO_3$ reacts with one gm mole of each of NaCl, NH_4CNS or KCNS. Since, equivalent weight of silver nitrate is equal to its molecular weight, the equivalent weights of NaCl, NH_4CNS or KCNS will also be equal to their molecular weight, *i.e.,*

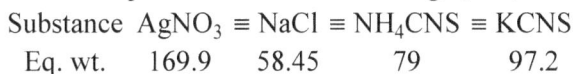

Substance	$AgNO_3$	$\equiv$	NaCl	$\equiv$	NH_4CNS	$\equiv$	KCNS
Eq. wt.	169.9		58.45		79		97.2

2.4.1   Determine the Strength of the given Solution of Sodium Chloride using $\dfrac{N}{20}$ Silver Nitrate Solution and Fluorescein as Indicator

Theory

It is an example of a precipitation titration involving reaction between silver nitrate and sodium chloride giving precipitate of AgCl.

$$AgNO_3 + NaCl \longrightarrow AgCl\!\downarrow + NaNO_3$$

or

$$Ag^+ + Cl^- \longrightarrow AgCl\!\downarrow$$

The equivalence point is indicated by use of fluorescein, an adsorption indicator. Fluorescein anion is coloured because of the quinonoid form (I).

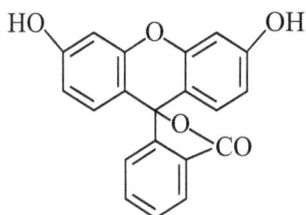

Fluorescein

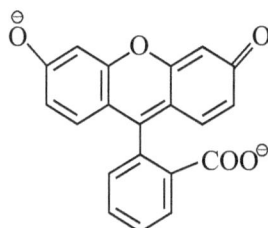

(I) Quinonoid form

The precipitated silver chloride adsorbs chloride ion on its surface as long as excess of chloride ions are present. At the end point, when a slight excess of silver nitrate is added (from the burette) the Ag^+ get adsorbed on the surface of AgCl which is in turn holds the indicator ion (I) due to secondary adsorption. The precipitate acquires a pink red colour due to this secondary adsorption of the indicator as shown below:

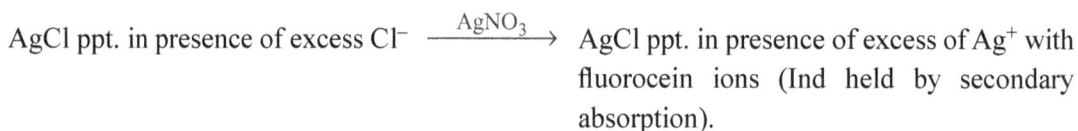

AgCl ppt. in presence of excess Cl^- $\xrightarrow{\text{AgNO}_3}$ AgCl ppt. in presence of excess of Ag^+ with fluorocein ions (Ind held by secondary absorption).

Procedure

Following steps are followed:

(*i*) Prepare a standard solution $\left(\dfrac{N}{20}\right)$ of sodium chloride by accurately weighing about 0.3 g of pure sodium chloride and dissolving in water. The solution is made upto 100 mL by addition of distilled water in a 100 mL volumetric flask.

(*ii*) Pipette out the standard sodium chloride solution (20 mL) in a conical flask. Add deionised water (about 30 mL) and fluorescein indicator (1 mL). Titrate with silver nitrate solution with constant shaking till the white ppt. of AgCl (that is formed during the titration) just acquires a pink red colour. Repeat to obtain three concordant readings.

(*iii*) Finally repeat the titration with the given sodium chloride solution till three concordant readings are obtained.

Notes:

1. Eosine (tetraiodofluorescein) can also be used as the indicator.

2. It is important to shake the conical flask continuously throughout the titration so that the indicator is adsorbed only at the equivalence point.

3. At the end point, the AgCl precipitate primarily adsorbs the silver ions and nitrate ions may be held by

secondary adsorption. Since, the fluoresceinate ions are more strongly adsorbed than the nitrate ions, the secondary adsorption layer is of fluorescinate and not of nitrate ions.

Calculations

Mass of sodium chloride $= y\, g/100$ mL

Normality (N) of standard NaCl $= \dfrac{y \times 10}{58.5}$

Normality (N_1) of silver nitrate

$$N_1 V_1 = NV$$

$$\text{(AgNO}_3\text{)} \quad \text{(Standard NaCl)}$$

$$N_1 = \frac{N \times 20}{V_1}$$

Normality (N_2) of given sodium chloride

$$N_2 V_2 = N_1 \times V_3$$

$$\text{(given NaCl)} \quad \text{(AgNO}_3\text{)}$$

$$N_2 = \frac{N_1 \times V_3}{20}$$

Amount of the given sodium chloride $= N_2 \times 58.6\ g/$litre.

2.4.2 Determine the Strength of a given Solution of Silver Nitrate by Volhard's Method

Theory

Volhard's method involves estimation of silver nitrate by titrating against ammonium or potassium thiocyanate solution. Ferric alum [ferric ammonium sulphate, $Fe_2(SO_4)_3.(NH_4)_2SO_4.24H_2O$] is used as an indicator.

Treatment of silver nitrate with ammonium or potassium thiocyanate in presence of nitric acid, gives a white precipitate of silver thiocyanate

$$AgNO_3 + NH_4CNS \longrightarrow NH_4NO_3 + AgCNS\downarrow$$

When all Ag^+ ions have been precipitated, a slight excess of ammonium thiocyanate (added from the burette) gives a red colour due to the formation of ferric thiocyanate.

$$Fe^{+3} + 3CNS^- \longrightarrow Fe(CNS)_3$$

$$\text{(Red)}$$

The titration is carried out in presence of nitric acid in order to avoid hydrolysis of ferric alum (used as the indicator).

The precipitated silver thiocyanate has a tendency to adsorb Ag^+ which may lead to premature formation of ferric thiocyanate (red colour before the end point). This may be avoided either by constantly shaking the solution in the titration flask or by adding a few drops of chloroform or nitrobenzene to the titration solution.

Procedure

Following steps are followed:

(*i*) Prepare a standard solution (~$N/20$) by accurately weighing ~0.85 g of silver nitrate and dissolving in deionised water to make 100 mL of the solution in a 100 mL volumetric flask.

(*ii*) Take the standard silver nitrate solution (20 mL) in a conical flask. Add dilute nitric acid (5 mL) and ferric alum (1 mL) to it. Titrate against ammonium thiocyanate (from the burette) with continuous stirring. A white precipitate of AgCNS is formed (a red colour that may be formed before the end point disappears on shaking). Continue the titration till the white precipitate settles down and the supernatant solution first acquires a permanent light red colour. Repeat to obtain three concordant readings.

(*iii*) Finally repeat the titration with the given silver nitrate solution as explained in (*ii*) above.

Calculations

Mass of silver nitrate in 100 mL solution = y g

Mass of silver nitrate in 1000 mL solution = $y \times 10$ g/litre

Normality (N) of standard silver nitrate = $\dfrac{y \times 10}{169.90}$

Normality (N_1) of ammonium thiocyanate

$$NV = N_1 V_1$$
$$\text{(AgNO}_3\text{)} \quad \text{(NH}_4\text{SCN)}$$
$$N_1 = \frac{NV}{V_1}$$

Normality (N_2) of given silver nitrate solution

$$N_2 V_2 = N_1 \times V_3$$
$$\text{(AgNO}_3\text{)} \quad \text{(NH}_4\text{SCN)}$$
$$N_2 = \frac{N_1 V_3}{V_2}$$

Amount of the given nitrate solution = $N_2 \times 169.90$ g / litre.

2.4.3 Determine the Strength of a given Solution of Sodium Chloride by Mohr's Method using Potassium Chromate as the Indicator

Theory

Mohr's method involves addition of $AgNO_3$ solution to a solution of NaCl containing potassium chromate. The Ag^+ first reacts with Cl^- to give a white precipitate of AgCl and not Ag_2CrO_4 as the solubility product of silver chloride is less than that of silver chromate.

$$Ag^+ + Cl^- \longrightarrow AgCl\downarrow$$

As the end point or the equivalence point approaches, Ag^+ may also react prematurely with CrO_4^{2-} (present in the solution as indicator) to form a red precipitate of $AgCrO_4$. This precipitate, however, dissolves on shaking as long as Cl^- ions are present in solution to form AgCl.

$$2Ag^+ + CrO_4^{2-} \longrightarrow Ag_2CrO_4\downarrow$$

(Red ppt.)

$$Ag_2CrO_4 + 2Cl^- \longrightarrow 2AgCl\downarrow + CrO_4^{2-}$$

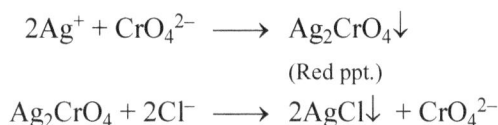

When all chloride ions have reacted with $AgNO_3$, a slight excess of $AgNO_3$ now added reacts with potassium chromate to give a red precipitate of silver chromate.

The reaction between $AgNO_3$ and NaCl can be quantitatively carried out in neutral medium, in this method of estimation. Since, in alkaline medium silver hydroxide gets precipitated leading to erroneous results and in acidic medium some chromate is converted to dichromate.

$$2CrO_4^{2-} + 2H^+ \longrightarrow 2HCrO_4^- \longrightarrow Cr_2O_7^{2-} + H_2O$$

Thus, the concentration of chromate ions is reduced and even at the equivalence point, a red precipitate of Ag_2CrO_4 is not formed as its solubility product is not achieved on addition of a slight excess of $AgNO_3$.

Procedure

Prepare a standard solution ($N/20$) of sodium chloride by dissolving ~0.3 g of NaCl in 100 ml distilled water in a 100 mL volumetric flask.

Pipette out the standard sodium chloride solution (20 mL) in a conical flask. Add potassium chromate solution (4–5 drops, 1%) and titrate with silver nitrate (initially a white precipitate is formed) till the precipitate acquires a light pink colour. Repeat to obtain three concordant readings.

Finally pipette out the given sodium chloride solution (20 mL) and carry out the titration as described above.

Notes:

1. The titration should be carried out rapidly, without exposing the precipitate of AgCl to light, as it tends to darken on exposure to light.

2. The titration flask should be shaken continuously to avoid premature precipitation of $AgCrO_4$.

Calculations

Mass of sodium chloride $= y$ g/100 mL

Normality (N) of standard sodium chloride solution $= \dfrac{y \times 10}{58.5}$

Normality (N_1) of $AgNO_3$ solution

$$N_1V_1 = NV$$

(AgNO₃)   (NaCl)

$$N_1 = \frac{NV}{V_1}$$

Normality (N_2) of given NaCl solution

$$N_2V_2 = N_1 \times V_3$$

(Given NaCl)   (AgNO₃)

$$N_2 \times 20 = N_1 \times V_3$$

$$N_2 = \frac{N_1 \times V_3}{20}$$

Amount of the given NaCl solution = $N_2 \times 58.5$ g/litre.

2.5   COMPLEXOMETRIC TITRATIONS

Introduction

Complexometric titrations involves the formation of a stable, soluble and stoichiometric complex between a metal ion and the complexing agent (called ligand) to give a neutral compound. The complexing agent (or the ligand) contains atleast one lone pair of electrons that it can donate to the central metal atom. The number of co-ordination bonds is the co-ordination number of the central metal.

For example, the coordination number of Fe(II) in the complex ion $[Fe(CN_6)]^{4+}$ is six (six CN^- groups are attached).

Ligands such as H_2O, CN^-, NH_3, containing only one donor atom (*i.e.*, O, N) are called *monodentate,* since they attach themselves to the central atom through only one coordination bond. *Polydentate* ligands contain two or more than two donor atoms in the same molecule and are accordingly called bi-dentate, tridentate etc. The complexes formed with polydentate ligands are called *chelates* and the corresponding ligands are called the chelating agents.

It is found that tertiary amines containing carboxylic groups are very good complexing agents that produce stable complexes with many metal ions. One of the most useful chelating agent is ethylenediamine tetra acetic acid (EDTA). It is a polyprotoic acid having the following structure:

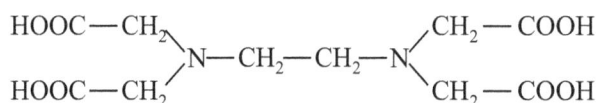

$$HOOC-CH_2 \diagdown \atop HOOC-CH_2 \diagup N-CH_2-CH_2-N \diagup ^{CH_2-COOH} _{\diagdown CH_2-COOH}$$

Its most common salt used in analysis is

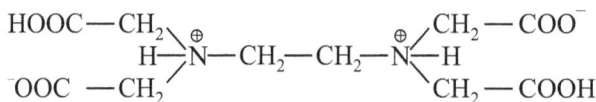

$$HOOC-CH_2 \diagdown \atop {}^-OOC-CH_2 \diagup H-\overset{\oplus}{N}-CH_2-CH_2-\overset{\oplus}{N}\diagup ^{CH_2-COO^-} _{\diagdown CH_2-COOH}$$

Here two hydrogen ions are lost more readily.

Thus, it has six potential co-ordination sites (four oxygens and two N-atoms). Hence, it usually forms octahedral complexes, though there may be exceptions. It usually forms a 1:1 complex with a metal.

Four simplicity EDTA is given the formula H_4Y and its sodium salt Na_2H_2Y that gives H_2Y^{2-} as the complexing ion. Its reaction with the bivalent cation M^{2+} may be written as:

$$M^{2+} + H_2Y^{2-} \rightleftharpoons MY^{2-} + 2H^+$$

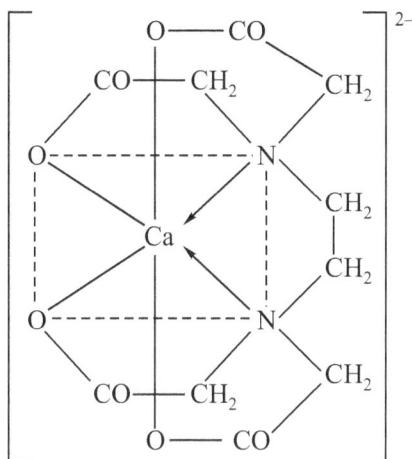

Complex of EDTA with Ca^{2+}

Advantages of using of EDTA

Use of EDTA has a number of advantages some of which are given below:

1. EDTA complexes with a large number of metal ions.
2. The complexes are formed instantaneously.
3. Complexes have a known constant stoichiometry of 1:1.
4. Complexes of bivalent metal ions are stable in alkaline medium, whereas those of trivalent metal ions are stable even in the acidic medium.
5. Most of the complexes formed are soluble in water.
6. EDTA is available in a high state of purity.
7. The solution of EDTA is stable for months if kept in plastic or borosilicate glass containers.
8. EDTA is a primary standard and so can be used directly to estimate the strength of a metal ion solution.

Types of EDTA Titrations

The EDTA titrations are of different types. Some of these are given below:

1. **Direct Titrations:** The sample is adjusted to the required pH by adding the appropriate buffer. In order to avoid the precipitation of hydroxide of the metal (or any other interferring ions present), an auxilary complex forming substance like a tartarate or citrate is added (if required). A suitable indicator is added and the titration is performed by adding EDTA solution till the colour changes at the end point. The titration can also be performed by using conductometric method.

   It is, however, found that sometime it is not possible to keep the metal ions in solution because of high pH or non-availability of a suitable indicator or if the rate of formation of the EDTA complexes is too slow or the complexe formed is not stable, it is necessary to look for other methods to perform the EDTA titration. Some of such methods are given below.

2. **Back Titration:** In these titrations, an excess of EDTA solution is added to the metal ion solution and an indicator is added. The excess amount of EDTA is then back titrated with a suitable metal salt solution.

3. **Displacement Titration:** The titrations are performed for such metal ions that form stronger EDTA complexes than those of magnesium or zinc. They then displace the less strongly bound metal ion from its EDTA complex. The metal ion thus displaced can be quantitatively estimated by performing an EDTA titration. Calcium is titrated by this method. Hence, the metal cation M^{n+} to be determined may be titrated with Mg or Zn complex of EDTA. The following reaction occurs:

$$M^{n+} + MgY^{2-} \rightleftharpoons (MY)^{(n-4)+} + Mg^{2+}$$

The amount of Mg^{2+} or Zn^{2+} ions free is equivalent to the cation present and can be titrated with a standard solution of EDTA and a suitable indicator.

4. **Alkalimetric Titration:** When a solution of disodium ethylenediamineteraacetate (Na_2H_2Y) is added to metal ion solution, a complex is formed with the liberation of two equivalents of hydrogen.

$$M^{+n} + H_2Y^{2-} \rightleftharpoons (MY)^{(n-4)+} + 2H^+$$

The H^+ thus set free can be titrated with a standard alkali by the usual acid-base titration.

Indicators

The Indicators used in complexometric titrations are usually organic compounds that change colour when they form complex with metals. Such indicators are called **metallochromic indicators**. It is essential that the metal-indicator complex be less stable than the corresponding metal-EDTA complex so that EDTA can remove the metal from the metal-indicator complex quite fast.

However, metallochromic indicators are very sensitive and hence, need to be added in concentration of 10^{-5} to 10^{-6} M. Hence, the amount of metal bound to the indicator is generally negligible (0.1% of the total metal to be determined).

Some of the important indicators used are:

(*i*) **Eriochrome Black T (Solochrome Black T):** It is sodium-1(-1 hydroxy-2-naphthylazo)-6-nitro-2 napthol-4-sulphonate.

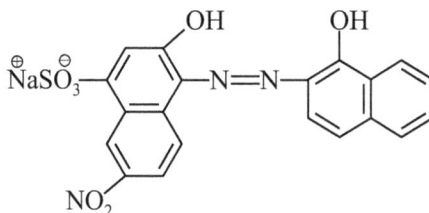

It shows different colours at different pH. For example, red (below pH 5.5), blue (pH 7 to pH 11), yellowish orange (above pH 11.5) and between pH 7 to pH 11, the indicator changes colour from blue to red at the end point.

(*ii*) **Murexide:** It is the ammonium salt of purpuric acid with the following structure:

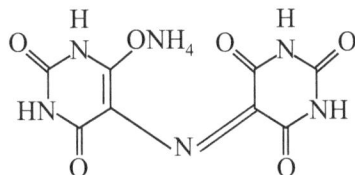

Murexide shows red violet colour (pH 9), violet (pH 9 to 11) and blue colour (above pH 11). The aqueous solution of murexide is not very stable. A fresh solution should be used every time.

It forms stable chelates with Co^{2+}, Ni^{2+}, Cu^{2+} and Ca^{2+}.

(*iii*) **Xylenol Orange:** [3, 3′-bis(N, N-di(carboxymethyl)-aminomethyl-o-cresol sulphonephthalein].

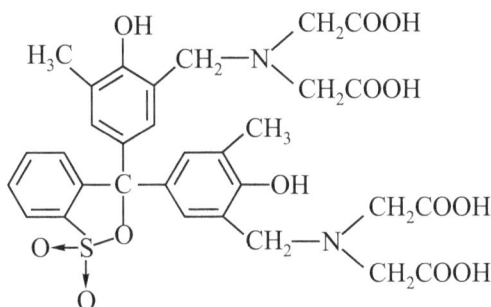

Xylenol orange gives yellow colour in acidic solution and forms red coloured stable complex with Zn^{2+}, Cd^{2+}, Co^{2+} and Pb^{2+} etc.

2.5.1 Determine the Strength of a given Solution of Magnesium Sulphate using EDTA $\left(\dfrac{N}{40}\right)$ and Eriochrome Black T as Indicator

Theory

The volumetric estimation of magnesium is carried out by EDTA (ethylenedlamine tetraacetic acid). Its disodium salt (Na_2H_2Y) is used for the preparation of standard solution.

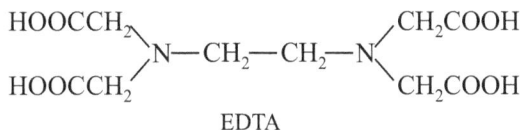

EDTA

The reaction between EDTA and Mg^{2+} takes place at pH 10, which is maintained by using buffers (*e.g.,* $NH_4Cl + NH_3$). At pH 10, the EDTA which is present as its tetracarboxylate ion forms a hexadentate ligand with Mg^{2+}. The four $-COO^-$ groups and the electron pairs on two nitrogen atoms are involved in the complex formation. The reaction may be represented as

$$H_2Y^{2-} + Mg^{2+} \longrightarrow MgY^{2-} + 2H^+$$

where H_2y^{2-} represents the dicarboxylate ion of EDTA and Mgy^{2-}, the complex between Mg^{2+} and Y^{4-} (the tetracarboxylate ion of EDTA).

The above reaction can be represented as:

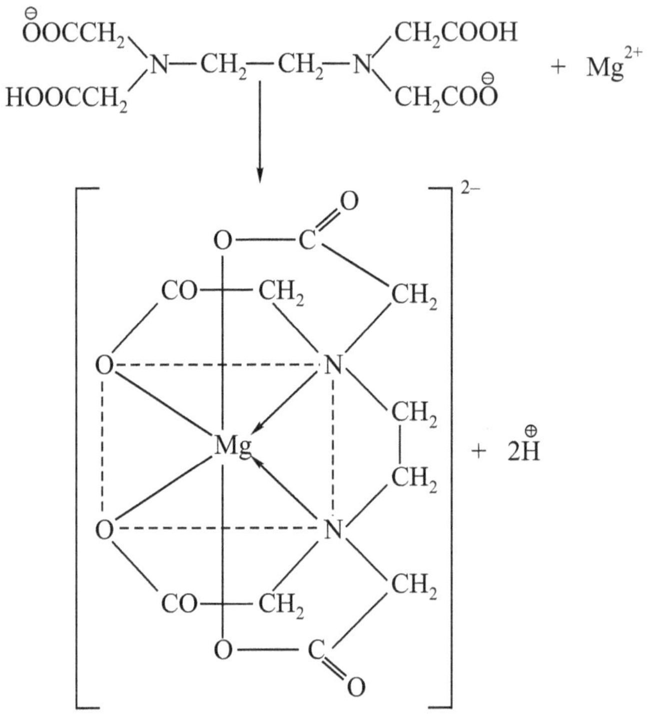

Metal sensitive indicators like Eriochrome black *T* are used for detection of the end point. The free indicator at pH 10 is blue in colour, it is red when present as a complex with the Mg^{2+}.

At pH 10, the indicator (represented as Ind) is present as its dianion and its reaction with Mg^{2+} may be represented as

$$Mg^{2+} + [H.Ind]^{2-} \rightleftharpoons [Mg. Ind]^{-1} + H^+ \qquad\qquad (i)$$
$$\text{Blue} \qquad\qquad\qquad \text{(I) Red}$$
$$Mg. Ind + EDTA \longrightarrow Mg. EDTA + Ind \qquad\qquad (ii)$$
$$\text{(II)}$$

When to a known volume of magnesium sulphate solution, a small amount of the indicator (which is a weak ligand) and the buffer is added, the solution acquires a red colour due to the formation of the Mg^{2+} indicator complex (*I*). On titration of the Mg-Indicator complex solution with EDTA (which is a strong ligand), forms the complex (*II*). Near the end point, the Mg^{2+} present in the magnesium indicator complex reacts completely with EDTA liberating the free indicator, which is blue in colour.

Hence, at equivalence point, the solution changes colour from red to blue.

Procedure

Following steps are involved:

(*i*) Prepare a standard solution of magnesium sulphate (M/40) (MgSO$_4$.7H$_2$O) by accurately weighing ~0.65 g of magnesium sulphate and dissolving in distilled water to obtain 100 ml of the solution.

(*ii*) Pipette out 10 ml of the standard magnesium sulphate in a conical flask. Add to it deionized water (10 mL), buffer (2mL, NH$_4$Cl.NH$_3$) and a pinch of Eriochrome black *T*. Titrate the resultant pink solution with EDTA till it changes to blue. Repeat to obtain three concordant readings.

(*iii*) Finally carry out the titration with the given magnesium sulphate solution.

Calculations

$$\text{Mass of MgSO}_4.7\text{H}_2\text{O in 100 mL} = y \text{ g}$$

$$\text{Molarity of standard MgSO}_4.7\text{H}_2\text{O} = \frac{y \times 10}{246.68} = M$$

$$\text{Molarity } (M_1) \text{ of EDTA is } M_1 \times V_1 = M \times V$$

$$\text{(EDTA)} \quad \text{(Standard MgSO}_4.7\text{H}_2\text{O)}$$

$$M_1 = \frac{M \times 10}{V_1}$$

Molarity (M_2) of given magnesium sulphate is

$$M_2 \times 10 = M_1 V_3$$

$$\text{(Given MgSO}_4.7\text{H}_2\text{O)} \quad \text{(EDTA)}$$

$$M_2 = \frac{M_1 V_3}{10}$$

Amount of given magnesium sulphate = M_2 × molecular weight = 246.68 × M_2g / litre

2.5.2 Determine the Strength of a given Solution of Zinc Sulphate using EDTA $\left(\dfrac{M}{40}\right)$ and Eriochrome Black T as Indicator

Theory

Zinc sulphate is estimated volumetrically by complexometric titration with EDTA. Eriochrome black T also known as Solochrome black *T* is used as the indicator.

A few drops of Eriochrome black *T* are added to a known volume of zinc sulphate containing a buffer (the pH of the solution should be 10). The solution acquires a red colour due to the formation of a complex between Zn^{2+} and indicator.

At pH 10, the indicator (Ind.) is present as a dianion (H.Ind^{2-}) and its reaction with Zn^{2+} is represented as:

$$\text{Zn}^{2+} + \text{H.Ind}^{2-} \longrightarrow [\text{Zn.Ind}]^{-1} + \text{H}^{\oplus}$$

$$\qquad\quad\text{Blue} \qquad\qquad\qquad \text{Red}$$

The pH of the solution is maintained at 10 because at this pH EDTA [H_4Y] forms complex with Zn^{2+} most efficiently when it is present as ethylene diamine tetracarboxylate ion [Y]$^{4-}$, EDTA itself is insoluble in water, its disodium salt is used for preparation of the solution for titration).

The dissociation constants (pK_a) of the four hydrogens of EDTA are 2.0, 2.7, 6.2 and 10.3, respectively.

When the red coloured Zn-indicator complex solution in the titration flask is titrated with the disodium salt of EDTA, Zn^{2+} forms a complex with its tetra carboxylate ion.

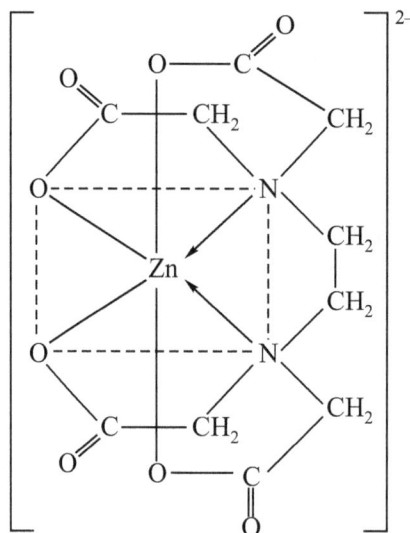

(Complex of Zn^{2+} with ethylene diamine tetra carboxylate ion).

Near the end point, Zn^{2+} which had formed a complex with the indicator dianion reacts completely with EDTA liberating the free indicator which is blue in colour. So the end point is indicated by change of colour from red to blue. The Zn^{2+} indicator complex is less stable than the Zn^{2+}-EDTA complex.

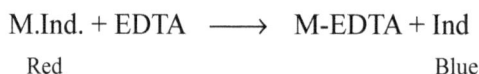

$$\text{M.Ind.} + \text{EDTA} \longrightarrow \text{M-EDTA} + \text{Ind}$$
$$\qquad \text{Red} \qquad\qquad\qquad\qquad\quad \text{Blue}$$

Procedure

(*i*) Prepare a standard solution of zinc sulphate (M/40) by dissolving ~0.72 g of $ZnSO_4$ in 100 mL water.

(*ii*) Pipette out the standard zinc sulphate solution (10 mL) in a conical flask, add deionized water (10 mL), buffer solution of NH_4Cl–NH_3 (2 mL) and a pinch of solid Eriochrome black *T*. The solution acquires a pink colour. Titrate with the EDTA solution till the solution becomes blue (the solution should have no reddish shade). Repeat to obtain three concordant readings.

(*iii*) Finally carry out titration with the given $ZnSO_4$ solution.

Notes:

1. Zinc acetate dihydrate or zinc ammonium sulphate hexahydrate can also be used for preparing a standad solution instead of zinc sulphate.

2. EDTA is insoluble in water, its disodium salt is used for preparation of solution $Na_2C_{10}H_{14}O_8N_2.2H_2O$, Mol. wt = Eq. wt. = 372.25).

3. pH of the solution is maintained at 10 using a buffer.

4. With an increase in pH, some metals tend to form soluble hydroxides, thus interferring in the estimation.

5. Deionized, water should be used in the titration and for preparation of solutions.

6. Polythene containers should be used for storage of EDTA solution.

7. Titration should be carried out slowly near the end point.

8. The last traces of reddish shade should disappear at the end point.

Calculations

Mass of zinc sulphate dissolved in 100 ml standard solution = y g

$$\text{Molarity } (M) \text{ of zinc sulphate solution } = \frac{y \times 10}{287.54}$$

$$\text{Molarity } (M_1) \text{ of EDTA} = M_1 \times V_1 = M \times 10$$
$$\text{(EDTA)} \qquad \text{(Standard ZnSO}_4\text{)}$$

$$M_1 = \frac{M \times 10}{V_1}$$

Molarity (M_2) of given zinc sulphate solution

$$M_2 V_2 = M_1 V_3$$
$$\text{(given ZnSO}_4\text{)} \quad \text{(EDTA)}$$

$$M_2 = \frac{M_1 V_3}{10}$$

Amount of given zinc sulphate solution = M_2 × Molecular weight = M_2 × 287.54 g/litre.

2.5.3 Determine the Amount of a given Solution of Calcium Carbonate using $\frac{N}{40}$ EDTA and Eriochrome Black *T* as Indicator

Theory

The titration between Ca^{2+} and EDTA (Na_2H_2Y) is a complexometric titration and is carried out at pH 10 (using $NH_4Cl - NH_3$ buffer).

Erichrome black T is used as the indicator, but the end point is not sharp. If however, a small amount of Mg^{2+} (as magnesium chloride) or Mg EDTA complex is added during the titration, the colour changes sharply from red to blue at the end point.

When EDTA is added form the burette to the titration flask containing the Ca^{2+}, Mg. EDTA complex and Mg.Ind. Complex, it first reacts with free Ca^{2+} ions.

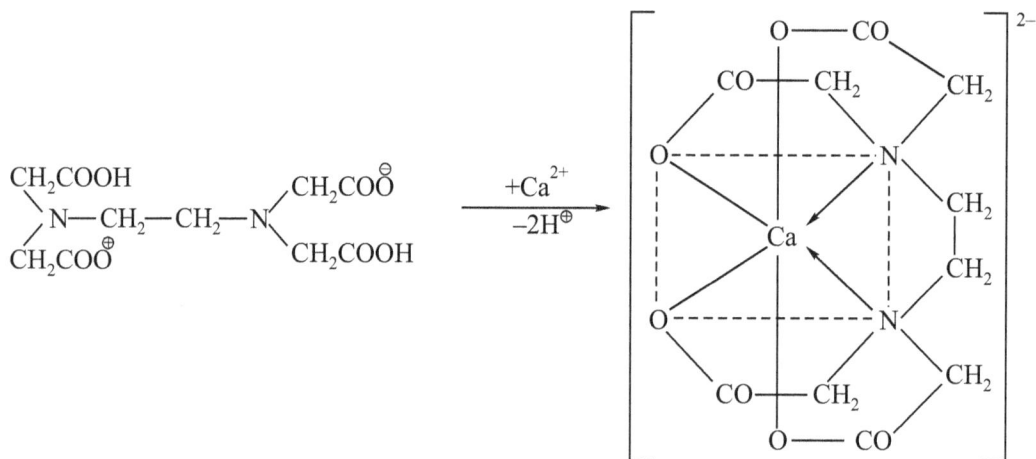

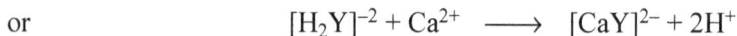

or $[H_2Y]^{-2} + Ca^{2+} \longrightarrow [CaY]^{2-} + 2H^+$

and then reacts with any free Mg^{2+} present in the solution.

$$Mg^{2+} + [H_2Y]^{2-} \longrightarrow [MgY]^{2-} + 2H^+$$

and finally at the end point, the EDTA reacts with the Mg.Ind complex (red) liberating free indicator (blue in colour) between pH $7 - 118$).

$$[Mg.Ind]^{-1} + [H_2Y]^{2-} \longrightarrow [MgY]^{2-} + [H.Ind]^{-2} + H^+$$

The Mg.Ind complex is more stable than Mg.EDTA complex as well as Ca.Ind complex. So when to a solution containing Ca^{2+} and Mg. EDTA complex, a small amount of the indicator (Eriochrome Black T) is added, Mg. Ind complex is formed and not Ca.Ind complex.

Procedure

Following steps are followed:

(i) Prepare a standard calcium carbonate solution by weighing accurately 0.25 g of $CaCO_3$ and dissolving in minimum amount of hydrochloric acid (6N). The solution is neutralised with few drops of dilute sodium hydroxide solution and volume made upto 100 mL with distilled water in a 100 mL volumetric flask.

(ii) Pipette out 10 mL of standard $CaCO_3$ solution in a conical flask, add 2 mL of the buffer solution (Buffer of NH_4Cl–NH_3. 142 mL of Conc. NH_3, sp.gr.088–0.90 and 17.5 g AR NH_4Cl and diluted to 250 mL) and 1 mL of Mg. EDTA complex solution [0.1M, prepared by mixing equal amounts of 0.4 M EDTA and $MgSO_4$, the solution neutralised with NaOH (pH 8–9, Check with phenolphthalin) and diluting to 0.1 M solution with distilled water] and a pinch of solid Erichrome black T indicator. The solution is slowly titrated with EDTA solution (Na_2H_2y) until the colour changes from wine red to blue. Repeat the titration to obtain three concordant reading.

(*iii*) Finally repeat the titration with given solution of $CaCO_3$.

Calculations

$$\text{Mass of } CaCO_3 = y \text{ g}/100 \text{ mL}$$

Molarity (M) of standard calcium carbonate solution $= \dfrac{y \times 10}{100.08}$

Molarity (M_1) of EDTA

$$M_1 \times V_1 = M \times 10$$
$$\text{(EDTA)} \quad \text{(Standard } CaCO_3 \text{ solution)}$$
$$M_1 = \dfrac{M \times 10}{V_1}$$

Molarity (M_2) of given $CaCO_3$ solution

$$M_2 V_2 = M_1 V_3$$
$$\text{(Given } CaCO_3) \quad \text{(EDTA)}$$
$$M_2 = \dfrac{M_1 V_3}{10}$$

Amount of the given $CaCO_3 = M_2 \times 100.08$ g/litre.

EXERCISE

1. Explain the term volumetric analysis. What are its applications. What are different types of volumetric analysis?

2. Write a note on Law of equivalence.

3. What do you understand by the term equivalent mass of a substance? Explain by taking an example of equivalent mass of a acid and a base.

4. Explain the terms molarity, normality, strength.

5. Explain acidimetry and alkalimetry. What indicators are used in acidimetry and alkalimetry.

6. Can you prepare a standard solution of hydrochloric acid, potassium permangnate why?

7. Using acidimetry and alkalimetry how will you determine (*i*) the amount of Na_2CO_3 and $NaHCO_3$ in a solution (*ii*) the amount of Na_2CO_3 and NaOH in a solution.

8. What are redox titrations? What indicators are used in these titrations?

9. How will you determine the equivalent weight of $KMnO_4$ and the strength of $KMnO_4$ in a solution?

10. How will you determine the strength of Mohr's salt soluting using (*i*) diphenyl amine as internal indicator and (*ii*) Potassium ferricyanide as external indicator.

11. What is iodometry? How will you determine the strength of a $CuSO_4$, solution?

12. How will you determine volumetrically the strength of tartaremetic solution and arsenious oxide solution?

13. What are precipitation titrations? What indicators are used in these titrations.

14. How will you determine the strength of silver nitrate solution by Volhard's method and Mohr's method?

15. What are complexometric titration?

16. How will you determine the strength of a given solution of magnesium sulphure, zinc sulphate and calcium carbonate using complexometric titrations.

○○○

3. Instrumental Methods of Volumetric Analysis

Contents

Instrumental Methods of Volumetric Analysis

INTRODUCTION

We have so far studied volumetric analysis by measuring the volumes of solution taking part in a chemical reaction (Chapter 2) using a burette by the process known as titration. It is, however possible to carry out volumetric analysis using certain instruments. Such methods and called 'Instrumental methods of volumetric analysis'. These methods include.

Potentiometric titrations, conductometric titrations, spectrophotometric titrations and pH-metric titrations.

In all these methods, the determination of end point is carried out using instruments.

3.1 POTENTIOMETRIC TITRATIONS

In potentiometric titration, a suitable electrode, known as indicator electrode is immersed in the solution to be titrated. This electrode acts as an 'indicator'. The indicator electrode is paired with a calomel electrode (a reference electrode). The two electrodes are connected to an electronic voltmeter. The addition of titrant from a burette causes change in concentration of ions resulting in change in e.m.f. of the indicator electrode. A sharp change in the electrode potential indicates the equivalence point.

The potential of a metal electrode in a solution of its own at 25°C is given by the equation.

$$E_{25} = E^\circ + \frac{0.0591}{n} \log C_M^{n+} \qquad \qquad ...(i)$$

where E° is the standard potential of the metal, n is the valency of the metal and C_M^{n+} is its concentration.

The concentration in fact is $- \log [M^{n+}]$ and this is equal to pM^{n+}

$$E_{25} = E^\circ - \frac{0.591}{n} pM^{n+} \qquad \qquad ...(ii)$$

For hydrogen electrode E_{25} will be

$$E_{25} = E_H^\circ - \frac{-0.0591}{n} pH^{n+} \qquad \qquad ...(iii)$$

where E_H° is the potential of normal hydrogen electrode and is arbitrarily taken as zero at all temperatures. This is because E_H°, the normal potential of hydrogen ions of unit activity in equilibrium with hydrogen gas at one atmospheric pressure is taken arbitrarily as zero.

So the equation (*iii*) becomes

$$E_{25} = -0.591 \text{ pH} \qquad \qquad ...(iv)$$

Principal of Potentiometric Titration

The equivalence point (end point), as has already been stated, in a potentiometric titration is detected by measuring the change in potential during the titration. The usual apparatus used for potentiometric titration is shown in Fig. 3.1.

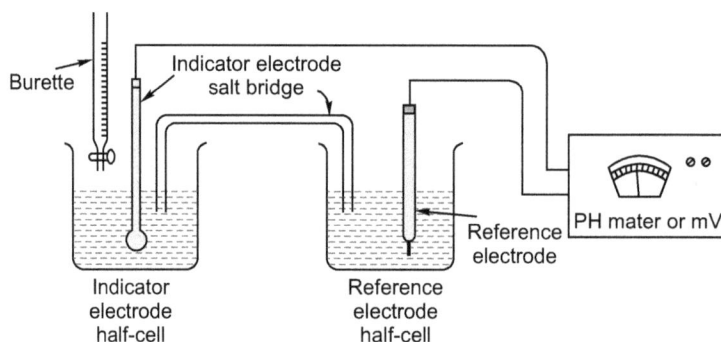

Fig. 3.1. Apparatus for potentiometric Titration.

The potential between the reference cell (whose potential is known) and the indicator half-cell (whose potential varies with the concentration of the solution) is measured at the beginning, after each addition of titrant and more closely near the end point (or equivalence point), when the readings change by larger values. During the addition of the titrant, the solution is stirred (using a magnetic stirrer).

The detection of the equivalence point is made on the basis of a plot between the potential and the volume of the titrant added (Fig. 3.2).

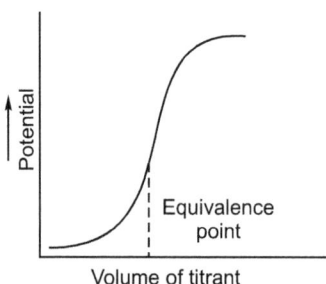

Fig. 3.2. Determination of equivalence point.

After the titration, a graph is plotted (Volume of titrant *V*s the potential) (Fig. 3.2), the equivalence point is determined by inspecting the graph. It is the point at which the curve is steepest (Fig. 3.2). In case there is uncertainty in determining the equivalence point, it is overcome by plotting a change in potential with change in volume of the reagent ($\Delta E/\Delta V$) Vs the average volume of the added reagent (Fig. 3.3). The equivalence point corresponds to the maxima in the curve, which is obtained by extraplotting the experimental points.

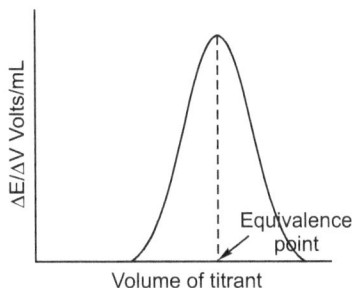

Fig. 3.3. A better procedure for determination of equivalence point.

The Indicator Electrode

The **indicator electrode** which is used in potentiometric titration must be appropriate. The choice of the indicator electrode depends on its characteristics. The potential of the indicator electrode should be related to the concentration of the species to be determined *via* the Nernst equation. Also the response to various concentrations should be rapid and reproducible. A number of indicator electrodes (like hydrogen electrode, antimony electrode and gas electrode) are available. Most commonly, a glass electrode is used (Fig. 3.4).

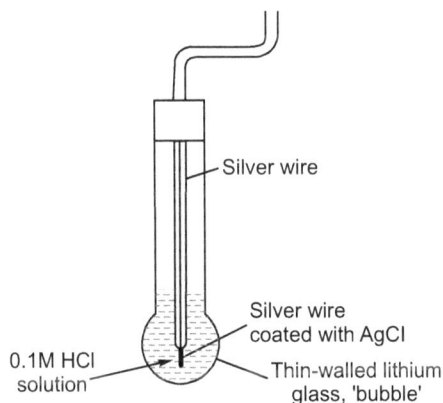

Fig. 3.4. Glass electrode.

Reference Electrodes

The **reference electrode** must have constant potential during potentiometric titration. Calomel electrode is most commonly used. It consists of metallic mercury and solid mercurous chloride (calomel) is contact with, and in equilibrium with aqueous solution of KCl. The half cell reaction of calomel electrode is represented as

$$Hg_2Cl_2(s) + 2e^- \rightleftharpoons 2Hg(s) + 2Cl^-$$

The electromotive force is given as

$$E_{Hg_2Cl_2}, Hg = E^{\circ}_{Hg_2Cl_2, Hg} - \frac{0.0591}{2} \log [Cl^-]^2$$

or

$$E_{Hg_2Cl_2, Hg} = E^{\circ}_{Hg_2Cl_2, Hg} - 0.059 \log [Cl^-]$$

Thus, as seen, the potential of a reference calomel electrode is dependent only on the concentration the chloride ion. Since, mercurous chloride is insoluble, the concentration of chloride ion depends only on the amount of KCl used with preparation of the electrode.

Three types of calomel electrodes are known. These include:

(i) Saturated calomel electrode (SCE). It consists of solid mercurous chloride and saturated solution of potassium chloride with metallic mercury

$$E^{\circ} = + 0.2415 \text{ volts.}$$

(ii) Normal calomel electrode (NCE). It consists of aqueous medium containing KCl (1F).

$$E^{\circ} = 0.280 \text{ volts.}$$

(iii) Decinormal calomel electrode (DCE). It consists of KCl (0.1F)

$$E^{\circ} = + 0.004 \text{ volts.}$$

Out of the above calomel electrodes, saturated calomel electrode (SCE) (Fig. 3.5) is most commonly used. As the solubility of KCl depends on the temperature, it is necessary that the temperature should be controlled for the operation of the electrode. Saturated calomel electrodes (commonly available) consists of two concentric tubes, the inner tube contains an amalgamated platinum wire which is connected to an external circuit. The inner tube is also filled with a paste of metallic mercury, mercurous chloride and potassium chloride. The outer tube contains a saturated solution of KCl having a few crystals of KCl in order to maintain saturation condition. The inner tube has a small pin hole which provides contact of the electrolyte with the outer tube. A porous asbestos, sealed into the tip of the outer tube provides contact with the solution of the sample.

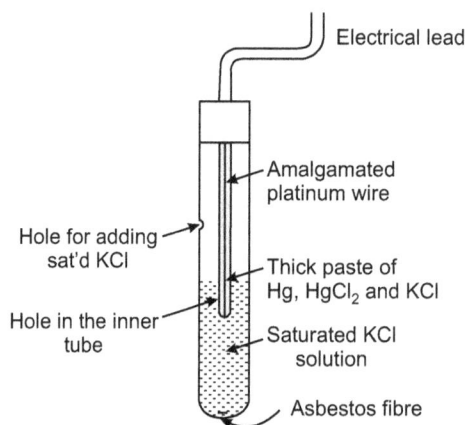

Fig. 3.5. Saturated calomel electrode.

It has been found (Pinkhoff and Treadwell), that a compensation electrode can be used in place of a reference electrode. The potential of the compensation cell is exactly equal to the E.M.F. of the indicator electrode at the end point. The end point, in such a stepup is located by a sudden reversal of polarity. No potentiometric assemble is needed. In place of a galvanometer, a simple capillary electrometer is used. When titration starts, mercury in the electrometer moves in one direction till the end point is reached and mercury stops to move. On further addition of the reagent, the mercury moves in the opposite direction. A disadvantage associated with this procedure is that a separate electrode system is needed for every titration.

Apparatus used for Potentiometric Titrations

The apparatus used for potentiometric titration consists of a single cell for housing both the electrodes instead of separate compartments, which are connected by a salt bridge (Fig. 3.6). It is more convenient to use, a single cell (vessel) for housing both the electrodes. The vessel for titration is a type of a beaker.

The indicator electrode and the saturated calomel electrodes are suspended in the container. A standard titrant is added in small amounts using a burette. Each addition of the titrant should preferably be of the same volume. A magnetic stirrer is used to stir the solution. The titration vessel should be kept closed having an arrangement for bubbling in inert gas and its outlet.

After each addition of the titrant, the solution should be stirred (in order to attain homogencity) and electrode potential Vs SCE measured. The end point is finally determined as shown in Figs. 3.2 and 3.3.

A diagrammatic representation of a simple apparatus for potentiometric titrations is shown in Fig. 3.6.

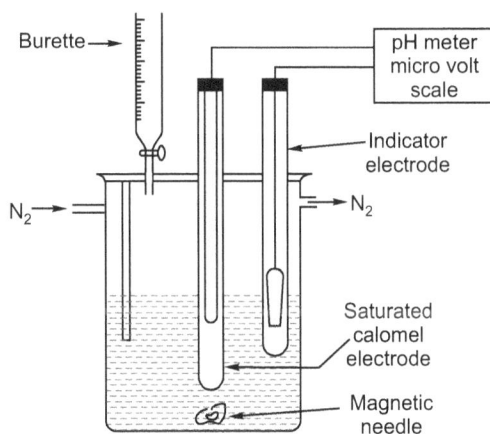

Fig. 3.6. A simple apparatus for potentiometric titration.

Applications of Potentiometric Titrations

Potentiometric titration find application in Neutralisation titrations (Acidimetry and alkalimetry), oxidation-reduction titrations (Redox titrations), precipitation titrations and complexometric titrations.

Acidimetry and Alkalimetry

The **acidimetry** and **Alkalimetry** commonly known as acid base titrations are invariably accompanied by changes in the concentration of H^+ and OH^- ion. The apparatus used for such potentiometric titrations is shown in Fig. 3.7.

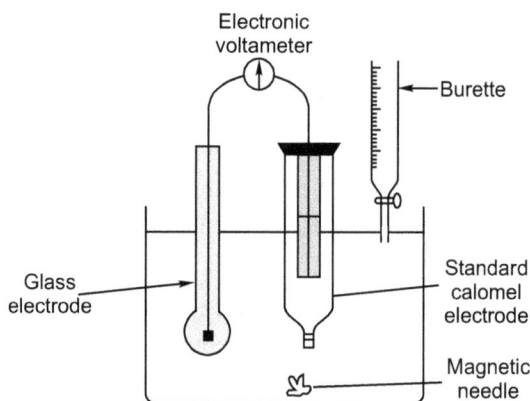

Fig. 3.7. Apparatus for potentiometric acid-base titrations.

The glass electrode is immersed in acid solution whose strength is to be determined. The glass electrode is coupled with a standard calomel electrode. The cell thus formed is connected to a potentiometer or a electronic voltameter. On addition of alkali from the burette, the pH of the solution changes. The e.m.f. of the cell also changes with pH of the solution is accordance of the reaction

$$E = E^° + 0.591 \text{ pH}.$$

Standard alkali solution is then added in small volumes from the burette, solution stirred and e.m.f. of the cell recorded. The e.m.f. is plotted against the volume of the alkali added. The shape of the curve for titration of a strong acid against a strong alkali (*e.g.*, HCl versus NaOH) is as shown in [Fig. 3.8(*a*)]. The equivalence point is indicated by the steepest portion of the curve. However, the steepness of the curve is less marked in case the solutions are very dilute or weak acids or bases are involved. In the latter case, it is difficult to judge the end point. In such cases, the slope of the curve, $\Delta E/\Delta V$ is plotted against volume of the alkali. The end point is indicated by the maxima of the curve [Fig. 3.8(*b*)].

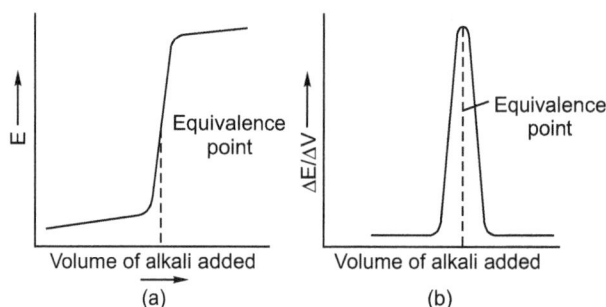

Fig. 3.8. Potentiometric titration curve of an acid and a base.

The potentiometric titrations are particularly useful where visual end point in the normal volumetric titration is difficult due to turbidity or presence of coloured species.

Aqueous medium gives satisfactory results in case of acids and bases having dissociation constants more than 10^{-8}. However, for weaker acids and bases, the reactions are incomplete resulting in changes near the end point. In such cases, non-aqueous solvents are used. For example, phenol ($K_\alpha = 10^{-10}$) can be titrated against a base in non-aqueous medium.

Redox Titrations

Redox titrations are also known as oxidation-reduction titrations. A typical example is titration of ferrous ions (Fe^{2+}) with ceric ions (Ce^{4+}). In this case Fe^{2+} is oxidised to Fe^{3+} and Ce^{4+} is reduced to Ce^{3+}.

$$Fe^{2+} + Ce^{4+} \longrightarrow Fe^{3+} + Ce^{3+}$$

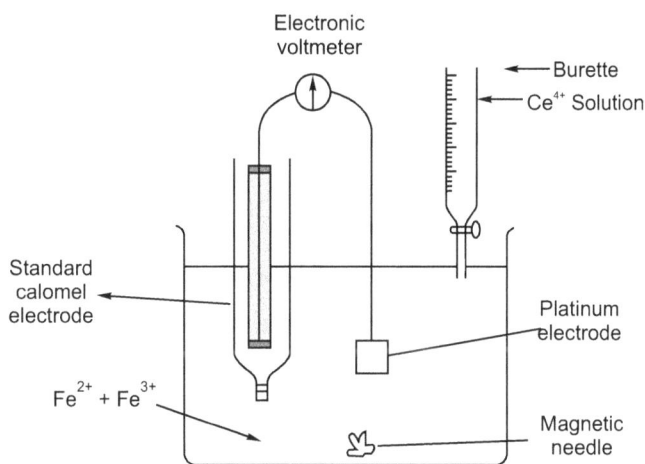

Fig. 3.9. Apparatus for potentiometric titration of Fe^{2+} with Ce^{4+}.

The apparatus used for titration is shown in Fig. 3.9. The indicator electrode is a strip of shining platinum which is dipped in the solution of Fe^{2+}. The indicator electrode is connected to a standard calomel electrode. A burette is used to add Ce^{4+} solution and after each addition, the cell potential is recorded.

The potential of the platinum electrode depends on the $[Fe^{3+}]/[Fe^{2+}]$ ratio. The potential of the cell, E also changes with the change of the ratio $[Fe^{3+}]/[Fe^{2+}]$. The cell potential changes with the addition of Ce^{4+} ions from the burette. Fig. 3.10 shows how the potential of the cell changes as titration proceeds. At the equivalence point, there is sharp rise in potential.

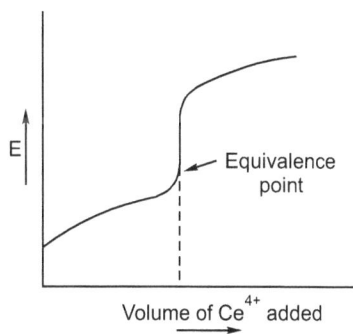

Fig. 3.10. Potentiometric titration curve of Fe^{2+} ions and Ce^{4+} ions.

Potentiometric titrations are particularly useful in case the solutions are coloured and indicator can not be used.

Precipitation Titrations

Precipitation titrations involve the formation of an insoluble substance when two reacting solutions are mixed together. A typical example is titration of a solution of sodium chloride against silver nitrate solution. The apparatus used for precipitation titration is shown in Fig. 3.11. A silver electrode is dipped in sodium chloride solution of unknown strength and is coupled with a calomel electrode through a KNO_3 salt bridge, so that KCl solution does not diffuse into the sample solution. In case, the calomel electrode comes in direct contact with a solution containing excess silver ions, there is a possibility that silver ions may seep through the sintered base of calomel electrode and react to form an insoluble layer of silver chloride.

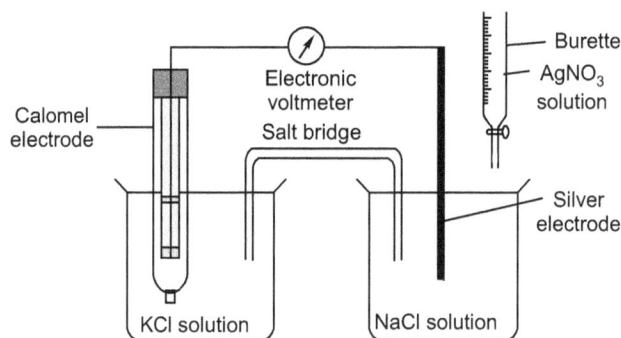

Fig. 3.11. Potentiometric Titration of sodium chloride against silver nitrate solution.

The change in cell potential is due to changes in the concentration of Ag^+ ions around the silver electrode

$$Ag^+ + e \longrightarrow Ag$$

Initially the concentration of Ag^+ is zero. As $AgNO_3$ is added from the burette, AgCl gets precipitated. At this stage, the solution will contain a small concentration of Ag^+ ions formed because of slight dissociation of AgCl. The concentration of Ag^+ ions increases slightly as Cl^- ions are removed in order to maintain the solubility product $K_{sp} = [Ag^+] [Cl^-]$. After the equivalence point, the concentration of Ag^+ ions increases and so the silver electrode potential sharply rises due to the presence of excess Ag^+ ions. The volume of $AgNO_3$ solution which is used to reach the equivalence point is shown in Fig. 3.12.

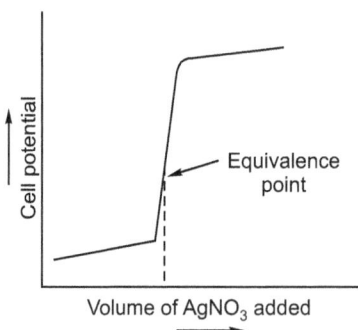

Fig. 3.12. Potentiometric titration curve for NaCl and $AgNO_3$.

Complexometric Titrations

Complexometric titrations involve the formation of a stable, soluble and stoichiometric complex between a metal ion and a complexing agent. Most of the potentiometric complexometric titrations have been carried out with EDTA. As an example the determination of calcium ions in an unknown solution by potentiometric titration. A simple apparatus (Fig.3.6) for potentiometric titration is used. It consists of a mercury indicator electrode and a saturated calomel electrode (Reference electrode) suspended in a titration vessel containing Ca^{2+} ions. A small concentration (10^{-4} M) of mercury (II) EDTA is introduced in the titration vessel. The mixture is finally titrated against standard EDTA solution. After each addition of the titrant, the potential of mercury indicator electrode is measured. Plotting the potential of the indicator electrode against the volume of titrant added is plotted (Fig. 3.13).

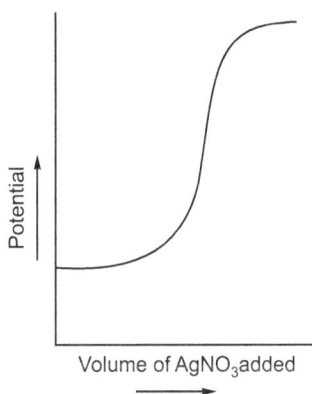

Fig. 3.13. Titration curve for the titration of Ca^{2+} ions with EDTA. Ammonia buffer (pH 10) is used.

Using potentiometric EDTA titrations, the strength of Mg^{2+} and Zn^{2+} in a solution can also be determined.

Automatic Potentiometric Titrations

Potentiometric titrations can easily be performed by using automatic equipment as well as for recording the titration curve. A typical instrument for automatic titration gives a titration curve of potential vs. reagent volume or in some cases $\Delta E/\Delta V$ or $\Delta^2 E/\Delta V^2$ against volume. It is easy to determine the end point on examination of the curve. The titration stops automatically when the potential reaches a pre-determined value. The instrument has a burette system with a selenoid operator valve for controlling the flow. The titrant is added very slowly so that the end point is not run over.

Advantages of Potentiometric Titrations

Potentiometric titrations have several advantages. The apparatus used for potentiometric titration is easily available and is dependable. By refering to the titration curves, it is easy to find the equivalence point. This method is particularly useful for coloured solutions where indicator does not work.

3.2   CONDUCTOMETRIC TITRATIONS

Introduction

Conductometric Titrations are used for those titrations in which there is sharp change in conductivity at the end point. The determination of end point of a titration by measurement of conductance are called conductometric titrations. This technique is useful for determining the end point in an acid base titration and precipitation titration. In conductometric titrations, advantage is taken of the fact that at a constant temperature, the conductance of a solution depends on the number of ions present and also their mobility. The procedure involves addition of a titrant (from a burette) to a measured volume of the solution to be titrated (taken in a conductance cell). The conductance (corresponding to various additions of titrant) are plotted against the volume of the titrant added. In this way, two linear curves are obtained, the intersection of there two curves is the end point.

Terms used in Conductometric Titrations

Various terms used in conductometric titrations are:

(i) **Conductivity:** An electrolyte solution conducts electric current *via* the movement of ions to the electrodes. The power of the electrolytes to conduct electric current is called **conductivity** or **conductance**.

(ii) **Specific Conductance:** The electrolytes (like metallic conductors) obey **Ohms law**, according to which, the current (I) flowing through a metallic conductor is given by the expression

$$I = \frac{E}{R}$$

where E is the potential difference (in volts) at two ends and R is the resistance measured in ohms.

The resistance R of a conductor is directly proportional to its length, l and is inversely proportional to the area (A) of its cross section.

$$R \propto \frac{l}{A}$$

or $$R = \rho \times \frac{l}{A} \qquad \qquad ...(i)$$

Where ρ (rho) is a constant proportionality and is called **specific resistance**. Its value depends on the material of the conductor. From equation (i) we can write

$$\rho = R \times \frac{A}{l}$$

If $$l = 1 \text{ cm and } A = 1 \text{ sq. cm, then}$$

$$\rho = R$$

(iii) **Specific Conductivity:** The reciprocal of specific conductance is called specific conductivity and is defined as the conductance of one centimeter cube of a solution of an electrolyte.

The specific conductance is denoted by the symbol κ (kappa). Thus,

$$\kappa = \frac{1}{\rho} = \frac{1}{R} \times \frac{l}{A}$$

Specific conductance is generally expressed in reciprocal ohms (R.O.) or mhos or ohm^{-1}.

(*iv*) **Equivalent Conductance:** The conductance of an electrolyte obtained by dissolving one gram equivalent of it in V_{mL} water is called **equivalent conductance**. It is denoted by $\wedge$ and is equal to the product of specific conductance, κ and the volume V in mL containing one gram equivalent of the electrolyte at the dilution V.

Thus, $\wedge = \kappa \times V$

Equivalent conductance is expressed in ohm^{-1} cm^2 equi^{-1}.

As the solution becomes more dilute, the equivalent conductance increases till a limiting value is reached. This value is known as equivalent conductance at infinite dilution (zero concentration).

(*v*) **Molar Concentration:** It is defined as the conductance of all ions produced by one mole (one gram molecular weight) of an electrolyte when dissolved in a certain volume V_{mL}.

It is denoted by μ and its value is obtained by multiplying specific conductance, κ, by volume in mL containing one mole of the electrolyte.

$$\mu = \kappa \times V$$

where V is the volume of the solution in mL containing one mole of the electrolyte. Its unit is ohm^{-1} cm^2 mol^{-1}.

Molar concentration can be calculated by using the relation:

$$\mu = \frac{\kappa \times 1000}{m}$$

Where m is the number at moles of electrolyte present in 1000 mL of the solution.

The specific conductance decreases on dilution, while equivalent conductance and molar conductance increase.

Applications of Conductivity

Some of the applications of Conductivity are:

(*i*) **Determination of the concentration of a solution containing a single strong electrolyte.**

In case of a solution of a single strong electrolyte there is a linear increase in conductance (containing about 20 percent of a solute), the analysis is based on calibration curves. Typical curves for some electrolytes are given in Fig. 3.14.

(*ii*) **Determination of solubility of sparingly soluble salts like AgCl, BaSO$_4$, PbSO$_4$ etc.**

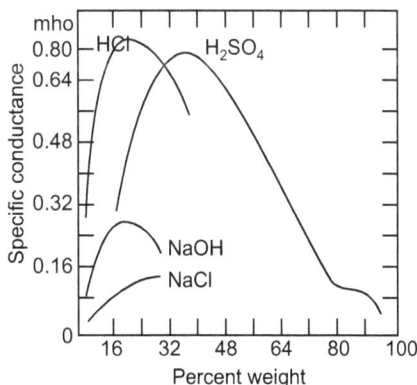

Fig. 3.14. Conductivity concentrations curves for some electrolytes.

The solubility of a sparingly soluble salt like AgCl is determinant by suspending a weighed precipitate of AgCl in conductivity water, warming the mixture and then cooling to 25°C. The specific conductance of the filtrate is found by using a conductivity cell by the usual method. The solubility of AgCl is given by the expression:

$$S = \frac{k_v \times E \times 1000}{\lambda_\infty}$$

Where E is the equivalent weight of AgCl (143.5); k_v is specific conductance

$$\lambda_\infty = \lambda_{Ag^+} + \lambda_{Cl^-}$$
$$= 61.92 + 76.51$$
$$= 138.27 \text{ mhos.}$$

$$\therefore \qquad S = \frac{k_v \times 143.5 \times 1000}{138.27}.$$

Knowing the experimental value of specific conductance, the solubility of AgCl can be calculated.

Types of Conductometric Titrations

As already stated, conductometric titrations are of two types, *viz.* Acidimetry and alkalimetry and precipitation titrations.

Acidimetry-Alkalimetry

Conductometric acidimetry-Alkalimetry titrations are of four types. These include:
- Titration of strong acid against a strong base,
- Titration of weak acid and against a strong base,
- Titration of strong acid against a weak base,
- Titration of weak acid against a weak base.

Strong Acid *vs* Strong Base

An example of conductometric titration of a strong acid against a strong base is titration of hydrochloric acid solution against a solution of sodium hydroxide.

The method consists in taking the acid solution (20 mL) in a conductance cell in a thermostat and adding 1 mL sodium hydroxide solution (from a burette) at a time. The solution after each addition of alkali is stirred and its conductance determined. The conductance is plotted against the volume of alkali added. It will be found that the points lie on two almost straight lines.

The intersection of the two interpolated lines is the end point. The volume of alkali corresponding to the end point is the volume of alkali required to neutralise 20 mL of the acid solution (Fig. 3.15).

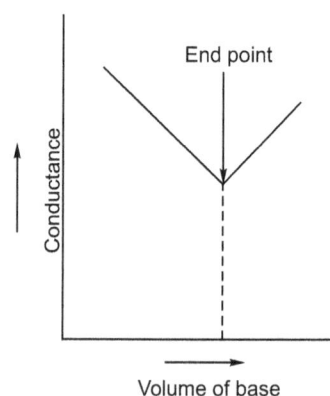

Fig. 3.15. Conductometric titration curve for strong acid and strong base.

The conductance of the solution before the addition of alkali is due to the presence of H^+ and Cl^- ions. Since H^+ ions possess greater mobility than Cl^- ions, the greater part of the conductance is due to it (H^+). On addition of alkali solution, the H^+ ions are removed by combining with OH^- ions forming water (which is feebly ionised)

$$H^+ + Cl^- + Na^+ + OH^- \longrightarrow Na^+ + Cl^- + H_2O \text{ (Feebly ionised)}$$

Due to the above reaction, the conductance of the solution decreases and continues to fall after each addition of NaOH solution till the end point is attained. After the end point, further addition of NaOH solution results in increase in conductance as OH^- ions are no longer removed in the chemical reaction.

The end-point of the titration is the point of minimum conductance.

In order to get accurate results, it should be ascertained that the volume change should be as small as possible. This is possible if the titrant is about ten times as strong as the acid solution. However, if this is not so, a correction to the readings has to be applied.

Thus, the actual volume $= \dfrac{v + V}{V} \times$ Observed conductance where v is the volume of the titrant and V is the original volume of the solution to be titrated.

A cell used for conductometric titration is given in Fig. 3.16. It has platinum electrodes which are sealed in position for accurate reproducible results. The surface of each electrode has a coating of a thin layer of platinum black. This increases the effective surface area. The cell is stored in distilled water in contact with the electrodes. This procedure prevents the deterioration of the electrodes. The electrodes are melted to thick platinum wire that are fused into glass tubes in which they make contact with mercury. The cell is made of pyrex or some other resistant type glass.

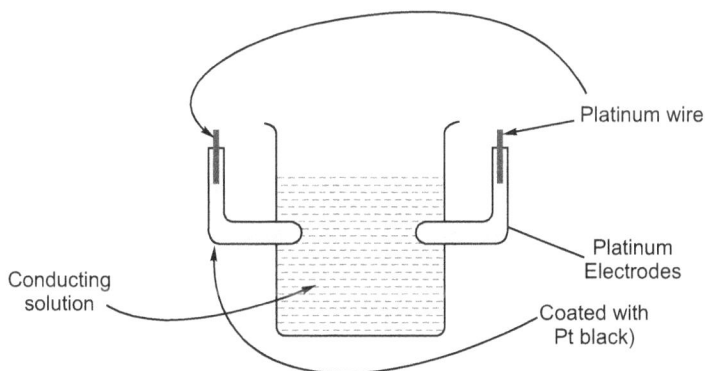

Fig. 3.16. Typical cell for conductometric measurement.

The conductometric cell is kept in a thermostat since the conductance varies will temperature.

Weak Acid *vs* a Strong Base

Conductometric titration of a weak acid (*e.g.*, acetic acid) against a strong base (*e.g.*, sodium hydroxide) is carried out as described in Fig. 3.16. In this case, the initial conductance of the solution is low as the weak acid (CH_3COOH) is poorly dissociated. Addition of sodium hydroxide gives sodium acetate, which is highly ionised. Initially, the acetate ions tend to suppress the ionisation of acetic acid still further due to common ion effect. However, after some time, there is increase in conductance due to the conducting power of highly ionised salt (NaOAc).

$$CH_3COOH + Na^+ + OH^- \longrightarrow CH_3COO^- + Na^+ + H_2O \quad \text{(Feebly ionised)}$$

As soon as the end point is reached, further addition of NaOH causes sharp increase in conductance due to the fast moving OH^- ions. The end point is the point of intersection of the two curves (Fig. 3.17).

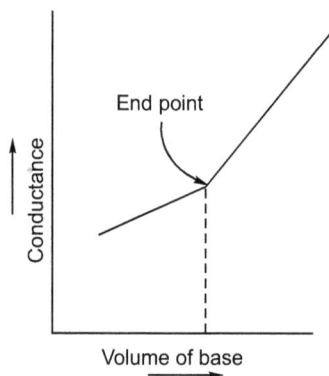

Fig. 3.17. Curve for titration of a weak acid against a strong base.

Strong Acid *vs* Weak Base

The conductometric titration of a strong acid (*e.g.*, hydrochloric acid) against a weak base (*e.g.*, ammonium hydroxide) is carried out as described earlier (Fig. 3.16).

The curve obtained for the titration of a strong acid against a weak base is shown in Fig. 3.18. In this case, the conductance of the solution first decreases due to the fixing up of the fast moving H^+ ions and their replacement by slow moving NH_4^+ ions.

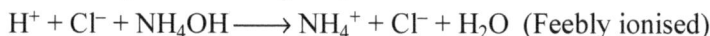

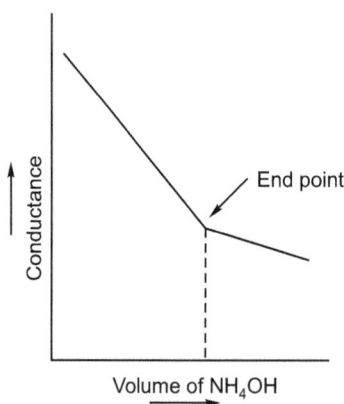

$$H^+ + Cl^- + NH_4OH \longrightarrow NH_4^+ + Cl^- + H_2O \quad \text{(Feebly ionised)}$$

Fig. 3.18. Curve for titration of strong acid against a weak base.

Weak Acid *vs* Weak Base

The conductometric titration of a weak acid (*e.g.*, acetic acid) against a weak base (*e.g.*, ammonium hydroxide) is also carried out as described earlier (*see Fig. 3.16*). In this type of titration (*viz.*, weak acid *vs* weak base), the conductometric method is most suitable as indicators (used in the normal volumetric analysis) do not give sharp end-point. The conductometric titration curve is shown in Fig. 3.19. In this case, the initial conductance of the solution (CH_3COOH) is poor due to poor dissociation of weak acid (CH_3COOH) the conductance increases as CH_3COONH_4 is formed. After the end point is reached, the conductivity remains almost constant since the weak base (NH_4OH) is a weak electrolyte. In this titration the end-point is sharp (Fig. 3.19).

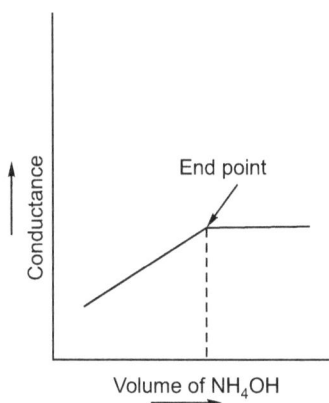

Fig. 3.19. Curve for titration of acetic acid against ammonium hydroxide.

Precipitation Titrations

An example of conductometric precipitation titration is the reaction of potassium chloride and silver nitrate. In this titration the end-point is determined accurately (Fig. 3.20). The accuracy of the determination is increased by using dilute solutions and addition of alcohol, which reduces the solubility of the precipitate and also prevents adsorption.

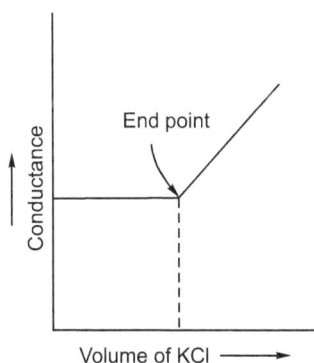

Fig. 3.20. Titration of KCl against $AgNO_3$.

In the above conductometric titration, the change in the conductance on the addition of $AgNO_3$ to KCl solution is not much, as the mobility of K^+ ion and Ag^+ ion is of the same order. The curve (Fig. 3.20) is nearly horizontal.

$$Ag^+ + NO_3^- + K^+ + Cl^- \longrightarrow K^+ + NO_3^- + \underset{(ppt.)}{AgCl} \downarrow$$

After the end point, the conductance increases sharply due to the increase in the number of free ions in the solution.

Advantages of Conductometric Titrations

Some of the advantages of conductometric titrations include:
- Coloured solution, where indicators do not work satisfactorily can be titrated by this procedure,
- This method is particularly useful for the titration of weak acids against weak bases which do not give a sharp change of colour with indicator in the usual volumetric titrations,
- Since the end-points are determined graphically, accurate and reproducible results are obtained,
- In polybasic acids, conductometric titrations can be used. In such cases, volumetric titrations do not give correct end points,
- Results obtained are much more accurate than in volumetric titrations,

It is, however, necessary to keep the temperature constant throughout the experiment. In acid-alkali titration, the titrant should be about ten times stronger than the solution to be titrated so that the change of volume is as little as possible.

3.3   SPECTROPHOTOMETRIC TITRATIONS

Introduction

We have known that in normal volumetric titration, the equivalence point of a reaction is detected visually by using an indicator as in acid-base titrations etc. or by the colour of the reactant (*e.g.*, potassium permanganate). Such titrations give results within one percent accuracy. In case, in some titrations, the colour change is gradual or the colour change does not contrast sharply, accurate results are not obtained. Such problems can be overcome if spectrophotometric titrations are carried out.

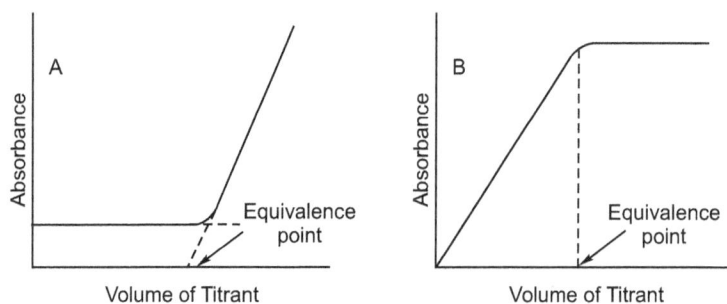

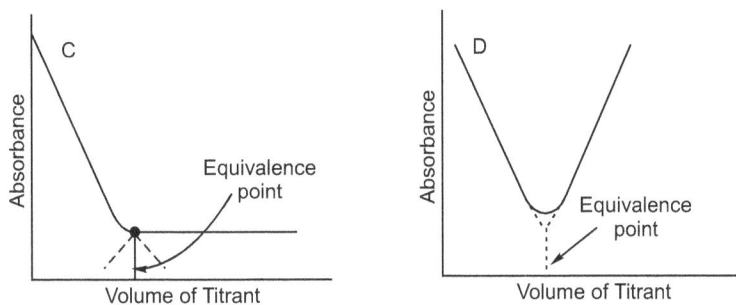

Fig. 3.21. Various titration curves. Determination of equivalence points.

In spectrophotometric titrations the end point is determined by using a spectrophotometer. The method consist in placing the titration vessel directly in the path of light of the instrument. The absorbance of the solution is determined after adding the titrant and a plot of absorbance as a function of volume of the titrant is made. Typical titration curves are given in Fig. 3.21.

Procedure of Titration

Spectrophotometric titration is carried out using a specially made titration cell (Fig. 3.22) (5 to 100 mL capacity). The cell fits into the cell compartment of the spectrophotometer. It is made of perspex sheet (which is opaque to *UV* light) and has two openings in order the accommodate circular quartz windows, which are arranged in such a way so that the beam of monochromatic light passes through their centres to the photoelectric cell. In the cell, there are two small openings, one for inserting the tip of a microburette and another for a micro stirrer. The whole cell except the quartz windows is covered with paper so that any extraneous light is excluded.

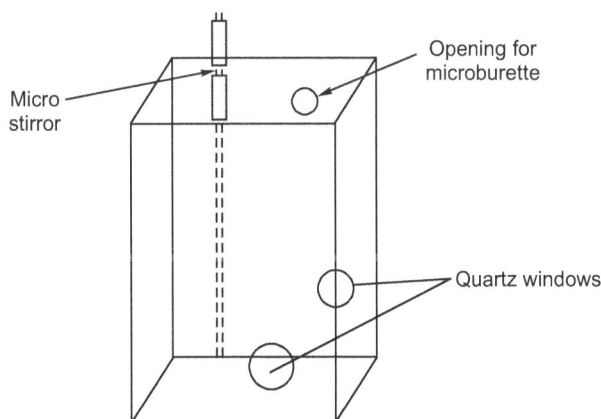

Fig. 3.22. Titration cell.

The solution to be titrated is taken in the cell, which is placed in the path of the light of the spectrophotometer. The light is adjusted to the wavelength at which the experiment has to be carried out. The instrument is then set to read zero absorbance (in case the reactant is colourless) or to some other starting value in case the reagent is coloured. After the addition of a known volume of titrant to the stirred solution in the titration cell, the absorbance is read. This is repeated at a number of points before the end point is reached and continued after the end point. Finally, a graph is plotted of the absorbance against the volume of the titrant added. Using this graph the equivalence point is determined.

The absorbance is known to depend on the concentration and so the effect of dilution must be taken into account. The error due to dilution is overcome by the following procedure:

(*i*) It is best if the titrant be more concentrated than the solution to be titrated. In this way there is negligible change in volume.

(*ii*) Alternatively a simple correction factor is added in the calculation

$$A = A' - \frac{V+v}{V}$$

where A and A' are the corrected and measured absorbance respectively. V is the original volume of the solution and v is the volume of the titrant added.

Applications

Following are given some of the applications of spectrophotometric titrations.

(*i*) **Acid-base Titration:** Titration of phenol with NaOH. The absorbance due to the formation of phenolate ion is followed.

(*ii*) **Redox Titrations:** Titration of Ce^{3+} with Co^{3+}. Formation of Ce^{4+} is followed.

(*iii*) **Complexometric Titrations:** Titration of Fe^{3+} with EDTA. Disappearance of Fe sulphosalicylic acid complex is followed.

Cu^{2+} can be titration with EDTA. Formation of Cu-EDTA complex is followed.

(*iv*) **Precipitation Titrations:** Titration of SO_4^{2+} ions with Ba^{2+} ions. Appearance of turbidity is followed.

3.4 pH-METRIC TITRATIONS

Introduction

In the usual titrations (*see chapter 2*), the titration is carried our using a indicator and observing the end point visually (by change in the colour). However, in pH-metric titrations, the end point is found by noting the pH of the solution (using a pH meter) and plotting it against the volume of the titrant added.

As an example, in acidimetry-alkalimetry, the titration of an acid (or a base) is carried out by a standard solution of a base (or an acid) as the case may be. The reaction involved is a nutralisation reaction

$$\text{Acid} + \text{Base} \longrightarrow \text{Salt} + \text{Water}$$

At equivalence point, the pH of the solution will be equal to, more or less than 7 depending on the hydrolysis of formed salt. Following cases may arise.

(*i*) Titration of a strong acid with a strong base. In this case, the salt formed is not hydrolysed by water and so the pH at the end point will be 7.

$$NaOH + HCl \longrightarrow NaCl + H_2O$$

(ii) Titration of a weak acid with a strong base. The salt formed in this case is hydrolysed by water to give a basic solution and so the pH at the end point will be greater than 7.

$$CH_3COOH + NaOH \longrightarrow CH_3COONa + H_2O$$

(iii) Titration of a strong acid with a weak base. In this case, the salt formed is hydrolysed with water to give acidic solution and so the pH at the end point will be less than 7.

$$HCl + NH_4OH \longrightarrow NH_4Cl + H_2O$$

pH-Metric Titration of a Strong Acid with a Strong Base

As an illustration, we consider the titration of a strong acid (HCl) against a strong base (NaOH). Following procedure is followed:

(i) Prepare a standard solution of 0.1 M oxalic acid.

(ii) Take the given 0.1 M NaOH solution (20 mL) in a conical flask. Titrate against standard oxalic acid (added from the burette) by using phenolphthalein as indicator in order to find the exact normality of the given NaOH solution.

(iii) Titrate NaOH solution of unknown strength with HCl using phenolphthalein as indicator to determine the exact normality to the given HCl solution volumetrically.

(iv) Switch on the pH meter. Calibrate the pH meter by immersing the glass electrode of the pH meter in a suitable buffer solution with known pH.

(v) Take 25 mL HCl solution in a beaker and dip the calibrated glass rod into it. Rinse and fill the burette with standardised NaOH solution [step (i)].

(vi) Add the NaOH solution (5 drops each time) dropwise, shake the solution and note the pH of the resultant solution. Continue noting pH after each addition of 5 drops of NaOH solution and determine the exact volume of the NaOH used at the end point by plotting pH against the volume of base added (Fig. 3.23).

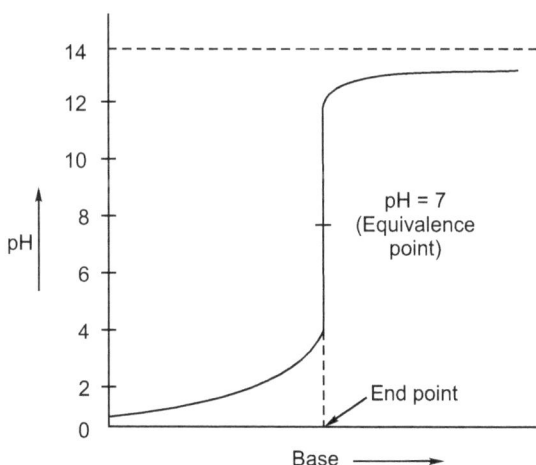

Fig. 3.23. Plot of Titration of pH against base added (strong acid *Vs* strong base).

The calculations are done in a similar way as described in chapter 2 (Section 2.2.1).

Other pH metric titrations are done using the same procedure. The end point in case of titration of a weak acid, 1 M CH$_3$COOH and strong base, 1 M NaOH solution and titration of a strong acid ($\dfrac{M}{100}$ HCl and weak base, $\dfrac{N}{100}$ NH$_4$OH) using a pH meter is as given in Fig. 3.24 and Fig. 3.25 respectively.

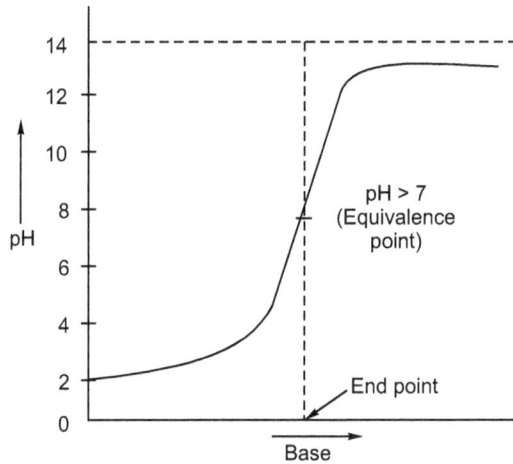

Fig. 3.24. Determination end point in the titration of a weak acid with strong base.

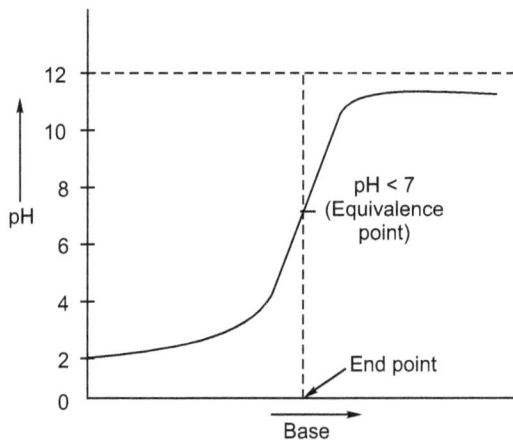

Fig. 3.25. Determination of end point in the titration of a strong acid with a weak base.

EXERCISES

1. Explain what are potentiometric titrations and the principle involved.
2. What are the advantages of potentiometric titrations over usual indicator volumetric procedures?
3. Write notes on (*a*) Indicator electrode and (*b*) Reference electrode.
4. Discuss the applications of potentiometric titrations.
5. How will you determine the concentration of Fe^{2+} using potentiometric titrations?
6. What are conductometric titrations? Explain.
7. Define conductivity, specific conductivity, equivalent conductance and molar concentration.
8. Explain with the help of a graph, the conductometric titration of (*i*) strong acid with strong base, (*ii*) weak acid with strong base and (*iii*) weak acid with weak base.
9. What are the advantages of conductometric titrations?
10. How are precipitation titrations conducted using conductometric titrations?
11. Describe the procedure for spectrophotometric titrations.
12. In spectrophotometric titration how the error due to dilution is overcome.
13. Give applications of spectrophotometric titrations.
14. Explain the use of pH metric titrations. How these are performed and end point determined?

○○○

PART III

Gravimetric Analysis
Estimation of More than one Constituent Present Together in a Solution
Instrumental Methods of Quantitative Analysis

4. Gravimetric Analysis

Contents

4

Gravimetric Analysis

4.1 INTRODUCTION

Gravimetric analysis constitutes a method of quantitative analysis. It involves quantitative estimation of various metals or anions associated with them. The method involves precipitation of an insoluble salt of an element or a compound which has to be quantitatively estimated from a solution, filtration, washing, drying, heating and finally weighing it or converting it into a pure stable compound of a definite composition and weighing it. As an example, for the estimation of barium, the method involves precipitation of barium from solution as insoluble $BaSO_4$, filtering, washing, drying and finally weighing. On the other hand, in the estimation of iron, it is precipated from the solution as hydrated ferric oxide (ferric hydroxide) which is of variable composition and so it is heated to give the oxide, Fe_2O_3 which is finally weighed.

Steps Involved in Gravimetric Estimations

Following are the steps involved in gravimetric estimations:

    (*i*) Cleaning of the crucible.

    (*ii*) Heating the crucible to its constant weight.

    (*iii*) Precipitation of the insoluble salt of the element or the compound to be estimated.

    (*iv*) Digestion of the precipitate.

    (*v*) Filtering and washing of the precipitate.

    (*vi*) Drying the precipitate.

    (*vii*) Incineration and heating of the precipitate.

    (*viii*) Weighing, heating and cooling till a constant weight is obtained.

    (*ix*) Calculations.

Cleaning of the Crucible

The crucibles used are of two types, *viz.*, silica crucible and sintered glass crucible.

    For cleaning a new silica crucible (Fig. 4.1), it is washed with conc. HCl and then with water.

However, for cleaning an old (or used) silica crucible, the solid residue is removed first by shaking and then by heating with conc. HNO_3 or aqua regia and finally washing with water.

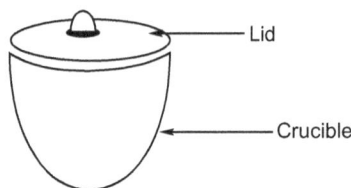

Fig. 4.1. A silica crucible.

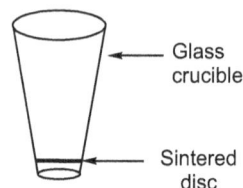

Fig. 4.2. A sintered glass crucible.

A sintered glass crucible (Fig. 4.2) is made of pyrex glass and is fitted with a sintered disc. These are numbered 1, 2, 3, 4 etc. The type of sintered glass to be used depends on the particle size of the precipitate. Thus, G_4 is used for filtration of fine particle size precipitate and G_1 is used for larger particle size precipitate.

The new sintered glass crucible is cleaned by washing with conc. HCl and then with water. However, a used or old sintered crucible is cleaned by removing the residue (if any) by gentle shaking and then by heating in a beaker containing conc. HNO_3 or aqua regia and finally washing with water.

Heating the crucible to constant weight

The cleaned silica crucible is dried by heating it on a clay pipe triangle (Fig. 4.3) with a Bunsen flame for about 30 min. It is allowed to cool in air (2-3 min) and then placed in a dessicator, (Fig. 4.4), allowed to cool and weighed. The process of heating, cooling and weighing is repeated till the crucible attains a constant weight. The constant weight crucible is kept in a dessicator.

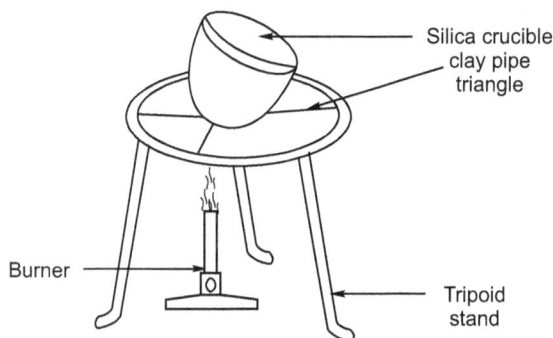

Fig. 4.3. Heating a silica crucible.

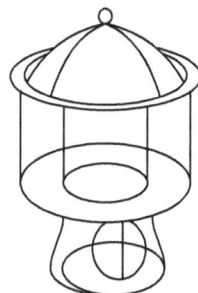

Fig. 4.4. A dessicator.

The cleaned sintered glass crucible is dried by heating (150-200°C) in an air oven, cooling by keeping in a dessicator and weighed. The process of heating, cooling and weighing is repeated till the crucible attains a constant weight.

A dessicator (Fig. 4.4) is a closed glass container in which dry atmosphere is maintained by keeping a dessicant like silica gel, anhydrous calcium chloride etc.

Precipitation

For precipitation of the insoluble salt of the element or compound, the given solution (known volume) is taken in a beaker, diluted with water, solution heated (if necessary) and treated with a precipitating reagent (the reagent is added slowly to the stirred solution) till the precipitation is complete. As an example a solution containing a barium salt is treated with dilute H_2SO_4 to give $BaSO_4$ precipitate.

Digestion of the Precipitate

The precipitate along with the solution obtained above is heated on a steam bath (~ 30 minutes— a few hours). This helps to increase the size of particles and the undesired impurities get dissolved.

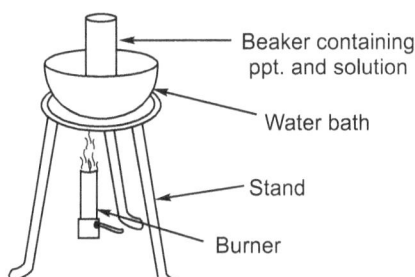

Fig. 4.5. Digestion of the precipitate.

Filtering and Washing of the Precipitate

After digestion of the precipitate, it is filtered either by using a filter paper or a sintered glass crucible and then washed with a suitable liquid in order to remove undesirable substances including the excess of the reagent present.

In case a filter paper is to be used, it is important to use a filter paper which leaves neglible amount of ash after incineration. Normally, Whatmann filter paper No. 42 (for fine particles), No. 41 (for gelatinous precipitate) and No. 40 (for bigger size particles) are used. A round filter paper (9-11 cm diameter) is used. It is first folded in half and then in quarters. It is then opened with three parts on one side and one part on the other side (Fig. 4.6) and then placed in a funnel (3″ diameter)

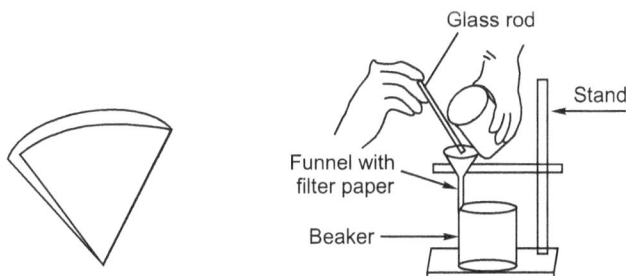

Fig. 4.6. Whatmann filter paper (folded).

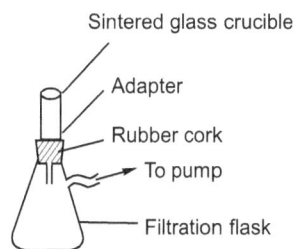

Fig. 4.7. Filtration of the precipitate.

Fig. 4.8. Filtration using sintered glass crucible.

and moistend with distilled water. The Whatmann filter paper should not project out of the funnel. The funnel alongwith the filter paper is then placed in a funnel stand and a beaker or a conical flask is kept below the stem of the funnel for collecting the filtrate. The tip of the stem of funnel should touch the side of the receiver. By this arrangement (Fig. 4.7), the speed of filtration increases. The supernatant liquid is then filtered leaving as much of the precipitate in the beaker as possible.

In case a sintered glass crucible is used for filtration, it is fitted using an adapter into a filtration flask (Fig. 4.8), which is attached to a suction pump. The liquid is transferred from the beaker into the sintered crucible using a glass rod.

The precipitate left in the beaker is washed with a wash liquid (in both the methods) (10-15 mL) and the supernatant liquid transferred to the funnel. The process is continued 3-4 times till the fitrate is free of the precipitating reagent. Finally all the precipitate is transferred from the beaker to the funnel or sintered glass crucible (as the case may be). Any precipitate sticking to the walls of the beaker is dislodged with the help of a 'policeman', which is a glass rod (10-12 cm long) with about 0.5 cm rubber tube pulled over one end of the glass rod (Fig. 4.9). Thus, all the precipitate along with the solvent is transferred into the funnel or sintered crucible. Generally, ammonium salt solution or dilute acid solution is used as wash liquid. Alcohol can also be used, and most of precipitates are insoluble in it.

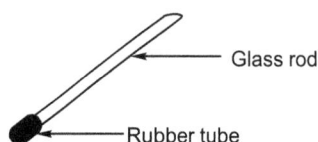

Fig. 4.9. A policeman.

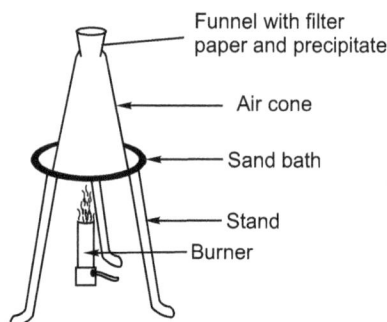

Fig. 4.10. Drying the precipitate.

Drying of the Precipitate

The precipitate (in the funnel) after filtration and washing is dried. In case filter paper is used for filtration, the funnel along with the precipitate and the filter paper is placed in a hot air drying cone (Fig. 4.10), which is kept over a sand bath and heated with a small flame. Adequate precaution is taken to ascertain that the filter paper does not get charred or catches fire and that the paper does not get so much dry that it tends to crack. In case, sintered glass crucible has been used for filtration, it is dried in an oven at a temperature depending on the nature of the ppt. (discussed in each experiment).

Incineration and Heating of the Precipitate

The sintered glass crucible containing the precipitate is heated in an oven at a temperature which depends on the nature of the precipitate (this aspect has been discussed in each estimation).

Alternatively, the dried filter paper along with the precipitation is folded to form a small packet and placed in a clean, dried silica crucible (whose constant weigh has been determined earlier). The crucible, kept over a clay pipe triangle which is placed on a tripoid stand is heated over a very small flame to carbonise the paper. Care is taken that the paper does not catch fire. After carbonisation, the flame is gradually increased and finally the crucible heated strongly (below the decomposition temperature of the precipitate) for about 30 minute. It is then cooled in air (4-5 minutes) and then in a dessicator and then weighed. The process of heating, cooling and weighing is repeated till the weight of the crucible along with the residual mass is constant.

As already stated, the filter paper during carbonisation should not catch fire. However, if it does, the flame below the crucible is removed, and the crucible covered with its lid in order to extinguish the fire.

It is found that in some cases, the precipitate may get partially reduced by the carbon of the filter paper. If this happens, the reduced form has to reconverted into the original form by suitable treatment. As an example, during the estimation of barium as sulphate, some of the precipitate gets reduced to the corresponding sulphide during heating. BaS is converted back to $BaSO_4$ by treatment with HCl and H_2SO_4.

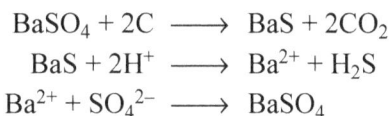

$$BaSO_4 + 2C \longrightarrow BaS + 2CO_2$$
$$BaS + 2H^+ \longrightarrow Ba^{2+} + H_2S$$
$$Ba^{2+} + SO_4^{2-} \longrightarrow BaSO_4$$

Calculation

Finally, calculations are done in order to estimate the amount of metal or a compound in the solutions. The method of calculation is discussed in detail in each experiment.

4.2 ESTIMATION OF COPPER

Copper can be estimated either as cuprous thiocyanate (CuSCN) or as copper oxide (CuO).

4.2.1 Estimation of Copper as Cuprous Thiocyanate

Copper can be estimated by precipitation as cuprous thiocyanate. The method consist in first reducing Cu^{2+} to Cu^+ by sulphurous acid (obtained from $NaHSO_3$ and HCl) and then treating with ammonium thiocyanate solution (10%)

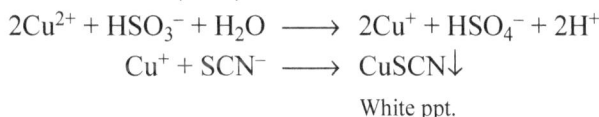

$$2Cu^{2+} + HSO_3^- + H_2O \longrightarrow 2Cu^+ + HSO_4^- + 2H^+$$
$$Cu^+ + SCN^- \longrightarrow CuSCN\downarrow$$
$$\text{White ppt.}$$

Procedure

Following steps are followed:

(i) To the given copper sulphate solution[1] (20 mL) taken in a 400 mL beaker is added few drops of dilute HCl and a slight excess of freshly prepared sulphurous acid solution[2] (35-40 mL).

(ii) The above solution is diluted with distilled water (125 mL), heated to nearly boiling and treated with slight excess of ammonium thiocyanate solution (10%) with stirring added for complete precipitation[3] of copper as cuprous thiocyanate. The solution is heated (1-2 min) and allowed to stand (4-5 hrs).

(*iii*) The precipitate of cuprous thiocyanate is filtered using a G_4 sintered glass crucible which has earlier been cleaned and heated to constant weight.

(*iv*) The precipitate is washed a number of times with wash liquid[4] and then with alcohol (20%, 5-6 times). Finally the whole of the precipitate is transferred into the crucible.

(*v*) The sintered glass crucible is heated at 110-115°C is an oven. It is cooled to room temperature (desiccator) and weighted.

(*vi*) The process of heating, cooling and weighing is repeated (3-4 times) till a constant weight is obtained.

Notes:

1. The copper sulphate solution, which is used for estimation is prepared by accurately weighing about 26 g of $CuSO_4.5H_2O$ (AR) and dissolving in distilled water in presence of acetic acid in a 1000 mL measuring flask and solution made up to the mark. 20 mL of this solution is used for estimation. This solution is prepared by the instructor.

2. Freshly prepared saturated sulphurous acid solution is prepare by adding 20 mL of 2 N HCl to 25 mL of 10% $NaHSO_3$ solution.

3. The complete precipitation of cuprous thiocyanate is indicated by the smell of SO_2 and a colourless solution.

4. The wash liquid is prepared by mixing distilled water (200 mL), ammonium thiocyanate (2 mL, 10%) and freshly prepared saturated sulphurous acid (10-15 drops).

All steps are followed as per details given in introduction (Section 4.1).

Calculations

Let the mass of CuSCN formed from 20 mL of the given solution of $CuSO_4 = w$ g.

$$121.62 \text{ g CuSCN} \equiv 63.54 \text{ g copper}$$

$\therefore \qquad w \text{ g of CuSCN} \equiv \dfrac{63.54}{121.62} \times w \text{ g of copper}$

$\therefore$ Mass of copper per litre of the given solution

$$= \dfrac{63.54 \times w \times 50}{121.62} \text{ g.}$$

4.2.2 Estimation of Copper as Copper Oxide (CuO)

Copper can also be estimated by precipitation as $Cu(OH)_2$ from a cupric salt solution with sodium hydroxide. The precipitated $Cu(OH)_2$ on heating gives copper oxide (CuO), which is weighed.

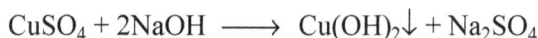

$$CuSO_4 + 2NaOH \longrightarrow Cu(OH)_2\downarrow + Na_2SO_4$$

$$Cu(OH)_2 \xrightarrow{\Delta} CuO + H_2O$$

Procedure

Following steps are followed:

(i) To the given copper sulphate solution[1] (20 mL) taken in a 400 mL beaker is added water (100 mL), conc. HCl (1 mL) and phenolphthalein (2 drops).

(ii) The above solution is heated to boiling and hot dilute solution of sodium hydroxide[2] (2%) is added slowly with stirring till the supernatant liquid acquires permanent pink colour.

(iii) The precipitate obtained in step (ii) is heated on a hot water bath till it turns brown and settles down.

(iv) The precipitate is filtered using Whatmann filter paper No. 40 by decantation. To the precipitate (in the beaker) is added hot water (15-20 mL), the mixture stirred, allowed to settle and filtered again by decantation. The process of washing with not water is repeated till the filtrate is free of SO_4^{2-} (test wih $BaCl_2$ solution) and alkali (test with litmus paper). The precipitate is finally transferred quantitatively from the beaker to the filter paper.

(v) The precipitate alongwith the filter paper and funnel is dried in a hot air conc. (Fig. 4.10).

(vi) After drying, the filter paper alongwith the dried precipitate is folded into a small packet, placed in a preheated constant weight silica crucible and heated on a small low flame to carbonise the filter paper. The crucible is cooled, conc. HNO_3 (2-3 drops) added (in order to oxidise Cu_2O or Cu that might have formed due to reduction of CuO by carbon of the filter paper). The crucible is heated on a low flame (3-4 min) and then strongly (800-900°C) for 20-30 min. It is cooled and weighed.

(viii) The process of heating, cooling and weighing is repeated till constant weight of the crucible along with CuO is obtained.

Notes:

1. The copper sulphate solution is prepared by the instructor as in Section 4.2.1 [note (i)].
2. Excess of NaOH should not be added as the ppted $Cu(OH)_2$ may dissolve. So phenolphthalein is added as an indicator.

Calculations

Let the mass of CuO from 20 mL of the given solution $= w$ g.

$$79.54 \text{ g } CuO \equiv 63.54 \text{ g copper}$$

$\therefore \qquad w \text{ g of } CuO \equiv \dfrac{63.54}{79.54} \times w \text{ g copper}$

$\therefore$ Mass of copper present per litre of the given solution

$$= \dfrac{63.54 \times w \times 50}{79.54} \text{ g.}$$

4.3 ESTIMATION OF IRON

Iron is estimated as iron oxide, which is obtained by precipitating iron as ferric hydroxide and heating. The solution of ferrous ammonium sulphate used for the estimation of iron, is first treated with conc. HNO_3 to oxidise Fe^{2+} to Fe^{3+}. Subsequent addition of NH_4OH gives hydrated oxide, $Fe_2O_3.X.H_2O$ (or $Fe(OH)_3$).

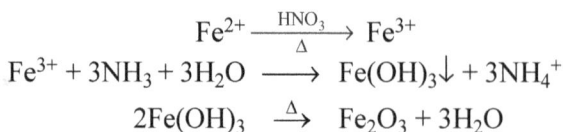

$$Fe^{2+} \xrightarrow[\Delta]{HNO_3} Fe^{3+}$$
$$Fe^{3+} + 3NH_3 + 3H_2O \longrightarrow Fe(OH)_3\downarrow + 3NH_4^+$$
$$2Fe(OH)_3 \xrightarrow{\Delta} Fe_2O_3 + 3H_2O$$

Procedure

Following steps are involved:

(i) To the given ferrous ammonium sulphate solution[1] (20 mL) taken in a 400 mL beaker is added distilled water (40-50 mL), dilute HCl (1 mL; 1:1) and conc. HNO_3 (1-2 mL). The solution is gently boiled to get a clear solution (check for complete oxidation of Fe^{2+} to Fe^{3+}; there should be no precipitation by treating a drop of the solution with a drop of potassium ferric cyanide in a groove tile). In case, oxidation is not complete, some more conc. HNO_3 (1 mL) is added and the solution heated.

(ii) The above solution is diluted with distilled water (140-150 mL), heated to boiling and ammonia solution[2] added slowly with stirring till a slight excess of ammonia is present. The solution is boiled and allowed to stand. The supernatant liquid should be colourless. In case the supernatant solution is yellow, some more ammonia solution (1:1) is added for completion of precipitation.

(iii) The formed precipitate is filtered using Whatmann filter paper No. 41 by decantation. To the residual precipitate is added ammonium nitrate solution (40-50 mL, 1%), solution stirred, allowed to stand and filtered by decantration. Washing of the precipitate continued (4-5 times) with ammonium nitrate solution till the filtrate does not give test for chloride ions. Finally transfer all the precipitate to the filter paper.

(iv) The precipitate alongwith the filter paper and the funnel is dried using a hot air cone (Fig. 4.10).

(v) The dried filter paper alongwith the precipitate is folded and heated gently in a preheated constant weight silica crucible. Adequate precaution is taken to ascertain that the filter paper does not catch fire.

(vi) Finally the crucible is heated to red heat, cooled and weighed. The process of heating, cooling and weighing is repeated till a constant weight is obtained.

Notes:

1. The required ferrous ammonium sulphate, $FeSO_4.(NH_4)_2SO_4.6H_2O$ solution is prepared by the instructor by weighing exactly about 50 g of ferrous ammonium sulphate (AR), dissolving in distilled water in presence of little dilute H_2SO_4 in a 1000 mL measuring flask and making the volume up to the mark. 20 mL of this solution is used for estimation.

2. An excess of NH_4OH solution is used for complete precipitation. Sodium or potassium hydroxide should not be used for precipitation, as the alkali metal ions are absorbed strongly on the surface of $Fe(OH)_3$ and cannot be removed by washing.

Calculations

Let the mass of Fe_2O_3 obtained from 20 mL of the given solution is w g.

$$159.70 \ g \ of \ Fe_2O_3 \equiv 111.69 \ g \ of \ iron$$

$$\therefore \qquad w \ g \ of \ Fe_2O_3 \equiv \frac{111.69}{159.70} \times w \ g \ of \ iron$$

Mass of iron per litre of the given solution

$$= \frac{111.69 \times w \times 50}{159.70} \ g$$

4.4 ESTIMATION OF BARIUM

Barium is estimated as barium sulphate by using dilute H_2SO_4 in presence of HCl

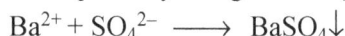

$$Ba^{2+} + SO_4^{2-} \longrightarrow BaSO_4\downarrow$$

Procedure

Following steps are involved:

(i) To the given barium chloride solution[1] (20 mL) taken in a 400 mL beaker is added conc. HCl[2] (about 0.5 mL) and water (about 100 mL).

(ii) The solution is heated to near boiling (80-90°C) and hot dilute H_2SO_4 (2N) added dropwise with stirring till the precipitation is complete.[3]

(iii) The above solution alongwith the precipitate is heated (30 min.) on a boiling water bath. The solution is cooled and the precipitate is allowed to settle down.

(iv) The precipitate is filtered using a Whatmann filter paper No. 42 by decantation. The ppt. is washed a number of times with hot water containing a trace of H_2SO_4 (1-2 drops of dil. H_2SO_4 in 1000 mL H_2O) and filtered each time by decantation till the filtrate is free of Cl^- and SO_4^{2-} ions. The absence of Cl^- and SO_4^{2-} is tested with $AgNO_3$ solution and $BaCl_2$ solution respectively.

(v) Finally all the precipitate is quantitatively transferred from the beaker on to the filter paper.

(vi) The filter paper alongwith the precipitate and the funnel is dried in a hot air cone (Fig. 4.10).

(vii) The dry filter paper alongwith the precipitate is folded, kept in a preheated constant weight silica crucible[4] and heated gently over a small flame to char the paper without inflaming it. The heating is continued for some more time till the carbon gets oxidised to CO_2.

(viii) The crucible containing the residue is cooled, conc. HCl[5] (one drop) and then conc. H_2SO_4[5] (one drop) added. The crucible is first heated gently then strongly[6] (30 min) using a mecker burner. The crucible is cooled and weighed.

(ix) The process of heating, cooling and weighing is repeated till constant weight.

Notes:

1. The required barium chloride solution is prepared by the instructor by exactly weighing about 12.50 g of barium chloride ($BaCl_2.2H_2O$; A.R.) and dissolving in distilled water in a 1000 mL measuring flask and the solution made upto the mark. 20 mL of this solution is used for estimation.

2. The precipitation of $BaSO_4$ is carried out in presence of HCl. This helps to increase the particle size of the precipitate and makes filtration easy. Large amounts of HCl are avoided as soluble $BaHSO_4$ may be formed by reaction between formed $BaSO_4$ and HCl.

3. The completion of precipitation is ascertained by allowing the precipitate to settle and adding some more dil. H_2SO_4 to the clear supernatant solution–no more precipitation should be observed.

4. In place of silica crucible, sintered glass G_4 crucible can also be used. In this case it is not necessary to add conc. HCl and conc. H_2SO_4 during heating (see note 5).

5. The precipitated $BaSO_4$ may be reduced to some extent by the carbon of the filter paper to BaS

$$BaSO_4 + 4C \xrightarrow{\Delta} BaS + 4CO$$

Addition of HCl converts BaS to $BaCl_2$

$$BaS + 2HCl \longrightarrow BaCl_2 + H_2O$$

and addition of H_2SO_4 converts $BaCl_2$ to $BaSO_4$

$$BaCl_2 + H_2SO_4 \longrightarrow BaSO_4\downarrow + 2HCl$$

6. The ppt. of $BaSO_4$ should not be heated to very high temperature (1400°C) in order to prevent its decomposition

$$BaSO_4 \xrightarrow{1400°C} BaO + SO_2$$

Calculations

Let the mass of $BaSO_4$ obtained from 20 mL of the given solution $= w$ g.

$$233.43 \ g \ BaSO_4 \equiv 137.33 \ g \ of \ barium$$

$$\therefore \qquad w \ g \ of \ BaSO_4 \equiv \frac{137.33}{233.43} \times w \ g \ of \ barium$$

Mass of Ba present per litre of the given solution

$$= \frac{137.33 \times w \times 50}{233.43} \ g.$$

4.5   ESTIMATION OF SULPHATE

Sulphate is estimated as barium sulphate by precipitating it from the solution by adding $BaCl_2$ solution in the presence of a little HCl

$$SO_4^{2-} + Ba^{2+} \longrightarrow BaSO_4\downarrow$$

Procedure

(*i*) To the given copper sulphate solution[1] (20 mL) taken in a 400 mL beaker is added distilled water (about 100 mL) and conc. HCl (1 mL).

(*ii*) The above solution is heated to near boiling (70-80°C) and a hot solution of $BaCl_2$ (10%) added dropwise with stirring till the precipitation is complete.[2]

(*iii*) Rest of the steps are discussed in the estimation of barium as $BaSO_4$ (Section 4.4). The only difference is that the precipitate is washed with hot water to remove undesired ions and not with water containing a trace amount of H_2SO_4.

Notes:

1. The required copper sulphate solution is prepared by the instructor by accurately weighing about 14 g of copper sulphate ($CuSO_4.5H_2O$; A.R.) and dissolving in distilled water in a 1000 mL measuring flask and made up to mark 20 mL of this solution is used for estimation.

2. The completion of precipitation is ascertained by allowing the ppt. of $BaSO_4$ settle down and adding few drops of $BaCl_2$ solution–there should be no more ppt. formation.

Calculations

Let the mass of $BaSO_4$ obtained from 20 mL of the given solution $= w$ g.

$$233.43 \text{ g of } BaSO_4 \equiv 96.06 \text{ g } SO_4^{2-}$$

$$\therefore \qquad w \text{ g of } BaSO_4 \equiv \frac{96.06}{233.43} \times w \text{ g of } SO_4^{2-}$$

Mass of SO_4^{2-} ions present per litre of the given solution

$$= \frac{96.06 \times w \times 50}{233.43} \text{ g.}$$

4.6   ESTIMATION OF SILVER

Silver is estimated as silver chloride by precipitating it from the given solution by dilute HCl in presence of HNO_3.

$$AgNO_3 + HCl \longrightarrow AgCl\downarrow + HNO_3$$

Silver chloride, being photosensitive, special precautions are taken. The precipitate is kept in dark.

Procedure

Following steps are followed:

(*i*) To the given silver nitrate solution[1] (20 mL) taken in a 400 mL beaker is added distilled water (about 50 mL) and conc. HNO_3[2] (0.5 mL).

(*ii*) The above solution is heated (about 70°C) on a water bath and hot dilute HCl (0.2 N) is added dropwise with stirring till the precipitation is complete.[3]

(*iii*) The beaker containing the precipitate is wrapped from out side with carbon paper (in order to prevent light from entering). The beaker is subsequently heated on a water bath (in order to coagulate the ppt.) for about 1 hour. During this time the beaker in kept covered with a watch glass, which is covered with black paper on the top side. This procedure protects the ppt. of AgCl from light.

(*iv*) The precipitate of AgCl is filtered using a sintered glass crucible (G_4)[4] by decantation, washing the precipitate with dilute nitric acid (water containing a few drops of dilute HNO_3) till the washings are free of chloride ions. Finally whole of the precipitate is quantitatively transferred from the beaker to the sintered glass crucible.

(*v*) The sintered glass crucible along with the precipitate is first dried in an oven at 100°C and then at about 130°C. The crucible is cooled (desiccator) and weighed.

(*vi*) The process of heating in an oven and cooling is repeated till the crucible has a constant weight.

Notes:

1. The silver nitrate solution for estimation is prepared by the instructor by exactly weighing about 18 *g* of $AgNO_3$(AR) dissolving in distilled water in presence a little HNO_3 (AR) in a 1000 mL measuring flask and the solution made up to the mark. 20 mL of this solution is used for estimation.

2. AgCl is precipitated from a solution of $AgNO_3$ solution containing HNO_3, which prevents the formation of colloidal AgCl and also reduces its solubility in the medium.

3. Complete precipitation of AgCl is ascertained by letting the ppt. of formed AgCl settle down and adding few drops of dilute HCl – no more precipitate is obtained.

4. In place of sintered glass crucible, Whatmann filter paper No. 42 can also be used. In this case, filtration is done by decantation, washing the residue with dilute HNO_3 (0.1 *N*) as in the above case till the washings are free of Cl^- ions. The precipitate is then quantitatively transferred from the beaker to the filter paper and dried on a hot air cone. The precipitate from the filter paper is transferred on the glazed paper and the ppt. covered with a funnel (the funnel is covered with carbon paper from outside in order to protect the ppt. of AgCl from light). The filter paper is folded into a small packet and placed in a preheated constant weight silica crucible. The crucible is heated gently with a small flame till the paper chars. Finally the ppt. of AgCl from glazed paper is transferred to the cooled silica crucible and then heated to 110-120°C in an oven for about 30 minute. The crucible is allowed to cool to room temperature and weighted. The process of heating, cooling and weighing in repeated till a constant weight of the crucible is obtained.

Calculations

Let the mass of AgCl obtained from 20 mL of the given $AgNO_3$ solution = *w* g

$$143.34 \text{ } g \text{ of AgCl} \equiv 107.88 \text{ } g \text{ silver}$$

∴ $$w \text{ } g \text{ of AgCl} \equiv \frac{107.88}{143.34} \times w \text{ } g \text{ of silver}$$

Mass of silver present per litre of the given solution

$$= \frac{107.88 \times w \times 50}{143.34} \text{ } g.$$

4.7   ESTIMATION OF CHLORIDE

Chloride is estimated as silver chloride by treatment of the solution containing chloride ions with $AgNO_3$ solution is presence of a little HNO_3.

$$Cl^- + Ag^+ \longrightarrow AgCl\downarrow$$

Procedure

Following steps are followed:

(i) To the given sodium chloride solution[1] (20 mL) taken in a 400 mL beaker is added distilled water (80-100 mL) and conc. HNO_3[2] (1 mL).

(ii) The above solution is heated to about 70°C in a water bath. $AgNO_3$ solution (2%) is added dropwise with stirring till the precipitation is complete.[3] The beaker is wraped with carbon paper and heated to 70 – 80°C and the ppt. allowed to settle down.

(iii) The ppt. obtained in step (ii) is filtered by decantation using Whatmann filter paper No. 42. The residue in the beaker is washed with water containing a little HNO_3 till the washing is free of Ag^+. Finally all the precipitate is transferred quantitatively from the beaker to the filter paper.

(iv) Subsequent steps are same given in note 4 of Section 4.6.

Notes:

1. The sodium chloride solution to be given for estimation is prepared by the instructor by accurately weighing about 5 g of sodium chloride (A.R.) in distilled water in a 1000 mL measuring flask and making up the volume to the mark. 20 mL of this solution is used for the estimation of chloride ion.

2. The precipitation of AgCl is carried out is presence of HNO_3 which helps to prevent co-precipitation of other ions (if present) and also to form easily filterable ppt.

3. The completion of precipitation is ascertained by letting the ppt. of AgCl settle and adding few drops of dil $AgNO_3$ to the supernatant liquid – no precipitation.

Calculations

Let the mass of AgCl obtained from 20 mL of the given NaCl solution = w g.

$$143.34 \text{ g AgCl} \equiv 35.46 \text{ g chloride ions}$$

$$w \text{ g of AgCl} \equiv \frac{35.46}{143.34} \times w \text{ g of chloride ions}$$

Mass of chloride ions present per litre of the solution

$$= \frac{35.46 \times w \times 50}{143.34} \text{ g.}$$

4.8 ESTIMATION OF ZINC

Zinc is estimated as oxide by precipitating as zinc carbonate by the addition of Na_2CO_3 solution to the zinc sulphate solution and then igniting the formed $ZnCO_3$ to give the oxide (ZnO).

$$ZnSO_4 + Na_2CO_3 \longrightarrow ZnCO_3\downarrow + Na_2SO_4$$

$$ZnCO_3 \xrightarrow{\Delta} ZnO + CO_2$$

Procedure

Following steps are followed:

(i) To the given solution of zinc sulphate[1] (20 mL) taken in a 250 mL beaker is added distilled water (50-60 mL). The solution is heated to almost boiling and a solution of sodium carbonate[2] (10%) added dropwide with stirring until a permanent turbidity is obtained. Phenolphthalein (2-3 drops) is then added followed by addition of more sodium carbonate solution till a permanent light pink colour is obtained. The solution is heated (3-4 min) and the ppt. formed is allowed to settle.

(*ii*) The above ppt. is filtered by decantation using Whatmann filter paper No. 41. The ppt. is washed with hot water and filtered by decantation. Washing and filtration by decantation is continued till the washing is free of CO_3^{2-} and SO_4^{2-}.

(*iii*) Whole of the precipitate in the beaker is quantitatively transferred to the filter paper and the precipitate alongwith the filter paper and funnel dried on a drying cone.

(*iv*) The precipitate in the filter paper is transferred to a glazed paper and covered with a funnel.

(*v*) The filter paper is folded and put in a **weighed crucible**, moistened with ammonium nitrate solution[3] and heated slowly at low temperature to carbonise the paper. The crucible is cooled and the ppt. of $ZnCO_3$ (on the glazed paper) transferred to the crucible.

(*vi*) The crucible is heated strongly for 10-15 min. By this heating $ZnCO_3$ is converted into ZnO. The crucible is cooled and weighed.

(*vii*) The process of heating, cooling and weighing is repeated till the crucible attains a constant weight.

Notes:

1. The zinc sulphate solution used for estimation is prepared by the instructor by exactly weighing about 56 g of zinc sulphate (A.R.) in distilled water in a 1000 mL measuring flask and making upto the mark.

2. Excess of Na_2CO_3 is avoided (this is checked by using phenolphthalein as indicator) as the ppt. of $ZnCO_3$ may dissolve due to the formation of sodium zincate.

3. Ammonium nitrate convets any metallic zinc formed by the reduction of ZnO with carbon of the filter paper to ZnO.

$$ZnO + C \longrightarrow Zn + CO$$
$$Zn + NH_4NO_3 \longrightarrow ZnO + N_2 + 2H_2O$$

Calculations

Let the mass of ZnO obtained from 20 mL of the given solution of $ZnSO_4 = w$ g

$$81.37 \, g \text{ of ZnO} \equiv 65.37 \, g \text{ zinc}$$

$$\therefore \qquad\qquad w \, g \text{ of ZnO} \equiv \frac{65.37}{81.37} \times w \, g \text{ zinc}$$

Mass of zinc present per litre of the given solution

$$= \frac{65.37 \times w \times 20}{81.37} \, g.$$

4.9   ESTIMATION OF LEAD

Lead is estimated either as lead sulphate or lead chromate.

4.9.1 Estimation of Lead as Lead Sulphate

Lead is precipitated as lead sulphate by addition of dilute H_2SO_4 to a solution of lead nitrate or lead acetate in presence of rectified spirit.

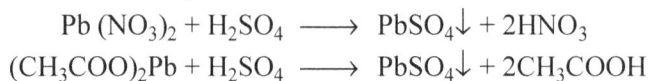

$$Pb(NO_3)_2 + H_2SO_4 \longrightarrow PbSO_4\downarrow + 2HNO_3$$
$$(CH_3COO)_2Pb + H_2SO_4 \longrightarrow PbSO_4\downarrow + 2CH_3COOH$$

Procedure

Following steps are followed:

(*i*) To the given lead nitrate (or lead acetate) solution[1] (20 mL) taken in a 250 mL beaker is added distilled water (20-25 mL) and conc. H_2SO_4 (2 mL). The solution is heated to almost boiling, cooled and ethyl alcohol[2] (20 mL) added. The solution is stirred and allowed to stand for about one hour. Complete precipitation of lead sulphate is confirmed.[3]

(*ii*) The ppt. of $PbSO_4$ is filtered by using Whatmann filter paper[4] No. 40 by decantation. The precipitate (in the beaker) is washed with water containing ethyl alcohol (95 mL water containing 5 mL ethyl alcohol) and filtered by decantation. The washing and filtration by decantation is continued till the washings are free of SO_4^{2-} ions.

(*iii*) The precipitate alongwith the filter paper and funnel is dried in an air cone. The filter paper is folded into a small packed, placed in a weighed silica crucible and heated slowly.

(*iv*) To the crucible is added a drop of conc. HNO_5' heated gently and a drop of conc. H_2SO_4 added[5]. The curcible is heated on a low flame and then strongly at about 500°C. The crucible is allowed to cool and weighed.

(*v*) The process of heating, cooling and weighing is repeated till a constant weight is obtained.

Notes:

1. The lead nitrate (or lead acetate) solution for the estimation is prepared by the instructor by exactly weighing about 13 g of lead nitrate (or lead acetate) (*AR*) and dissolving in water in presence of a little glacial acetic acid in a 1000 mL measuring flask and making up the volume to the mark. 20 mL of this solution is used for estimation.

2. The precipitate of $PbSO_4$ is insoluble in ethyl alcohol and so the precipitation and washing of the precipite is carried out in presence of ethyl alcohol.

3. The complete precipitation of lead as $PbSO_4$ is ascertained by allowing the ppt. to settle down and adding few drops of dil. H_2SO_4 there should not be any turbidity formation.

4. In place of Whatmann filter paper, sintered glass crucible can also be used. In this case treatment with HNO_3 and H_2SO_4 are avoided (see step 5).

5. The $PbSO_4$ formed may get reduced to PbS by the carbon of the filter paper during heating. Heating with HNO_3 and H_2SO_4 converts the formed PbS into $PbSO_4$.

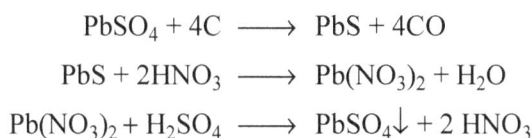

$$PbSO_4 + 4C \longrightarrow PbS + 4CO$$
$$PbS + 2HNO_3 \longrightarrow Pb(NO_3)_2 + H_2O$$
$$Pb(NO_3)_2 + H_2SO_4 \longrightarrow PbSO_4\downarrow + 2 HNO_3$$

Calculations

Let the weight of $PbSO_4$ obtained from

$$20 \text{ mL of the solution} = w \text{ g}$$

$$303.28 \text{ g of } PbSO_4 \equiv 207.19 \text{ g lead}$$

$$\therefore \qquad w \text{ g of } PbSO_4 \equiv \frac{207.19}{303.28} \times w \text{ g of lead.}$$

Mass of the lead present in 1000 mL of the solution

$$= \frac{207.19 \times w \times 50}{303.28} \text{ g.}$$

4.9.2 Estimation of Lead as Lead Chromate

Lead can also be estimated as lead chromate by adding a solution of sodium or potassium chromate to a solution of lead nitrate or lead acetate.

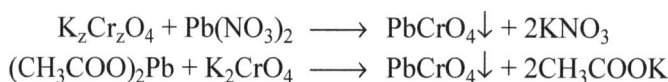

$$K_zCr_zO_4 + Pb(NO_3)_2 \longrightarrow PbCrO_4\downarrow + 2KNO_3$$
$$(CH_3COO)_2Pb + K_2CrO_4 \longrightarrow PbCrO_4\downarrow + 2CH_3COOK$$

Procedure

Following steps are followed:

(*i*) To 20 mL of lead salt solution[1] taken in a 250 mL beaker is added water (20 mL).

(*ii*) The above solution is heated to almost boiling and a solution of potassium chromate. (4%) added dropwise with constant stirring till the supernatant liquid is yellow orange in colour. The resultant mixture is heated (5 min) and then cooled.

(*iii*) The precipitate obtained in step (*ii*) is filtered by decantation using a sintered glass crucible G_3. The precipitate is washed with water containing acetic acid (95 mL water and 5 mL acetic acid) till the washings are free of chromate ions (no precipitate is obtained by adding lead acetate to the filtrate).

(*iv*) Finally the precipitate is quantitatively transferred from the beaker to the sintered glass crucible and washed with ractified spirit (50%).

(*v*) The crucible alongwith the precipitate in heated in an oven at 120°C for about 30 min. It is cooled and weighted.

(*vi*) The process of heating, cooling and weighing is repeated till a constant weight is obtained.

Note:

1. The lead salt solution is prepared by the instructor as in Section 4.9.1 (note 1).

Calculations

Let mass of lead chromate from 20 mL of the given solution $= w \text{ g}$

$$323.22 \text{ g of lead chromate} \equiv 207.19 \text{ g lead}$$

$$\therefore \qquad w \text{ g of lead chromate} \equiv \frac{207.19}{323.22} \times w \text{ g lead}$$

Mass of lead present per litre in the given solution

$$= \frac{207.19 \times w \times 50}{323.22} \ g.$$

4.10   ESTIMATION OF NICKEL

Nickel is estimated as its dimethyl glyoxime complex by treating the given nickel ammonium sulphate solution with 1% alcoholic solution of dimethyl glyoxime in presence of NH_4OH.

$$NiSO_4(NH_4)_2SO_4. \ 6H_2O \ + \ 2CH_3{-}C{=}NOH \atop CH_3{-}C{=}NOH \ + \ 2NH_4OH$$

Complex
(scarlet red)

$$\longrightarrow \ \cdots \ + \ 2(NH_4)_2SO_4 \ + \ 8H_2O$$

Procedure

Following steps are followed:

(*i*) To the given nickel ammonium sulphate solution[1] (20 mL) contained in a 400 mL beaker is added distilled water (100 mL). The solution is heated to about 70°C and dimethyl glyoxime solution[2] (25-30 mL, 1%) is added. This is followed by immediate gradual addition of ammonia solution[3] (1 : 1) with constant stirring until the precipitation is complete[4].

(*ii*) The above solution is heated for 30 min. on a boiling water bath and then cooled.

(*iii*) The scarlet red complex formed in step (*ii*) is filtered using a constant weight preheated sintered glass crucible (G_3) by decantation. The residual complex is washed with cold water and filtered by decantation till the washings are free of SO_4^{2-} ions.

(*iv*) The sintered glass crucible alongwith the complex is dried in an oven at 110-120°C (about 45 minutes), cooled and weighed.

(*v*) The process of heating, cooling and weighing is repeated till the crucible + complex has a constant weight.

Notes:

1. The nickel ammonium sulphate, $NiSO_4.(NH_4)_2SO_4.6H_2O$(AR) solution is prepared by the instructor by accurately weighing about 18 g of the nickel ammonium sulphate and dissolving in distilled water alongwith few mL of dilute HCl in a 1000 mL measuring flask. The solution is made upto the mark by distilled water. 20 mL of this solution is used for estimation.

2. A slight excess of dimethyl glyoxime solution is used so that any cobalt, zinc or copper present as impurities form soluble salts with the reagent and remain in solution.

3. The complex is soluble in acids or in alcoholic solution containing more than 50% alcohol but is insoluble in dilute ammonia solution and so the pecipitation is carried out in presence of dilute ammonia solution.

4. The completion of precipitation is ascertained by testing the supernatant solution by adding a few drops of dimethyl glyoxime solution–no precipitation should occur.

Calculations

Let the mass of nickel dimethyl glyoximate obtained from 20 mL of the given solution $= w$ g.

288.91 g of nickel dimethyl glyoximate (complex)

$$= 58.69 \text{ g of nickel}$$

$\therefore \qquad\qquad w \text{ g of the complex} = \dfrac{58.69}{288.91} \times w \text{ g of nickel}$

Mass of nickel present per litre of the given solution

$$= \dfrac{58.69 \times w \times 50}{288.91} \ g.$$

4.11   ESTIMATION OF CHROMIUM

Chromium is estimated by precipitating it as hydroxide (or hydrated oxide) and heating to give chromium oxide, Cr_2O_3. The precipitating reagent used is ammonia solution.

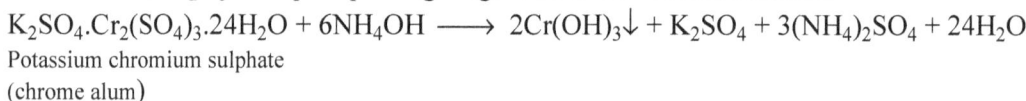

$$K_2SO_4.Cr_2(SO_4)_3.24H_2O + 6NH_4OH \longrightarrow 2Cr(OH)_3\!\downarrow + K_2SO_4 + 3(NH_4)_2SO_4 + 24H_2O$$
Potassium chromium sulphate
(chrome alum)

$$2Cr(OH)_3 \xrightarrow{\ \Delta\ } Cr_2O_3 + 3H_2O$$

Procedure

Following steps are involved:

(i) To the given chrome alum solution (20 mL) taken in a 400 mL beaker is added water (about 200 mL), solid ammonium chloride (0.5 g) and methyl red (2-3 drops).

(ii) The above solution is heated to boiling and hot ammonia solution (1:1) added dropwise with constant sitirring till the precipitation is complete and the solution attains a yellowish colour. The solution is boiled for 1-2 minutes.

(iii) The precipitate of chromium hydroxide obtained in step (ii) is filtered[1] using a Whatmann filter paper No. 41. The ppt. is washed with ammonium nitrate solution (1 %) till the washings are free of Cl^- and SO_4^{2-} ions.

(iv) The precipitate alongwith the filter paper and funnel is dried in an hot air cone. The filter paper is folded and heated gently in a preweighed silica crucible till the paper chars. Finally, the crucible is heated for 20-30 min. over a mecker burner. The crucible is cooled in a dessicator and weighed.

(v) The process of heating, cooling and weighing is repeated till a constant weight of the crucible and ppt. is obtained.

Note:

1. Before filtration of the precipitate [step (*iii*)], about half a test tube of filter paper pulp is added to speed up the filtration.

   In an alternative method, chromium is estimated in sodium or potassium dichromate solution by reducing the Cr (VI) to Cr (III). It is then precipitated as hydroxide, heated and weighed as Cr_2O_3. The reduction of Cr (VI) to Cr (III) is carried out with ethyl alcohol is presence of H_2SO_4.

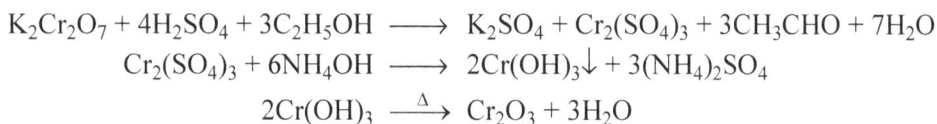

$$K_2Cr_2O_7 + 4H_2SO_4 + 3C_2H_5OH \longrightarrow K_2SO_4 + Cr_2(SO_4)_3 + 3CH_3CHO + 7H_2O$$
$$Cr_2(SO_4)_3 + 6NH_4OH \longrightarrow 2Cr(OH)_3\downarrow + 3(NH_4)_2SO_4$$
$$2Cr(OH)_3 \xrightarrow{\Delta} Cr_2O_3 + 3H_2O$$

Following precedure is followed:

To the given sodium dichromate or potassium dichromate solution (20 mL) in a beaker (400 mL capacity) is added dilute H_2SO_4 (4 mL) and C_2H_5OH (15 mL). The solution is heated, excess of alcohol and formed acetaldehyde removed. The solution is diluted with water (200 mL) and chromium estimated as given above [steps (*ii*) to (*v*)].

Calculations

Let the mass of Cr_2O_3 obtained from 20 mL of the given solution be w g.

$$151.99 \ g \ Cr_2O_3 \equiv 104.00 \ g \ chromium$$

$\therefore$
$$w \ g \ Cr_2O_3 \equiv \frac{104.00}{151.99} \times w \ g \ chromium$$

Mass of chromium present per litre of the given solution

$$= \frac{104.00 \times w \times 50}{151.99} \ g.$$

4.12 ESTIMATION OF ALUMINIUM

Aluminium is estimated as oxide. Thus, a solution of the given aluminium sulphate on treatment with ammonium hydroxide solution gives a ppt. of the hydroxide (hydrated aluminium oxide) which an heating gives Al_2O_3.

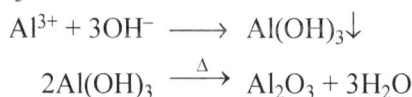

$$Al^{3+} + 3OH^- \longrightarrow Al(OH)_3\downarrow$$
$$2Al(OH)_3 \xrightarrow{\Delta} Al_2O_3 + 3H_2O$$

Procedure

Following steps are followed:

(*i*) To the given aluminium solution (20 mL) taken in a 400 mL beaker is added solid ammonium chloride (0.5 g) and methyl red (2-3 drops).

(*ii*) The above solution is heated to just boiling and hot ammonia solution (1:1) added slowly with constant stirring till the precipitation is complete and the solution acquires yellowish colour. The solution is boiled (1-2 min.) and half a test tube of filter paper pulp added.

(*iii*) The precipitate is immediately filtered using a Whatmann filter paper No. 41. The ppt. is washed with ammonium nitrate solution till the washings are free of Cl^- and SO_4^{2-} ions.

(*iv*) The ppt. alongwith the filter paper and the funnel is dried in a hot air cone. The filter paper is folded and heated gently in a pre-weighed silica crucible till the filter paper chars.

(*v*) Finally the crucible is heated (1200°C) over a Mecker burner for 20-30 min, cooled in a dessicator and weighed.

(*vi*) The process of heating, cooling and weighing is repeated till a constant weight is obtained.

Calculations

Let the mass of Al_2O_3 obtained from 20 mL of the given solution $= w$ g.

$$101.94 \text{ g of } Al_2O_3 \equiv 53.96 \text{ g Aluminium}$$

$$\therefore \qquad\qquad w \text{ g} + Al_2O_3 \equiv \frac{53.96}{101 \cdot 94} \times w \text{ g Aluminium}$$

Mass of aluminium present per litre of the given solution

$$= \frac{53.96 \times w \times 50}{101.94} \text{ g.}$$

4.13   ESTIMATION OF MAGNESIUM

Magnesium can be estimated as pyrophosphate ($Mg_2P_2O_7$) or as magnesium ammonium phosphate hexahydrate ($MgNH_4PO_4.6H_2O$). It can also be estimated as oxinate (8-hydroxyquinolate) $[(C_9H_6NO)_2Mg.2H_2O]$ or $Mg(C_9H_6.ON)_2$.

4.13.1   Estimation of Magnesium as Pyrophosphate ($Mg_2P_2O_7$) or as Magnesium Ammonium Phosphate Hexahydrate ($MgNH_4PO_4.6H_2O$)

Magnesium is precipitated as ($MgNH_4PO_4.6H_2O$) by treatment of the magnesium salt solution with excess of diammonium hydrogen phosphate $[(NH_4)_2HPO_4]$ followed by addition of excess ammonia solution.

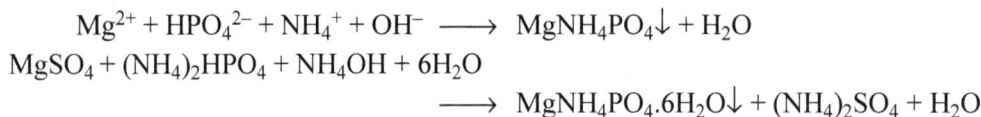

$$Mg^{2+} + HPO_4^{2-} + NH_4^+ + OH^- \longrightarrow MgNH_4PO_4\downarrow + H_2O$$
$$MgSO_4 + (NH_4)_2HPO_4 + NH_4OH + 6H_2O$$
$$\longrightarrow MgNH_4PO_4.6H_2O\downarrow + (NH_4)_2SO_4 + H_2O$$

The hexahydrate is washed with alcohol, anhydrous ether and dried in a dessicator for about 30 min and weighed. Alternatively the hexahydrate may be heated at 1000 − 1100°C to give magnesium pyrophosphate and weighed.

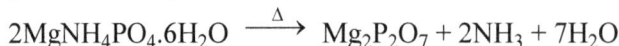

$$2MgNH_4PO_4.6H_2O \xrightarrow{\Delta} Mg_2P_2O_7 + 2NH_3 + 7H_2O$$

Procedure

Following steps are followed:

(*i*) To the given magnesium sulphate solution[1] (20 mL) taken in a 400 mL beaker is added conc. HCl (5 mL), water (130 mL) and a few drops of methyl red indicator. The solution becomes red.

(*ii*) To the above solution is added diammonium hydrogen phosphate solution (10 mL, 10%) and then conc. ammonia drop wise with stirring until the solution becomes yellow. The solution is stirred (5 min) and excess conc. ammonia solution (5 mL) added. The solution is allowed to stand (4-5 hrs).

(*iii*) Weighing as magnesium ammonium phosphate hexahydrate ($MgNH_4PO_4.6H_2O$).

The precipitate obtained in step (*ii*) is filtered using a constant weight sintered glass crucible (G_3) (the crucible before using is washed with alcohol and then with ether). The precipitate in the crucible is washed with dilute ammonia solution (1:9, $NH_3 : H_2O$) till the washings are free of Cl^- ions. Subsequently the ppt. is washed with ether (4 × 5 mL). The ppt. is dried by drawing air through the crucible for about 10 minutes. The crucible is kept in a dessicator for about 30 min. and then weighted.

(*iv*) Weighing as magnesium pyrophosphate $Mg_2P_2O_7$.

The ppt. of $MgNH_4PO_4.6H_2O$ [step (*ii*)] is filtered using a porcelain filtering crucible and washed with ammonia solution (0.8 M) till the washings are free of Cl^- ions. The crucible is dried by heating at 100–150°C in an oven and then at 1000–1100°C in an electric muffle furnace for 25-30 minutes. The crucible is cooled and weighed. The above process of heating at 1000–1100°C and cooling is repeated till a constant weight is obtained.

Alternatively, the precipitate of magnesium ammonium phosphate is filtered using a Whatmann filter paper by decantation. The precipitate is washed with ammonia solution (0.8 M) till the washing are free of Cl^- ions (test with $HNO_3 + AgNO_3$). Whole of the ppt. is quantitatively transferred from the beaker to the funnel. The ppt. alongwith the filter paper and funnel is dried using an air cone. and the ppt. is then transferred from the filter paper to a glazed paper. The filter paper is incerated in a preheated constant weight silica crucible at low temperature. Subsequently, the ppt. on the glazed paper in transferred to the crucible. The crucible is strongly heated, cooled and weighed. The process is repeated till a constant weight is obtained.

Note:

1. To solution to be used for estimation is prepared by the instructor by dissolving accurately weighed 24 g. $MgSO_4$ in one litre of solution.

Calculations

(*i*) As $MgNH_4PO_4.6H_2O$

Let the mass of $MgNH_4PO_4.6H_2O$ obtained from 20 mL of the given solution = w g.

$$245.28 \text{ g of } MgNH_4PO_4.6H_2O \equiv 24.31 \text{ g Mg}$$

$$w \text{ g of } MgNH_4PO_4.6H_2O \equiv \frac{24.31}{245.28} \times w \text{ g of Mg}$$

Mass of magnesium per litre in the given solution

$$= \frac{24.31 \times w \times 50}{245.28} \text{ g.}$$

(*ii*) As $Mg_2P_2O_7$.

Mass of $Mg_2P_2O_7$ obtained from 20 mL of the given solution = w g.

$$222.55 \text{ g of } Mg_2P_2O_7 \equiv 48.62 \text{ g Mg}$$

$$w \text{ g of } Mg_2P_2O_7 \equiv \frac{48.62}{222.55} \times w \text{ g Mg}$$

Mass of magnesium present per litre of the given solution

$$= \frac{48.62 \times w \times 50}{222.55} \text{ g.}$$

4.13.2 Estimation of Magnesium as Oxinate (8-hydroxyquinolate) $(C_9H_6NO)_2Mg.2H_2O$ or $Mg(C_9H_6ON)_2$

Magnesium is estimated as oxinate by treating magnesium solution with 8-hydroxyquinoline in the presence of o-cresol phthalein as indicator

8-Hydroxy quinoline
(oxine)

Magnesium 8-Hydroxy quinolate
(oxinate)

$$2C_9H_7NO + Mg^{2+} + 2H_2O \longrightarrow (C_9H_6NO)_2 Mg . 2H_2O + 2H^+$$

Procedure

Following steps are followed:

(i) To the given magnesium sulphate solution (20 mL) taken in a 400 mL beaker is added water (100 mL), ammonium chloride (2 g) and o-cresolphthalein[1] (5-6 drops).

(ii) To the above solution is added ammonium hydroxide solution (6 N) until a violet colour is obtained. The solution is heated to 70-80°C and 8-hydroxyquinoline solution (1% in 2N acetic acid) is added slowly till the supernatant solution has a distinct yellow colour (excess of reagent is avoided).

(iii) The above solution alongwith the precipitate is heated on a steam bath for about 10 minutes with constant stirring.

(iv) The precipitate is filtered using a sintered glass crucible (G_4). The ppt. is washed with dilute ammonia solution (1 : 40) until the washings are colourless.

(v) The precipitate alongwith the crucible is dried at 100-110°C until a constant weight is obtained. At this stage, the oxinate is present as dihydrate, $(C_9H_6NO_2)$ Mg. 2H_2O. The ppt. is dried at 150-160°C and is weighed as $(C_9H_6NO_2)$ Mg.

Note:

1. The precipitation is best carried out at pH about 9.5. The indicator (o-cresolphthalein) shows colour change at this pH.

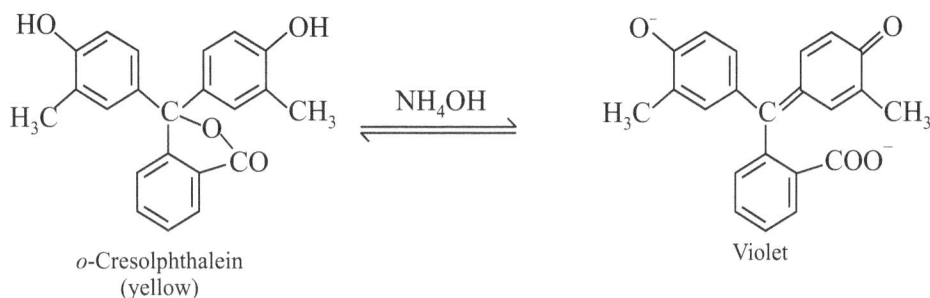

o-Cresolphthalein
(yellow)

Violet

Calculations

(*i*) Magnesium as $(C_9H_6NO_2)_2Mg.2H_2O$

Mass of magnesium oxinate dihydrate obtained from 20 mL of the given solution = w g.

346.63 g of magnesium oxinate dihydrate ≡ 24.31 g Mg

w g of magnesium oxinate dihydrate ≡ $\dfrac{24.31}{346.63} \times w$ g Mg

Mass of Mg present per litre of the given solution

$$= \dfrac{24.31 \times w \times 50}{346.63} \text{ g.}$$

(*ii*) Magnesium as oxinate $[(C_9H_6NO)_2Mg]$

Mass of magnesium oxinate = w g

310.61 g of magnesium oxinate ≡ 24.31 g Mg

w g of magnesium oxinate ≡ $\dfrac{24.31}{310.61} \times w$ g Mg

Mass of Mg present per litre of the given solution

$$= \dfrac{24.31 \times w \times 50}{310.61} \text{ g.}$$

EXERCISES

1. Describe various steps involved in gravimetric estimations.
2. For the gravimetric estimation of copper as cuprous thiocyanate or copper oxide, which is a better method. Give reason.
3. For the gravimetric estimation of iron, why the precipitated iron hydroxide has to be ignited to give Fe_2O_3.
4. Why in some estimations, Whatmann filter paper is used for filtration and in some other estimations sintered glass crucible is used?
5. For the estimation of zinc, the precipitated $ZnCO_3$ has to be converted into ZnO. Why?
6. Lead can be estimated as $PbSO_4$ or $PbCrO_4$. Which is a better method. Justify.
7. How is nickel estimated?
8. Magnesium is estimated as pyrophosphate or as magnesium ammonium sulphate or as oxinate. Which one is a better procedure.

οοο

5. Estimation of more than one Constituent Present Together in a Solution

Estimation of More than one Constituent Present Together in a Solution

5.1 INTRODUCTION

So far we have described the estimation of a single constituent present in a solution. It is, however, important in certain cases to carry out estimation of more than one constituent present in a solution. This has special use in the analysis of alloys, which is commonly done in post graduate classes and also in industrial establishments.

5.2 ESTIMATION OF COPPER AND NICKEL IN A GIVEN SOLUTION

Copper and nickel can be estimated gravimetically in the given solution. However, copper alone can be estimated volumetrically in the same solution (for details see Section 2.3.5).

Gravimetric Estimation of Copper and Nickel

Using 50 mL of the given solution containing Cu and Ni, Copper is precipitated as cuprous thiocyanate (as discussed in Section 4.2.1) and estimated. The filtrate and washings obtained from CuCNS are used for gravimetric estimation of nickel, for which following steps are involved.

(*i*) The filtrate (after removing CuCNS) and the washings are evaporated (sand bath) to about 100 mL. To the concentrated solution, conc. HNO_3 (15 mL) and conc. HCl (15 mL) added and the solution evaporated to near dryness on a sand bath (this treatment destroys excess thiocyanate present).

(*ii*) The residue is extracted with conc. HCl (3-4 mL) and distilled water (100 mL). The resulting solution is treated with ammonia solution till it becomes slightly ammonical.

(*iii*) Finally, nickel is precipitated as dimethyl glyoxime complex (*for details see Section 4.10*) and estimated.

Notes:

1. The solution containing Cu and Ni to be used for estimation is prepared by the instructor by accurately weighing about 26 g of $CuSO_4.5H_2O$ (A.R.) and dissolving is distilled water in a 1000 mL measuring flask. To this solution is added 18 g of nickel ammonium sulphate (A.R.) (accurately weighed). The

clear solution is made up to the mark with distilled water. 50 mL of the mixed solution is used for gravimetric estimation of Cu and Ni.

2. When ever a solution is to be concentrated by heating on a sand bath, the beaker is partially covered by a watch glass. After concentration the watch glass is washed by distilled water and the washings collected in the same beaker. This practice must be followed in all cases.

Calculations

The calculations are done as explained in Sections 2.3.5, 4.2.1 and 4.10.

Result

1. Cu (Volumetric estimation) = ... g/litre
2. Cu (Gravimetric estimation) = ... g/litre
3. Ni (Gravimetric estimation) = ... g/litre.

5.3   ESTIMATION OF COPPER AND ZINC IN A GIVEN SOLUTION

Copper and zinc can be estimated gravimetrically in the given solution. Copper alone can be estimated volumetrically in the same solution (For details see Section in 2.3.5).

Gravimetric Estimation of Copper and Zinc

The given solution of copper and zinc (40 mL) is diluted with distilled water to 200 mL in a 200 mL measuring flask. From this solution 40 mL is taken in a 400 mL beaker for gravimetric estimation of copper and zinc (the remaining solution is used for volumetric estimation of copper (see Section 2.3.5).

Copper is precipitated as CuCNS as described in Section 4.2.1. The filtrate and washings are used for gravimetric estimation of Zn.

For the gravimetric estimation of zinc, the combined filtrate and washings and evaporated to about 100 mL using a sand bath. To the remaining solution is added conc. HNO_3 (30 mL) and conc. HCl (15 mL). The solution is evaporated to almost dryness in order to destroy excess of thiocyanate.

The residual solution is dissolved in water and zinc estimated as zinc oxide as described in Section 4.8.

Note:

1. The solution containing copper and zinc is prepared by the instructor by weighing accurately about 26 g of $CuSO_4.5H_2O$(A.R.) and 7 g of zinc oxide in distilled water to which conc. HCl is added. The volume is made upto 1000 mL in a measuring flask. 40 mL of this solution is used for estimation.

Result

1. Cu (Volumetric estimation) = ... g per litre
2. Cu (Gravimetric estimation) = ... g/litre
3. Zn (Gravimetric estimation) = ... g/litre.

5.4   ESTIMATION OF COPPER AND MAGNESIUM IN A GIVEN SOLUTION

Copper and magnesium are estimated gravimatically in the given solution. Copper alone can be estimated volumetrically in the same solution (For details see Section 2.3.5).

Gravimetric Estimation of Copper and Magnesium

The given solution of copper and magnesium (40 mL) is diluted to 200 mL with distilled water in a 200 mL measuring flask. 40 mL and this solution is taken in a 400 mL beaker for gravimetric estimation of Cu and Mg. The remaining solution is used for volumetric estimation of copper.

Copper is estimated as cuprous thiocyanate as described in Section 4.2.1. The filtrate and washings are combined and evaporated to about 100 mL using a sand bath. To the concentrated solution is added conc. HNO_3 (30 mL) and conc. HCl (15 mL) and the solution evaporated nearly to dryness. This process destroys excess of thiocyanate. The remaining solution in the beaker is boiled with conc. HCl (4-5 mL) and diluted with distilled water. The resulting solution is used for the gravimetric estimation of magnesium as pyrophosphate ($Mg_2P_2O_7$) or as magnesium oxinate [$Mg(C_9M_6ON)_2$] as described in Sections. 4.13.1 and 4.13.2.

Note:

1. The solution containing copper and magnesium is prepared by the instructor by accurately weighing about 20 g copper sulphate. $CuSO_4.5H_2O$(A.R.) and about 30 g. $MgSO_47H_2O$ and dissolving both in distilled water in a 1000 mL measuring flask. The volume is made upto the mark. 40 mL of this solution is used for the estimation of Cu and Mg.

Result

  (*i*) Copper (Volumetric estimation) = ... g/litre
 (*ii*) Copper (Gravimetric estimation) = ... g/litre
(*iii*) Magnesium (Gravimetric estimation) = ... g/litre.

5.5   ESTIMATION OF COPPER AND BARIUM IN THE GIVEN SOLUTION

Copper and barium are estimated gravimatically in the given solution. Copper alone can be estimated volumetrically in the given solution (For details see Section 2.3.5).

Gravimetric Estimation of Copper and Barium

From the given solution (40 mL), nitric acid (if present) is removed and copper is estimated as cuprous thiocyanite (as described in Section 4.2.1). The combined filtrate and washings are evaporated (using sand bath) to about 100 mL. To the concentrated solution is added conc. HNO_3 (30 mL) and conc. HCl (15 mL) and the solution evaporated to almost dryness. The residue is extracted with water and few mL of conc. HCl added. From this solution barium is estimated as barium sulphate as described in Section 4.4.

Note:

1. The solution containing copper and barium is prepared by the instructor by accurately weighing about 18 g of copper metal (A.R.) (dissolved in HNO_3) and 15 g $BaCl_2.2H_2O$ (A.R.) and dissolved in water.

The two solution are mixed in a 1000 mL measuring flask and volume made upto the mark by distilled water. 40 mL of this solution is used for estimation.

5.6   ESTIMATION OF COPPER AND SILVER IN THE GIVEN SOLUTION

Copper and silver are estimated gravimatrically in the given solution. Copper and silver both can also be individually estimated volumetrically in the given solution (*for details see Section 2.3.5 and 2.4.2*).

Gravimetric Estimation of Copper and Silver

The given solution (40 mL) containing copper and silver is diluted with distilled water to 200 mL in a measuring flask. 40 mL of this solution is taken in a 400 mL beaker for gravimetric estimation of copper and silver. The remaining solution is used for the volumetric estimation of copper and silver.

The volumetric estimation of copper and silver is done as described in sections 2.35 and 2.4.2 respectively).

Gravimetically, silver is estimated as silver chloride as explained in Section 4.6. The combined filtrate and washings obtained after filtering silver chloride ppt. is treated with conc. H_2SO_4 (B.D.H.) (3-4 mL) and evaporated using a sand bath until white dense fumes of SO_3 are evolved (to destroy HNO_3 present). The residue is dissolved in distilled water, solution neutralised with ammonia solution (1:1) and the gravimetric estimation of barium carried out as described a Section 4.4.

Note:

1. The solution of copper and silver for estimation is prepared by the instructor by accurately weighing about 26 g of $CuSO_4.5H_2O$(A.R.) and 15 g of $AgNO_3$ (A.R.) and dissolving in distilled water and making up the volume to the mark in 1000 mL measuring flask. 40 mL of this solution is used for estimation of Cu and Ba.

Result

   (*i*) Silver (Volumetrically) = ... g/litre
  (*ii*) Silver (Gravimetrically) = ... g/litre
 (*iii*) Copper (Volumetrically) = ... g/litre
 (*iv*) Copper (Gravimetrically) = g/litre.

5.7   ESTIMATION OF SILVER AND NICKEL IN THE GIVEN SOLUTION

Silver and nickel are estimated gravimetrically in the given solution. Silver can also be estimated alone volumetrically (for detail see Section 2.4.2).

Gravimetric Estimation of Silver and Nickel

To the given solution (40 mL) containing silver and nickel is added distilled water and solution made up to 200 mL in a measuring flask. 40 mL of this solution is taken in a 400 mL beaker for estimating silver and nickel gravimetrically and the remaining solution is used for estimation silver volumetrically (*see Section 2.4.2*).

Silver is estimated as silver chloride as described in Section 4.6.

The filtrate and washing are combined and evaporated to about 100 mL using a sand bath. Nickel is estimated as its dimethyl glyoxime complex as described in section 4.10.

Note:

1. The solution containing silver and nickel for estimation is prepared by the instructor by weighing exactly about 15 g. $AgNO_3$ (A.R.) and 18 g nickel ammonium sulphate (A.R.) and dissolving in distilled water. The solution is made up the mark in a 1000 mL measuring flask. 40 mL of this solution is used for estimation.

Result

(*i*) Silver (Volumetrically) =   ... g/litre

(*ii*) Silver (Gravimetrically) = ... g/litre

(*iii*) Nickel (Gravimetrically) = ... g/litre.

5.8   ESTIMATION OF SILVER AND ZINC IN THE GIVEN SOLUTION

Silver and zinc are estimated gravimetrically in the given solution. Silver can also be estimated alone volumetrically (for details see Section 2.4.2).

Gravimetric Estimation of Silver and Zinc

To the given solution (40 mL) containing silver and zinc is added distilled water and solution made upto 200 mL in a 200 mL measuring flask. 40 mL of this solution is used for the estimation of Ag and Zn and the remaining solution is used for volumetric estimation of silver (*for details see Section 2.4.2*).

Silver is estimated gravimetrically as silver chloride as described in Section 4.6.

The filtrate and washings are combined and evaporated to about 100 mL using using a sand bath. Zinc is estimated as oxide in the concentrated solution gravimetrically as described in Section 4.8.

Note:

1. The solution containing silver and zinc is prepared by the instructor by accurately weighing about 15 g $AgNO_3$ (A.R.) and 7 g ZnO (A.R.) (moistened with water and dissolved in conc. HNO_3). The $AgNO_3$ and ZnO solution are mixed and the solution made up to 1000 mL in a measuring flask. 40 mL of this solution is used for estimation of Ag and Zn.

Result

Copper (Volumetrically) = ... g/litre

Copper (Gravimetrically) = ... g/litre

Zinc (Gravimetrically) = ... g/litre.

5.9   ESTIMATION OF SILVER AND MAGNESIUM IN THE GIVEN SOLUTION

Silver and magnesium are estimated gravimetrically in the given solution. Silver can also be estimated alone volumetrically in the given solution (For details see Section 2.4.2).

Gravimetric Estimation of Silver and Magnesium

To the given solution (40 mL) is added distilled water and the solution made upto 200 mL in a measuring flask. 40 mL of this solution is used for the gravimetric estimation of silver and magnesium and the remaining solution is used for volumetric estimation of silver (*see Section 2.4.2*).

Silver is estimated gravimetrically as silver chloride as described in Section 4.6.

The filtrate and washings obtained are mixed and concentrated to about 100 mL using a sand bath. Magnesium is estimated gravimetrically as oxinate in the concentrated solution as described in Section 4.13.2.

Note:

1. The solution containing silver and magnesium for estimation is prepared by the instructor by accurately weighing about 15 g $AgNO_3$ (A.R.) and 30 g $MgSO_4.7H_2O$ (A.R.) and dissolving in distilled water in 1000 mL measuring flask. The solution is made up to the mark. 40 mL of this solution is used for the estimation of Ag and Mg.

5.10   ESTIMATION OF IRON AND NICKEL IN THE GIVEN SOLUTION

Iron and nickel are estimated gravimetrically in the given solution. Iron alone can also be estimated volumetrically in the same solution (For details see Sections 2.3.2, 2.3.3 and 2.3.4).

Gravimetric Estimation of Iron and Nickel

To the given solution (40 mL) is added distilled water and solution made upto 200 mL in a measuring flask. 40 mL of this solution is used for the gravimetric estimation of Fe and Ni. The remaining solution is used for the volumetric estimation of iron by any of the methods as described in Sections 2.3.2, 2.3.3 or 2.3.4.

To the solution (40 mL) taken in a 400 mL beaker is added solid ammonium chloride (5 g). Iron is then precipitated as iron hydroxide and estimated as Fe_2O_3 as described in Section 4.3.

The filtrate and washings are combined and concentrated to about 100 mL using a sand bath. Nickel is estimated in the concentrated solution as nickel dimethyl glyoximate as described in Section 4.10.

Note:

1. The solution containing iron and nickel is prepared by the instructor by accurately weighing about 60 g ferrous ammonium sulphate, $FeSO_4.(NH_4)_2SO_4.6H_2O$ (A.R.) and 18 g of nickel ammonium sulphate (A.R.) in water and making up the volume to 1000 mL in a measuring flask. 40 mL of this solution is used for the estimation Fe and Ni.

Result

Iron (Volumetrically) = ... g/litre

Iron (Gravimetrically) = ... g/litre

Nickel (Gravimetrically) = ... g/litre.

5.11   ESTIMATION OF IRON AND MAGNESIUM IN THE GIVEN SOLUTION

Iron and magnesium are estimated gravimetrically in the given solution. Iron alone can be estimated volumetrically in the same solution (For details see Sections 2.3.2, 2.3.3 and 2.3.4).

Gravimetric Estimation of Iron and Magnesium

To the given solution (40 mL) is added distilled water and the solution diluted to 200 mL in a measuring flask. 40 mL of this solution is used for the estimate of iron and magnesium and the remaining solution is used for volumetric estimation of iron as discussed in Sections 2.3.2, 2.3.3 and 2.3.4.

To the solution (40 mL) taken in a 400 mL beaker is added solid ammonium chloride and then iron is precipitate as hydroxide and estimated as Fe_2O_3 as discussed in Section 4.3.

The filtrate and washings are combined and concentrated to about 100 mL by heating on a sand bath. Magnesium is finally estimated as oxinate as described in Section 4.13.2.

Note:

1. The solution containing iron and magnesium is prepaired by the instructor by exactly weighing about 60 g of $FeSO_4$. $(NH_4)_2SO_4.6H_2O$(A.R.) and 30 g of $MgSO_4.7H_2O$ (A.R.) and dissolving in 1000 mL distilled water in a measuring flask. 40 mL of this solution is used for the estimation of Fe and Mg.

Result

Iron (Volumetrically) = ... g/litre

Iron (Gravimetrically) = ... g/litre

Magnesium (Gravimetrically) = ... g/litre.

5.12   ESTIMATION OF COPPER, NICKEL AND ZINC IN THE GIVEN SOLUTION

Copper, nickel and zinc are estimated gravimetrically in the given solution. Copper alone can be estimated volumetrically in the same solution (For details see Section 2.3.5).

Gravimetric Estimation of Copper, Nickel and Zinc

To the given solution (40 mL) distilled water is added and the solution made up to 200 mL in a measuring flask. 40 mL of this solution is used for estimation of Cu, Ni and Zn gravimetrically. The remaining solution is used for the volumetric estimation of copper as discussed in Section 2.3.5.

In the solution (40 mL), copper is estimated gravimetrically as cuprous thiocyanate as described in Section 4.2.1.

The combined filtrate and washings (obtained after filtration of cuprous thiocyanate) are evaporated to about 100 mL (using a sand bath). To the concentrated solution is added conc. HNO_3 (25 mL) and conc. HCl (15 mL) and the solution evaporated nearly to dryness (to destroy excess of thiocyanate). The residue is dissolved in distilled water (about 30 mL) and solid ammonium chloride (5 g) added. This solution is used for the gravimetric estimation of **nickel** as **nickel** dimethyl glyoximate as discussed in Section 4.10.

The combined filtrate and washings (after filtering nickel dimethyl glyoximate) are evaporated to about 100 mL using a sand bath. To the concentrated solution is added conc. HNO_3(30 mL) and the solution evaporated to dryness. Conc. H_2SO_4 (1 mL) is added to the residue and the mixture evaporated till white fumes of SO_3 are obtained. The residue is dissolved in conc. HCl (5 mL) and diluted with distilled water. The resultant solution on warming becomes clear. Zinc is estimated gravimetrically (as ZnO) in the solution (For details see Section 4.8).

Note:

1. The solution containing Cu, Ni and Zn for estimation is prepared by the instructor by dissolving accurately weighted about 26 g $CuSO_4.5H_2O$ (A.R.), 78 g nickel ammonium sulphate, $NiSO_4$.$(NH_4)_2SO_4.6H_2O$(A.R.) and 7 g. ZnO (A.R.) and dissolving in distilled water and making up the volume to 1000 mL in a measuring flask. 40 mL of this solution is used for the estimation of Cu, Ni and Zn.

Result

Copper (Volumetric) = ... g per litre

Copper (Gravimetric) = ... g per litre

Nickel (Gravimetric) = ... g per litre

Zinc (Gravimetric) = ... g per litre.

5.13   ESTIMATION OF COPPER, NICKEL AND MAGNESIUM IN THE GIVEN SOLUTION

Copper, nickel and magnesium are estimated gravimetrically in the given solution and copper alone is estimated volumetrically in the same solution (*for details see section 2.3.5*).

Gravimetric Estimation of Copper, Nickel and Magnesium

To the given solution (40 mL) is added distilled water and the solution made up to 200 mL in a measuring flask. 40 mL of this solution is used for gravimetric estimation of Cu, Ni and Mg. The remaining solution is used for volumetric estimation of copper as discussed in Section 2.3.5.

Copper is estimated gravimetrically in the solution (40 mL) as cuprous thiocyanate as described in Section 4.2.1.

The combined filtrate and washings are evaporated to about 100 mL (using a sand bath). To the concentrated solution is added conc. HNO_3 (25 mL) and conc. HCl (15 mL) and the solution evaporated the nearly dryness (in order to destroy excess thiocyanate). The residue is dissolved in distilled water (about 30 mL) and solid ammonium chloride (5 g) added. The resulting solution is used for the gravimetric estimation of nickel as nickel dimethyl glyoximate as discussed in Section 4.10.

The combined filtrate and washings (after filtering nickel dimethyl glyoximate) are evaporated to to about 100 mL using sand bath. To the concentration solution is added conc. HNO_3(30 mL) and the solution evaporated to dryness. Conc. H_2SO_4 (1 mL) is added to the residue and the mixture evaporated till white fumes of SO_3 are obtained. The residue is dissolved in conc. HCl (5 mL) and diluted with distilled water. The resultant solution is warmed to get a clear solution. Finally, magnesium is estimated gravimatically as oxinate as described in Section 4.13.2.

Note:

1. The solution containing Cu, Ni and Mg for estimation is prepared by the instructor by dissolving accurately weighted about 26 g $CuSO_4.5H_2O$(A.R.), 18 g $NiSO_4$ $(NH_4)_2SO_4.6H_2O$(A.R.) and 30 g. $MgSO_4.7H_2O$ (A.R.) in distilled water and solution made upto 1000 mL in a measuring flask. 40 mL at this solution are used for estimation of Cu, Ni and Mg.

Result

Copper (Volumetric) = ... g/litre

Copper (Gravimetric) = ... g/litre

Nickel (Gravimetric) = ... g/litre

Magnesium (Gravimetric) = ... g/litre.

5.14 ESTIMATION OF SILVER, COPPER AND NICKEL IN THE GIVEN SOLUTION

Silver, copper and nickel are estimated gravimetrically in the given solution and silver alone is volumetrically estimated in the same solution. (*see Section 2.4.2*)

Gravimetric Estimation of Copper Silver and Nickel

To the given solution (40 mL) is added distilled water and the solution made up to 200 mL in a measuring flask. 40 mL of this solution is used for gravimetric estimation of Ag, Cu and Nl. The remaining solution is used for the volumetric estimation of silver (*for details see Section 2.4.2*).

Silver is estimated gravimetrically in the solution (40 mL) as silver chloride as discussed in Section 4.6.

The combined filtrate and washings are evaporated to about 100 mL using a sand bath. The remaining solution is used for the gravimetric estimation of copper as CuCNS as described in Section 4.2.1.

The filtrate and the washings (after filtration of CuCNS) are concentrated to about 100 mL using a sand bath. To the concentrated solution is added conc. HNO_3 (25 mL) and conc. HCl (15 mL) and the solution evaporated to nearly dryness (in order to destroy excess thiocyanate). The residue is dissolved in distilled water (about 30 mL) and solid ammonium chloride (5 g) added. The resulting solution is used for the gravimetric estimation of nickel as dimethyl glyoximate as discussed in Section 4.10.

Note:

1. The solution containing Cu, Ni and Mg required for estimation is prepared by the instructor by accurately weighing about 26 g. $CuSO_4.5H_2O$ (A.R), 15 g. $AgNO_3$(A.R.) and 18 g. $NiSO_4$ $(NH_4)_2SO_4.6H_2O$(A.R.) and dissolving in distilled water and conc. HNO_3 (50 mL) added. The solution is made upto 1000 mL by adding distilled water in a measuring flask. 40 mL of this solution is used for estimation of Ag, Cu and Ni.

Result

$$\text{Silver (Volumetrically)} = \ldots \text{g/litre}$$
$$\text{Silver (Gravimetrically)} = \ldots \text{g/litre}$$
$$\text{Copper (Gravimetrically)} = \ldots \text{g/litre}$$
$$\text{Nickel (Gravimetrically)} = \ldots \text{g/litre.}$$

5.15   ESTIMATION OF COPPER, SILVER AND ZINC IN THE GIVEN SOLUTION

Copper silver and zinc are estimated gravimetrically in the given solution and copper alone is estimated volumetrically in the same solution.

Gravimetric Estimation of Copper, Silver and Zinc

To the given solution (40 mL) is added distilled water and the solution made upto 200 mL in a measuring flask. 40 mL of this solution is used for the estimation of Cu, Ag and Zn. The remaining solution is used for the volumetric estimation of copper as discussed in Section 2.3.5. Since, the solution contains HNO_3 (see note 1), it is removed by evaporation to dryness on a sand bath and then decomposed with conc. H_2SO_4 (1 mL) till the white fumes is SO_3 are evolved. The residue is extracted which water and then copper is estimated volumetrically as stated above. In this estimation the mineral acid is replaced by acetic acid.

Silver is first estimated gravimetrically as silver chloride in 40 mL of the solution as described in Section 4.6.

The filtrate and washings (after filtering AgCl) are combined, solution concentrated to about 100 mL and copper estimated gravimetrically as CuSCN as discussed in Section 4.2.1.

The combined filtrate and washings (after filtering CuCNS) are evaporated to about 100 mL using a sand bath. To the concentrated solution is added conc. HNO_3 (30 mL) and the solution evaporated to almost dryness. Conc. H_2SO_4 (1 mL) is added to the residue and the mixture evaporated till white fumes of SO_3 are obtained. The residue is extracted with conc. HCl (5 mL) and diluted with distilled water. The resultant solution on warming becomes clear. Finally, zinc is estimated gravimetrically (as ZnO) in the solution (For details see Section 4.8).

Note:

1. The solution containing Cu, Ag and Zn required for the estimation is prepared by the instructor by accurately washing about 26 g. $CuSO_4.5H_2O$(A.R.), 15 g. $AgNO_3$(A.R.) and 7g.ZnO (A.R.) and dissolving in distilled water. To the solution is added conc. HNO_3(50 mL) and solution made upto

1000 mL by distilled water in a 1000 mL measuring flask. 40 mL of this solution is used for estimation of Cu, Ag and Zn.

Result

$$Cu \text{ (Volumetrically)} = ... \text{ g/litre}$$
$$Cu \text{ (Gravimetrically)} = ... \text{ g/litre}$$
$$Ag \text{ (Gravimetrically)} = ... \text{ g/litre}$$
$$Zn \text{ (Gravimetrically)} = ... \text{ g/litre.}$$

5.16   ESTIMATION OF SILVER, NICKEL AND ZINC IN THE GIVEN SOLUTION

Silver, nickel and zinc are estimated gravimetrically in the given solution and silver is estimated volumetrically in the same solution.

Gravimetric Estimation of Silver, Nickel and Zinc

To the given solution is added distilled water and solution made up to 200 mL is a measuring flask. 40 mL of this solution is used for the gravimetric estimation of Ag, Ni and Zn and the remaining solution is used for volumetric estimation of silver following the method described in Section 2.4.2.

First of all, silver is estimated as AgCl by following the procedure described in Section 4.6.

The filtrate and washings (obtained after filtering AgCl) are concentrated to about 100 mL on a sand bath and nickel is estimated gravimetrically. For this to the concentrated solution is added conc. HNO_3 (25 mL) and conc. HCl (15 mL) and the solution evaporated to near dryness. The residue is dissolved in distilled water (about 30 mL) and solid ammonium chloride (5g added). The resulting solution is used for the gravimetric estimation of nickel as dimethyl glyoximate as described in Section 4.10.

The filtrate and washings (obtained after filtering nickel complex) is concentrated to about 100 mL on a sand bath and the resulting solution used for gravimetric estimation of zinc on ZnO as described in Section 4.8.

Note:

1. The solution containing Ag, Ni and Zn is prepared by the instructor by dissolving accurately weighed about 15 g of $AgNO_3$ (A.R.) 18 g.$NiSO_4$ $(NH_4)SO_4.7H_2O$(A.R.) and 7 g of ZnO in distilled water and conc. HNO_3 (50 mL) and the solution is made up to 1000 mL with distilled water in a measuring flask. 40 mL this solution is used for the estimation of Ag, Ni and Zn.

Result

$$Ag \text{ (Volumetrically)} = ... \text{ g/litre}$$
$$Ag \text{ (Gravimetrically)} = ... \text{ g/litre}$$
$$Ni \text{ (Gravimetrically)} = ... \text{ g/litre}$$
$$Zn \text{ (Gravimetrically)} = ... \text{ g/litre.}$$

5.17   ESTIMATION OF SILVER, NICKEL AND MAGNESIUM IN THE GIVEN SOLUTION

Silver, nickel and magnesium are estimated gravimetrically in the given solution and silver is estimated volumetrically in the same solution using the procedure described in Section 2.4.2.

Gravimetric Estimation of Silver, Nickel and Magnesium

To the given solution (40 mL) is added distilled water and the solution made up to 200 mL in a measuring flask. 40 mL of this solution are used for gravimetric estimation of Ag, Ni and Mg and the remaining solution is used for the volumetric estimation of silver following the procedure described in Section 2.4.2.

Silver is first estimated in 40 mL of the solution as AgCl (for details see Section 4.6).

The filtrate and the washings (after filtration of AgCl) are concentrated on a sand bath to about 100 mL to the concentrated solution is added solid ammonium chloride and nickel is estimated gravimetrically as nickel dimethyl glyoxime as discussed in Section 4.10.

The combined filtrate and washings (after filtering nickel dimethyl glyoximate) are evaporated to about 100 mL using a sand bath. To the solution is added conc. HNO_3 (30 mL) and solution evaporated to dryness. To the residue is added conc. H_2SO_4 (1 mL) and the mixture evaporated till white fumes of SO_3 are obtained. The residue is dissolved in conc. HCl, and diluted with distilled water. The solution is warmed to get a clear solution and magnesium estimated gravimetrically as oxinate as described in Section 4.13.2.

Note:

1. The solution containing Ag, Ni and Mg for the estimation is prepared by the instructor by accurately weighing about 15 g $AgNO_3$(A.R.), 18 g $NiSO_3$, $(NH_4)_2SO_4.6H_2O$ and $30g.MgSO_4.7H_2O$(A.R.) in distilled water and conc. HNO_3 (50 mL) added. The solution is made up to 1000 mL in a measuring flask with distilled water. 40 mL of this solution is used for estimation of Ag, Ni and Mg.

Result

$$Ag \text{ (Volumetrically)} = \dots g/litre$$
$$Ag \text{ (Gravimetrically)} = \dots g/litre$$
$$Ni \text{ (Gravimetrically)} = \dots g/litre$$
$$Mg \text{ (Gravimetrically)} = \dots g/litre.$$

5.18   ESTIMATION OF IRON, NICKEL AND ZINC IN A GIVEN SOLUTION

Iron, nickel and zinc are estimated gravimetrically in the given solution and iron is estimated volumetrically in the same solution following the method described in Sections 2.3.2 or 2.3.3 or 2.3.4.

Gravimetric Estimation of Iron, Nickel and Zinc

To the given solution (40 mL) is added distilled water and the solution made up to 200 mL with distilled water in a 200 mL measuring flask. 40 mL of this solution used for gravimetric estimation of Fe, Ni and Zn. The remaining solution is used for the volumetric estimation of iron as described in Sections, 2.3.2, 2.3.3 or 2.3.4.

Iron is first estimated gravimetrically as Fe_2O_3 in 40 mL of the solution as described in section 4.3.

The filtrate and washings [obtained after filtering $Fe(OH)_3$] are concentrated to about 100 mL by using a sand bath. To the concentrated solution is added NH_4Cl (5 g) and nickel estimated gravimetrically as nickel dimethyl goximate as described in Section 4.10.

The combined filtrate and washings (after filtering Ni complex) are concentrated using a water bath to about 100 mL. To the solution is added conc. HNO_3 and the solution evaporated to dryness. To the residue is added conc. H_2SO_4 (1 mL) and the mixture evaporated till white fumes of SO_3 are obtained. The residue is dissolved in conc. HCl and diluted with distilled water. The solution is warmed to get a clear solution and zinc is estimated gravimatically as described in Section 4.8.

Note:

1. The solution containing Fe, Ni and Zn for the estimation is prepared by the instructor by dissolving accurately weighed about 60 g. $FeSO_4$ $(NH_4)_2SO_4.6H_2O$(A.R.), 10 g.$NiSO_4.(NH_4)_2SO_4.6H_2O$(A.R.) and 7g.$ZnO$(A.R.) (moistened with water and then dissolved in conc. HCl) in distilled water and making up the volume to 1000 mL with distilled water in a measuring flask. 40 mL of this solution are used for the estimation of Fe, Ni and zinc.

Result

$$Iron\ (Volumetrically) = ...\ g/litre$$
$$Iron\ (Gravimetrically) = ...\ g/litre$$
$$Ni\ (Gravimetrically) = ...\ g/litre$$
$$Zn\ (Gravimetrically) = ...\ g/litre.$$

EXERCISES

1.   How will you estimate the amount of:
    (a) Copper and nickel in a given solution.
    (b) Copper and zinc in a given solution.
    (c) Copper and magnesium in a given solution.
    (d) Copper and silver in a given solution.
    (e) Silver and nickel in a given solution.
    (f) Silver and zinc in a given solution.
    (g) Silver and nickel in a given solution.
    (h) Iron and nickel in a given solution.
    (i) Iron and magnesium in a given solution.
    (j) Copper, nickel and zinc in a given solution.

(*k*) Copper, nickel and magnesium in a given solution.

(*l*) Silver, copper and nickel in a given solution.

(*m*) Copper, silver and zinc in a given solution.

(*n*) Silver, nickel and zinc in a given solution.

(*o*) Silver, nickel and magnesium in a given solution.

(*p*) Iron, nickel and zinc in a given solution.

2. In case three metals *viz*., Fe, Ni and Zn are present in a solution which one is estimated volumetrically.

○○○

6. Instrumental Methods of Quantitative Analysis

Contents

Instrumental Methods of Quantitative Analysis

The instrumental methods have been used for the detection of metals (cations) (For details see chapter 1, Section 1.8 and 1.9). The methods used are Flame Atomic Emission Spectrometry (FAES) and Atomic Absorption Spectroscopy (AAS). Both these methods can also be used for the estimation of metals.

6.1 FLAME ATOMIC EMISSION SPECTROMETRY (FAES)

Introduction

As already stated (Section 1.8, chapter 1), in FAES, the solution of a metallic salt is aspirated into the flame. Following are given the sequence of events that take place:

- The solvent in the sample gets evaporated to give a residue.
- The residue decompose into neutral atoms.
- The neutral atoms get excited to form excited atoms, which are unstable.
- The unstable excited atoms revert back to the ground state and during this process emission of radiation takes place.
- The wavelength of the emitted radiation which is characteristics of the metal is measured and so the metallic element in solution is identified (see Table 1.16 in chapter 1). A simple FAES spectrometer (Fig. 1.1, Chapter 1) is used for this purpose.
- The intensity of the emitted radiation is helpful in finding the amount or the concentration of the metal present.

Simple FAES Spectrometer

The simple FAES spectrometer can also be used for estimation of the amount or concentration of a metal. For this, a calibration curve is plotted. The method consists in the preparation of a number of standard solutions of known concentration of an element which is to be estimated. As a first step distilled water is aspirated into the flame and the spectrometer should read zero. Subsequently, the standard solutions containing 10, 5, 2.5 and 1 ppm of the element are aspirated one by one in the flame and the intensities of emitted radiations measured in each case. Lastly, a calibration curve is plotted using emission *vs* concentration. The calibration curve should be a straight line.

Finally, the solution containing the metal (whose concentration is to be determined) is aspirated into the flame. From the measured intensity of emitted radiation and referring to the calibration curve, the concentration of the metal in the solution is determined.

Alternative procedure for finding the amount of a metal in solution involves using an internal standard. In this technique lithium is used as an internal standard. Various steps involved are given below:

- Standard solutions of lithium are first prepared by dissolving pure lithium carbonate (2.6617 g) in minimum amount of dilute HCl and the solution diluted to 500 mL with distilled water in a calibrated flask. This solution contains 1000 mg of lithium per mL. It is appropriately diluted to get solutions containing 20, 10, 5 and 2 ppm lithium solution.
- Each of the standard solutions of lithium (prepared above) are aspirated into the flame and the intensity of the emitted radiation recorded.
- The emitted intensity of the unknown solution is also recorded.
- The ratio of the two intensities are calculated and plotted against the concentration of the element to be determined to get a calibration curve.
- The unknown concentration of the metal is finally determined by referring to the calibration curve.

For the sake of convenience the preparation of standard solutions of potassium and calcium and given below.

Potassium chloride (A.R.) (0.9045 g) is dissolved in distilled water in a 500 mL standard flask. The is made upto 500 mL mark with distilled water. This solution contains 1000 mg potassium per mL. The solution is used to prepare solutions containing 10, 5 and 2 ppm of potassium ions.

Calcium carbonate (A.R.) (1.7486 g) is dissolved minimum amount of dilute HCl in a 500 mL standard flask. The solution is diluted with distilled water to 500 mL mark. This solution contains 1.400 mg calcium per mL and is used to prepare solutions of 100, 50, 25 and 10 ppm of calcium ions.

In place of a simple FAES spectrometer, an **Internal Standard Flame Photometer** can be used to get good results. In this flame photometer, an internal standard (lithium) is used. The sample solution containing internal standard (lithium in equal concentration) is sucked by an atomiser (which is operated by the fuel or the oxidant) and aspirated into the flame. The emitted radiation passes through a filter and then strikes a mirror. The mirror in turn divides the emitted radiation into two parts. One part is due to internal standard (lithium) and the other part is due to the presence of the element to be measured. Both the radiations fall on separate amplifiers and then on a common detector (a calibrated potentiometer), which records the ratio of the intensity of the element to be determind to that of the internal standard. A diagramatic representation of an internal standard flame photometer is given below (Fig. 6.1).

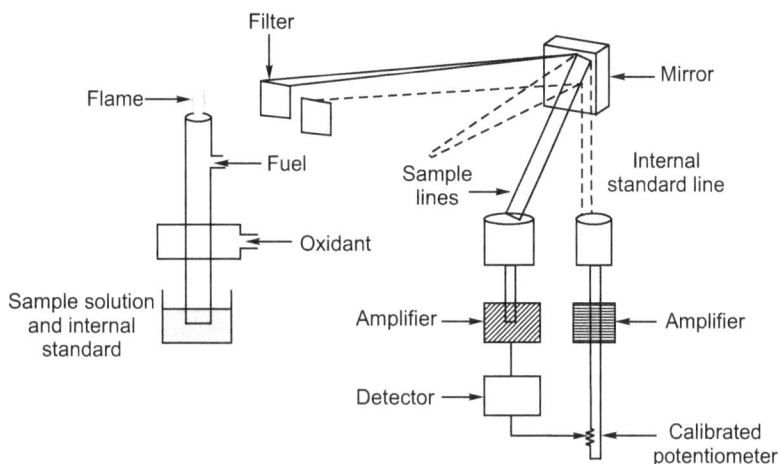

Fig. 6.1. Diagramatic representation of an internal standard flame photometer.

In standard flame photometer the errors due to difference in viscosity and surface tension are reduced considerably. So this photometer is much better than the simple flame photometer.

Interferences in Flame Atomic Emission Spectrometry

Some of the commonly encountered interferences in flame photometry are:

(*i*) **Spectral interference:** In cases, where two elements are present, both these exhibit different spectral characteristic, but their spectras may overlap and both the elements emit at same particular wavelength. It is not possible for the detector to differentiate between the sources of two radiations and will record total signal which gives erroneous results. Thus, as an example, iron line at 3247.28 Å overlaps with copper line at 3247.54 Å. Also iron line at 2852.13 Å overlaps with magnesium line at 2852.12 Å. Such interferences can be overcome by removing the interferring element. Alternatively, calibration curves can be prepared from solutions having similar quantities of interferring element.

(*ii*) **Cation-anion interference:** The radiation frequency emitted by an element is affected due to the presence of certain anions like oxalate, sulphate, phosphate and aluminate. As an illustration, calcium in presence of phosphate forms a stable salt and this results in depression of signal due to calcium (as calcium phosphate is stable and does not easily decompose). In a similar way, barium in presence of sulphate gives low emission intensity. This type of interference can be eliminated either by removal of the interferring anion or by using calibrated curves with the interferring anion at the same concentration in the standard solution.

(*iii*) **Cation-cation interference:** It is possible that there may be mutual interference of two cations. As an example there is mutual interference of aluminium with calcium and magnesium. Such interference can also be removed as in (*ii*) above.

(*iv*) **Interference due to formation of oxides:** Some elements (like alkaline earth elements) form stable oxides in presence of oxygen (which is used as an oxidant). This is responsible for removal of major fraction of the metal atoms and results in low intensity. Such interference can be removed by using higher temperature of flames resulting in dissociation of oxides to form free atoms.

Limitations of Flame Atomic Emission Spectrometry

Flame atomic emission spectroscopy has two important limitation. It is useful only for liquid samples and no information is obtained about the molecular form of the metal and only the total metal content in a sample can be determined.

Applications of FAES

An important application of FAES include **Quantitative analysis**. FAES is used for the estimation of alkali and alkaline metals. Besides the element like Al, Sb, As, Ba, Ca, Bi, Cu, Fe, Pb, Hg, Ni and Pd can also be estimated.

6.2   ATOMIC ABSORPTION SPECTROSCOPY (AAS)

Introduction

It has already been stated (Chapter 1, Section 1.9) that atomic absorption spectroscopy (AAS) is used for the detection of metallic element in solution. The method involves conversion of the sample to the atomic vapour followed by the measurement of the absorption of atomic vapour at some selected wavelength, which is characteristic of each individual element. The element, in this technique is first reduced to the elemental state followed by vaporisation and imposing the vapourised metal in the beam of the source. This procedure is specific as the atoms of only a particular element can absorb radiations of its characteristic wavelength. Thus, a metal can be detected and also estimated in presence of other metals. The method is extremely sensitive and can determine even trace element as low as 1 ppm. This method, however, cannot be used for elements like Ti, W, Mo, V etc. as oxides of these metals are formed in the flame.

Instrumentation

An AAS instrument consists of a glass tube containing an inert gas 'Argon' at several mm pressure, an anode and a hollow cathode, the inside of which is coated with the metal under investigation. A diagramatic representation of a anode and cathode in a glass tube is shown in chapter 1, Fig. 1.2.

In the flame automiser the solutions of the salt of the metal to be estimated is aspirated. The solution gets converted into the metal, which in turn gets decomposed in the flame and is reduced to the elemental state forming a cloy of atoms, which absorb a fraction of the radiation in the flame. In case, a flame automerizer is used, the emission by the metal in the flame is at the same wavelength as the absorption wavelength of the metal. This is because the same electronic transition is involved. The resultant energy is passed through a monochromator (which can be a prism or grating) (to eliminate any extraneous light resulting from the flame) and finally to the detector (commonly used detector is photomultiplier type). A diagramatic representation of AAS is shown in Fig. 1.4, Chapter 1.

The common fuel used in the burner of an AAS instrument are acetylene-air (temperature 2800°C), acetylene-oxygen or nitrous oxide (temperature 3000°C) and acetylene-air (temperature 2200°C). Depending on the metal to be estimated, the temperature in the oven is adjusted accordingly.

It has been found that much better results are obtained by using a graphite furnace in place of a flame automiser and the sample holder. The graphite furnace has hollow graphite cylinder placed in such a way that the beam of light from the hollow cathode lamp passes through it. The sample (small amount) is injected in the tube through a small hole at the top. When an electric current is passed through the tube, the metal sample is obtained in dry state and the absorption of the metal recorded as a single peak. An atomic absorption spectrometer using a graphite furnace is shown in Fig. 1.5, Chapter 1.

The graphite furnace is much superior compared to the flame detector and can detect various elements even in minute amounts. Table 6.1 given below gives the detection limits for some elements.

Table 6.1. *Comparative detection limits of some elements by Flame AAS and graphite furnace AAS (CL. Herbert, S.B. Khan and R.G. Smith Scheicher, American Laboratory, 65-79, Aug., 1979).*

Element	Detection limits (PPM)	
	Flame AAS	Graphite furnace AAS
Al	20	0.004
As	100	0.06
B	1000	–
Cd	1	0.008
Co	5	0.03
Cr	3	0.005
Cu	2	0.008
Fe	5	0.003
Mn	3	0.004
Mo	10	0.06
Ni	8	0.02
P	10	3.0
Pb	10	0.03
Pt	50	0.45
Se	100	0.10
Si	60	0.10
Ti	50	0.30
U	7000	–
V	20	0.15
Zn	0.6	0.0007

Determination of the Concentration of elements in ppm

The concentration of the elements in ppm is determind by referring to a calibration curve, which is drawn by spraying standard solutions of the element (to be determined) and plotting the absorbance against concentration (ppm). As an illustration a calibration curve for copper at 3428 Å is an shown in Fig. 6.2.

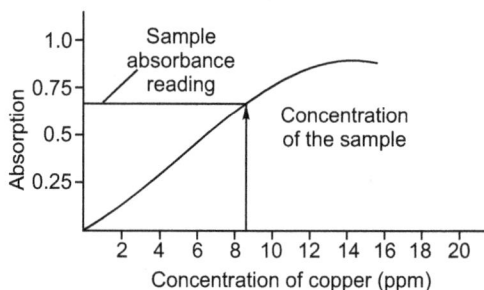

Fig. 6.2. Calibration curve for copper at 3428 Å.

Estimation of Various Elements

Following (Table 6.2) are given the various elements which can be estimated by atomic absorption spectroscopy.

Table 6.2. *Estimation of Various Elements by AAS Technique.*

Element	Hollow Cathode lamp mull of	Flame used for estimation	Detection limit (ppm)	Absorbance nm	Standard solution (1000 ppm) preparation for calibration curve (using measuring flask)
Lead	Lead	Acetylene-air	5	283.31	0.79978 Pb $(NO_3)_2$ (AR) in 500 mL distilled water
Mercury* (Poisonous)	Mercurty	Acetylene nitrous oxide	20	253.65	0.6767 g $HgCl_2$ (AR) in 500 mL distilled water
Silver	Silver	Acetylene-air	1	328.07	0.7877 g $AgNO_3$ (AR) dissolved in 500 mL distilled water
Bismuth	Bismuth	Acetylene-air	1	223.06	0.5000 g Bi metal in 25 mL conc HNO_3 and diluted to 500 mL with distilled water
Copper	Copper or Brass	Acetylene-air	1	324.75	1.9647 g $CuSO_4 . 5H_2O$ (AR) dissolved in 500 mL distilled water

Cadmium	Cadmium	Acetylene-air	1	228.80	1.5965 g of $CdSO_4$. 8/3 H_2O (AR) dissolved in 500 mL distilled water
Arsenic	Arsenic	Acetylene-nitrous oxide	1	193.70	0.6601 g AS_2O_3 (AR) dissolved in 25 mL conc. HCl and solution made upto 500 mL distilled water
Antimony	Antimony	Acetylene-air	50	206.83	Antimony metal (AR) (0.5000 g) dissolved in 10 mL conc. HCl. 5g tartaric acid (AR) and two drops of conc. HNO_3. Solution made up to 500 mL by distilled water
Tin	Tin	Acetylene-nitrous oxide	1	224.61	0.5000 g tin metal (AR) dissolved in 100 mL conc. HCl and 3mL conc. HNO_3 and solution made upto 500 mL by distilled water
Iron	Iron	Acetylene-air	1	248.33	0.5000 g iron (AR) dissolved in 10 mL conc. HCl and 3 mL conc. HNO_3. Heat the solution and cool and dilute to 500 mL with distilled water
Aluminium	Aluminium	Acetyle-nitrous oxide	1	309.28	8.791 g $KAl.(SO_4)_2$. $12H_2O$ (AR) dissolved in 500 mL distilled water
Chromium	Chromium	acetylene-nitrous oxide	1	357.87	1.9145 g $K_2Cr_2O_7$ (AR) dissolved in 500 mL distilled water
Cobalt	Cobalt	Acetylene-air	1	240.73	2.9691 g $Co(NO_3)_2$. $6H_2O$ (AR) in 500 mL distilled water
Nickel	Nickel	Acetylene-air	0.1	341.48	2.9766 g $Ni(NO_3)_2$ $6H_2O$ (AR) in 500 mL distilled water

...contd.

Manganese	Manganese	Acetylene-air	0.1	279.48	1.8015 g MnCl$_2$. 4H$_2$O (AR) dissolved in 25 mL conc. HCl and 25 mL H$_2$O. The solution made upto 500 mL with distilled water
Zinc	Zinc or Brass	Acetylene-air	0.1	213.86	2.6990 g ZnSO$_2$ 7H$_2$O (AR) in 500 mL distilled water
Barium	Barium	Acetylene-air	0.1	553.55	0.7684 g BaCO$_3$ (AR) in 8 mL 2 M HCl solution made upto 500 mL with distilled H$_2$O
Strontium	Strontium	Acetylene-air	0.1	460.75	0.8824 g SrCO$_3$ (AR) in 8 mL 2 M HCl. Solution made up to 500 mL with distilled water
Magnesium	Magnesium	Acetylene-air	0.1	285.21	0.5000 g Mg metal (AR) dissolved in 25 mL 6 M HCl. Solution made up to 500 mL with distilled water
Potassium	Potassium	Acetylene-air	0.1	766.49	0.9533 g KCl (AR) in 500 mL distilled water

Notes:

1. Low concentration of Cd, Cr and Pb can be determined by chelation with ammonium pyrrolidine dithiocarbamate followed by extraction of the formed complex with methyl isobutyl ketone and aspiration into acetylene-air flame.

2. Low concentration of Al and Be can be estimated by chelation with 8-oxyquinol, extraction with methyl isobutyl ketone and aspiration into acetylene-nitrous oxide flame.

3. Estimation of some element like As and Se can be done by conversion to their hydrides followed by aspiration into hydrogen-argon flame.

4. Mercury is an extremely toxic metal. Its intake is responsible for various neurological disorders ultimately leading to death. It is estimated by flameless atomic absorption method. The method involves reduction of Hg^{2+} to elemental Hg^0 by SnCl$_2$ solution at room temperature. The resulting mercury vapour is swept into an absorption cell, which is subjected to beam of radiant vapour from an Hg vapour lamp. The measurement of absorption at 253.7 nm gives the amount of mercury in water. For more details see Chapter 1, note c, page 91.

5. The AAS technique is useful for the determination of only one element at a time. However, now multielement hollow cathode source is available with a detection system having a spectral range of 2320 to 3281 Å. Using this technique, Michell has been able to determine simultaneously eight element (Zn, Cd, Ni, Co, Fe, Mn, Cu and Ag).

EXERCISES

1. Discuss the principle of flame atomic emission spectrometry (FAES).
2. What type of interferences one can encounter in FAES and how they are overcome.
3. Describe the sailent features of Atomic Absorption Spectroscopy. What is the principle involved in the estimation of elements by AAS.
4. How is mercury estimated in a water sample?

OOO

PART IV

Inorganic Preparations

7. Inorganic Preparations

Contents

7

Inorganic Preparations

7.1 SIMPLE SALTS

7.1.1 Cuprous Chloride, Cu_2Cl_2

Cuprous chloride is used as a catalyst in organic reactions and in petroleum industry. It is also used in denitration of cellulose, as a condensing agent for soaps, fats and oil. In gas analysis, it is used to absorb carbon monoxide.

Cuprous chloride can be prepared from Copper Sulphate by heating with copper and hydrochloric and

$$CuSO_4 + Cu + 2HCl \longrightarrow Cu_2Cl_2 + H_2SO_4$$

Chemicals:

Copper sulphate ($CuSO_4.5H_2O$)	3.0 g
Copper turning	2 g
Sodium chloride	1.6 g
Conc. HCl	2.2 mL
Sodium sulphite	1.2 g
Alcohol	12 mL

Procedure

A mixture of copper sulphate (3.0 g), sodium chloride (1.6 g) and conc. HCl (2.2 mL) is heated slowly in a round bottomed flask (50 mL capacity) on a wire guage. During heating, the mouth of the flask is covered with a funnel, the stem of the funnel going inside the flask. The heating is continued till the blue colour of $CuSO_4$ disappears. The mixture is poured into water (about 50 mL) to which sodium sulphite (1.2 g) had been added. A white precipitate of Cu_2Cl_2 is obtained. The precipitate of cuprous chloride is filtered (using suction). The residue is washed with water and alcohol. The sample is dried at 110°C in an oven, yield 2 g.

7.1.2 Cuprous Oxide, Cu_2O

Also known as red curpous oxide, it is a fungicide and is used in antifouling paints for marine use. Also used in photoelectric cells and as red pigment for glass, ceramic glazes, it is also used as a catalyst.

Cuprous oxide is prepared by the reduction of copper sulphate in alkaline medium with glucose.

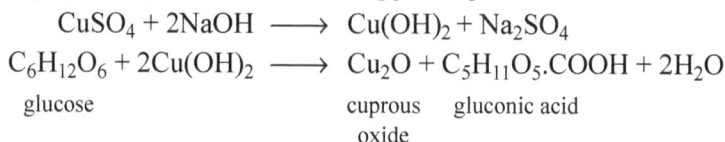

$$CuSO_4 + 2NaOH \longrightarrow Cu(OH)_2 + Na_2SO_4$$

$$\underset{\text{glucose}}{C_6H_{12}O_6} + 2Cu(OH)_2 \longrightarrow \underset{\substack{\text{cuprous} \\ \text{oxide}}}{Cu_2O} + \underset{\text{gluconic acid}}{C_5H_{11}O_5.COOH} + 2H_2O$$

Chemicals: Copper sulphate   3 g
                              Glucose   1.8 g
              Sodium hydroxide   20% solution (0.8 mL)

Procedure

A solution of $CuSO_4$ (3 g) in about 15 mL water in heated to boiling. To the solution is added glucose (1.8 g) followed by addition of sodium hydroxide solution (0.8 mL, 20%). The heating is continued till the blue colour of $CuSO_4$ disappears. The solution is cooled, supernatant liquid decanted and cuprous oxide filtered. The precipitate is washed with water, alcohol and dried on a porous plate, yield 1.5 g.

7.1.3 Ferrous Sulphate, $FeSO_4.7H_2O$

Also known as Green Vitriol, ferrous sulphate is used in the manufacture of iron and iron compounds and other sulphates. It is used as a reducing agent in chemical processes, as wood preservative and as weed killer.

Ferrous sulphate is prepared from 'Kipps Waste', a waste product of kipps apparatus obtained during the preparation of H_2S.

$$FeS + H_2SO_4 \longrightarrow FeSO_4 + H_2O$$

Chemicals: Kipps waste   50 mL
                    Iron filings   5 g
                    Dil. H_2SO_4 10 ml

Procedure

To the kipps waste (50 mL) taken in a China dish is added dil H_2SO_4 (about 10 mL) and iron filings (about 5 g). The mixture is heated slowly to boiling till the solution becomes saturated. The absence of ferric ions is ascertained by testing a drop of the solution with a drop of KCNS solution. When all the ferric ions have been reduced, the concentrated solution is filtered and allowed to cool. The separated green crystals of $FeSO_4.7H_2O$ are filtered and dried on a porous plate.

7.1.4 Lead Chromate, $PbCrO_4$

Also known as chrome yellow lead chromate is used as pigment in oil and water colours. It also finds use in printing fabrics, decorating china and porcelain.

Lead chromate is prepared by the reaction of potassium chromate with lead nitrate.

$$Pb(NO_3)_2 + K_2CrO_4 \longrightarrow PbCrO_4\downarrow + 2KNO_3$$

Chemicals: Lead nitrate   1 g

Potassium chromate   0.6 g

To lead nitrate solution (1 g in water) is slowly added with stirring a solution of potassium chromate (0.6 g) in water. The precipitated lead chromate is filtered, washed with water and dried yield, 1.2 g.

Note: Both lead and chromium salts are poisonous and adequate precaution should to taken during handling.

7.1.5 Lead Tetra-acetate, $Pb(CH_3COO)_4$

Lead tetra-acetate is a selective oxidizing agent in organic synthesis and is prepared by heating red lead with acetic acid in presence of acetic anhydride.

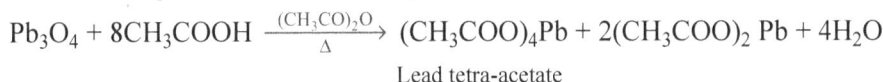

$$Pb_3O_4 + 8CH_3COOH \xrightarrow[\Delta]{(CH_3CO)_2O} (CH_3COO)_4Pb + 2(CH_3COO)_2\,Pb + 4H_2O$$

Lead tetra-acetate

The filtrate which contains lead acetate is treated with chlorine when it gets converted to lead tetra-acetate.

$$2(CH_3COO)_2Pb + Cl_2 \longrightarrow (CH_3COO)_4Pb + PbCl_2$$

Chemicals: Red lead   6 g

Glacial acetic acid   11 g

Acetic anhydride   7 g

Procedure

Powdered red lead (6 g) is added in small lots to a vigorously stirred solution of glacial acetic acid (11 g) and acetic anhydride (7 g) (at 55°C) contained in a three necked RB flask fitted with a stirrer and a thermometer. After the addition is complete the temperature is raised to about 80°C in order to complete the reaction. The thick dark coloured solution is cooled, filtered and residue washed with glacial acetic acid. The crude lead tetra-acetate thus, obtained is dissolved in glacial acetic acid containing a little acetic anhydride is treated with decolourising carbon, filtered and cooled. The separated lead tetra-acetate is filtered and dried in vacuum over KOH, yield about 3 g.

The mother liquor containing lead acetate [$Pb(CH_3COO)_2$] is heated to about 75° with stirring and a dry stream of chlorine gas passed through the solution. After the reaction is complete, decolourising carbon is added, temperature maintained at 75°C, yield 3-4 min and filtered using a pre heated Buchner funnel. On cooling the filtrate, crystals of lead tetra-acetate is obtained.

7.1.6 Magnesium Sulphate $MgSO_4.7H_2O$

Also known as Epsom salt, magnesium sulphate is used in manufacture of frosted paper, fire-proofing fabrics, dyeing and printing calicos and as fertilizer. It is also used as purgative. It occurs in nature as the mineral epsomite ($MgCO_3$).

Magnesium sulphate is prepared in the laboratory by treating magnesium carbonate with dilute H_2SO_4.

$$MgCO_3 + H_2SO_4 \longrightarrow MgSO_4 + H_2O + CO_2$$

Chemicals: Magnesium carbonate   2 g

Dil. H_2SO_4 2.5 mL.

Procedure

To a cooled solution of magnesium carbonate (2 g) in water is added dropwise with stirring, dil H_2SO_4 till there is no more effervescence. Excess of acid is avoided. In case the solution becomes acidic a little more $MgCO_3$ is added. The solution is heated to boiling and filtered. The filtrate is concentrated and cooled. The separated $MgSO_4.7H_2O$ is filtered by decantation and dried. Yield 1.5 g.

Note: In place of $MgCO_3$, the mineral epsomite can also be used.

7.1.7 Manganese Dioxide (Active), MnO$_2$

Active manganese dioxide is used as a reagent in organic synthesis. It is prepared from potassium permanganate by treating with manganese sulphate in presence of sodium hydroxide.

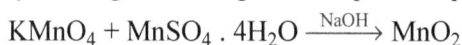

$$KMnO_4 + MnSO_4 . 4H_2O \xrightarrow{\text{NaOH}} MnO_2$$

Chemicals: Potassium permanganate   5 g

Manganese sulphate   5.4 g
tetrahydrate

Sodium hydroxide   40% 6 mL

Procedure

To a stirred solution of potassium permanganate (5g) in water (about 55 mL) are added dropwise simultaneously a solution of manganese sulphate tetrahydrate (5.4 g) in 10 mL water and sodium hydroxide solution (40%, 6 mL) during 1 hr. The mixture is stirred for 1 hr more. The precipitated MnO_2 is centrifuged, washed with water until the washings are colourless and free of sodium hydroxide. The product is dried at 110° for 24 hr and then powdered. Yield about 5.5 g.

7.1.8 Mercuric Oxide, HgO

Mercuric oxide is known to be useful in a number of organic transformations. The yellow variety is more reactive than the red variety. Yellow mercuric oxide is obtained by treating a solution of mercuric chloride with sodium hydroxide.

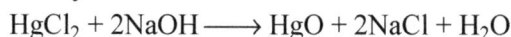

$$HgCl_2 + 2NaOH \longrightarrow HgO + 2NaCl + H_2O$$

7.1.9 Potassium Dichromate, K$_2$Cr$_2$O$_7$

Potassium dichromate is used in tanning leather, dyeing, painting and decorating porcelain. Also used as pigment and for water proofing fabrics. It is used as an oxidising agent in organic and inorganic synthesis.

Potassium dichromate is prepared by treating chromium acetate with bromine water in presence of alkali followed by acidification. The formed $Na_2Cr_2O_7$ is treated with KCl.

$$Cr(CH_3COO)_3 + 3NaOH \longrightarrow 3CH_3COONa + Cr(OH)_3$$
$$2Cr(OH)_3 + 8NaOH + 3Br_2 \longrightarrow 2Na_2CrO_4 + 4NaBr + 2HBr + 6H_2O$$
$$2Na_2CrO_4 + H_2SO_4 \longrightarrow Na_2Cr_2O_7 + Na_2SO_4 + H_2O$$
$$Na_2Cr_2O_7 + 2KCl \longrightarrow K_2Cr_2O_7 + 2NaCl$$

Chemicals:	Chromium acetate	3 g
	Sodium hydroxide	2.9 g
	Bromine water	6 mL
	Potassium chloride	2 g

Procedure

To a saturated solution of chromium acetate (3 g) in dilute HCl is added sodium hydroxide (2.9 g) and bromine water (6 mL). The mixture is heated. The green solution turns yellow indicating the formation of Na_2CrO_4. The solution is acidified with conc. H_2SO_4.

$$2Na_2CrO_4 + H_2SO_4 \longrightarrow Na_2Cr_2O_7 + Na_2SO_4 + H_2O$$

The solution is cooled and separated Na_2SO_4 filtered. The filtrate is concentrated, potassium chloride (2 g) added and solution heated.

$$Na_2Cr_2O_7 + 2KCl \longrightarrow K_2Cr_2O_7 + 2NaCl$$

The separated NaCl is filtered and filtrate cooled to give orange crystals of $K_2Cr_2O_7$. These are filtered and dried. Yield 3 g.

7.1.10 Potassium Permanganate, KMnO₄

Potassium permanganate is used for tanning leathers, purifying water and as an oxidizing reagent in organic synthesis and in redox titrations as an indicator.

Potassium permanganate is prepared by fusing manganese dioxide with KOH in presence of an oxidising agent like $KClO_3$ to give potassium manganate which on subsequent treatment with CO_2 yields $KMnO_4$.

$$3MnO_2 + 6KOH + KClO_3 \longrightarrow 3K_2MnO_4 + KCl + 3H_2O$$
$$\text{Pot. manganate}$$
$$3K_2MnO_4 + 2CO_2 \longrightarrow 2KMnO_4 + 2K_2CO_3 + MnO_2$$
$$\text{Pot. permanganate}$$

Chemicals:	Manganese dioxide	3.4 g
	Potassium hydroxide	4 g
	Potassium chlorate	1.6 g

Procedure

A mixture of powdered $KClO_3$ (1.6 g) and KOH (4 g) is heated gently in a nickel crucible on a sand bath until the mixture just melts. To the fused mass is added powdered MnO_2 (3.4 g) in small lots with stirring. Heating and stirring is continued till a green mass is obtained. The mixture is cooled, transferred to a beaker containing water (about 100-150 mL). Carbon dioxide gas is passed into the solution till a drop of the solution on placing on a filter paper does not give any green colour. The resultant solution is filtered, filtrate concentrated and cooled to give crystals of $KMnO_4$. Yield about 6 g.

7.1.11 Silver Oxide, Ag_2O

Also known as argentous oxide, silver oxide is a brown black powder and is used as a catalyst in the conversion of aldehydes to acids, hydroxy aromatic compounds to quinones and a host of other organic transformations.

Silver oxide is commercially available but can be prepared by treatment of a solution of $AgNO_3$ with KOH solution. The resulting precipitate is decanted, washed with distilled water until no NO^-_3 ions are detected in the supernatant liquid. The remaining Ag_2O is filtered and dried in air. In case oxidations are carried out in an aqueous solution, the Ag_2O is prepared in situ from $AgNO_3$ and NaOH or KOH.

7.1.12 Sodium Chloride

Sodium chloride commonly known as table salt or common salt is a essential nutrient factor. Intravenously it is used as isotonic solution to raise blood volume to combact dehydration.

Sodium chloride is present in large amounts in sea waters from which water is evaporated by the heat from the sun to give curde sodium chloride which is purified by passing hydrogen chloride gas into a saturated solution of commercial sample. On passing HCl gas, the concentration of chloride ions (Cl^-) increases resulting the precipitation of pure sodium chloride (common ion effect).

Chemicals: Common salt (NaCl)   15 g
                 Hydrogen chloride gas

Procedure

In the clear solution of common salt (NaCl) (15 g) in water is passed a current of hydrogen chloride gas till the precipitation of NaCl is complete. The clear liquid is decanted, and crystalline NaCl dried over a porous plate. Yield about 10 g.

The apparatus for generating HCl gas and passing HCl gas into the solution is shown below.

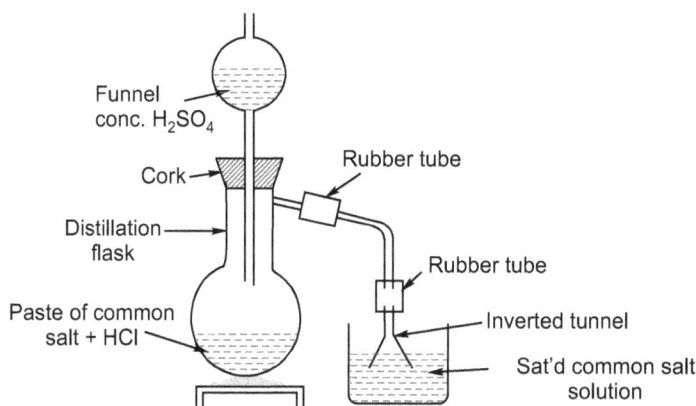

Fig. 7.1. Generation of HCl gas.

7.1.13 Sodium Perborate, BNaO$_3$

Sodium perborate is used for bleaching straw and other fibres. It is also used as a tropical antiseptic and in mouth wash.

Sodium perborate is prepared from sodium metaborate (BNaO$_2$, obtained by fusing equivalent molecular mass of borax and sodium carbonate) and hydrogen peroxide.

$$Na_2B_4O_7 + Na_2CO_3 \longrightarrow 4BNaO_2 + CO_2$$
$$\text{Borax} \qquad\qquad\qquad \text{Sodium}$$
$$\text{metabaorate}$$

$$BNaO_2 + H_2O_2 \longrightarrow BNaO_3 + H_2O$$
$$\text{Sodium}$$
$$\text{perborate}$$

Sodium perborate is also used as an oxidising reagent for the oxidation of primary aromatic amines to azo compounds or nitro compounds and sulfoxides to sulfones.

7.1.14 Sodium Sulphite, Na$_2$SO$_3$.7H$_2$O

Sodium sulphite is mainly used in photographic developers for fixing prints, bleaching wood, straw, silk and for generating SO$_2$. Also used for preserving meat, egg yolks etc.

Sodium sulphite is prepared by saturating sodium carbonate solution with SO$_2$.

$$Na_2CO_3 + H_2O + 2SO_2 \longrightarrow 2NaHSO_3 + CO_2\uparrow$$
$$2NaHSO_3 + Na_2CO_3 \longrightarrow 2Na_2SO_3 + H_2O + CO_2\uparrow$$
$$\text{Sodium sulphite}$$
$$\text{in solution}$$

Chemicals: Sodium carbonate   25 g

                           SO$_2$

Procedure

Sodium carbonate (25 g) is dissolved in about 60 mL water and the solution divided in two parts. One portion of this solution is saturated with SO$_2$ to form sodium bisulphite (the solution must smell of SO$_2$). To this solution is added the remaining sodium carbonate solution and the total solution evaporated in a silica dish and the concentrated solution allowed to stand. The separated crystals of sodium sulphite are separated from the mother liquor by decantation and dried between folds of filter paper. Yiled about 25 g.

7.1.15 Sodium Thiosulphate, Na$_2$S$_2$O$_3$.5H$_2$O

Commonly known as hypo, sodium thiosulphate is used to remove chlorine from solutions as per equation.

$$Na_2S_2O_3 + 4Cl_2 + 5H_2O \longrightarrow 2NaHSO_4 + 8HCl$$
$$Na_2S_2O_3 + 2HCl \longrightarrow 2NaCl + H_2O + 2S$$

Also used as fixer in photography, for extraction of silver ores and as mordant in dyeing.

Sodium thiosulphate is prepared by heating a solution of sodium sulphite solution with sulphur powder.

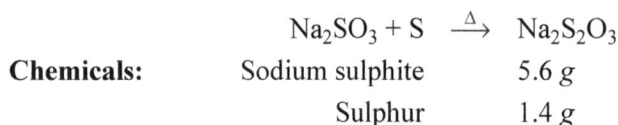

$$Na_2SO_3 + S \xrightarrow{\Delta} Na_2S_2O_3$$

Chemicals:	Sodium sulphite	5.6 g
	Sulphur	1.4 g

Procedure

A solution of sodium sulphite (5.6 g) in hot water (about 20 mL) and sulphur (1.4 g) is refluxed till almost all sulphur has reacted. It is ascertained that the solution is not alkaline (test a drop of the solution with a drop of phenolphthalein – there should not be any pink colour). In case the solution is alkaline, some more sulphur is added and the mixture refluxed for 2-3 hrs., for completion. The hot solution is filtered, filtrate concentrated and cooled. The separated sodium thiosulphate is filtered and dried at 40 – 50°C. Yield about 3 g.

7.2   DOUBLE SALTS

7.2.1 Chrome Alum, $K_2SO_4.Cr_2(SO_4)_3.24H_2O$

A double salt of potassium sulphate and chromium sulphate, chrome alum is obtained by the reduction of an acidifield solution of $K_2Cr_2O_7$ with alcohol or starch.

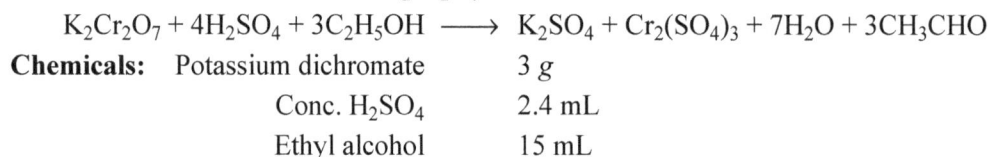

$$K_2Cr_2O_7 + 4H_2SO_4 + 3C_2H_5OH \longrightarrow K_2SO_4 + Cr_2(SO_4)_3 + 7H_2O + 3CH_3CHO$$

Chemicals:	Potassium dichromate	3 g
	Conc. H_2SO_4	2.4 mL
	Ethyl alcohol	15 mL

Procedure

Conc. H_2SO_4(2.4 mL) is added dropwise to a stirred suspension of $K_2Cr_2O_7$ (3 g) in water (15 mL) (temp. is kept < 40° during addition). The mixture is stirred till a homogeneous solution is obtained. In case there is a rise in temperature, few pieces of ice are added to keep the temperature < 40°C. To the above solution is added ethyl alcohol (15 mL) dropwise with stirring (temp. < 30°C). The mixture is cooled (ice-bath) and allowed to stand overnight. The clear supernatant liquid is decanted and the crystals of chrome alum dried. Yield about 3.6 g.

7.2.2 Chrome Red, $PbCrO_4.PbO$

A double salt obtained by mixing an aqueous solution of lead nitrate with potassium chromate and treating the formed lead chromate with sodium hydroxide solution.

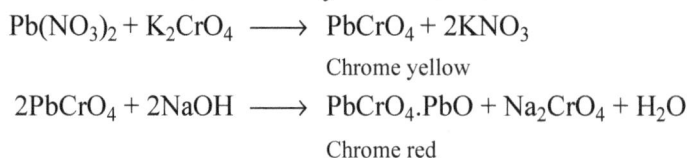

$$Pb(NO_3)_2 + K_2CrO_4 \longrightarrow PbCrO_4 + 2KNO_3$$

Chrome yellow

$$2PbCrO_4 + 2NaOH \longrightarrow PbCrO_4.PbO + Na_2CrO_4 + H_2O$$

Chrome red

Chemicals:

Lead nitrate	3 g
Potassium chromate	1.8 g
Sodium hydroxide	5%, 7.8 mL

Procedure

A solution of K_2CrO_4 (1.8 g) in water is added to a solution of lead nitrate (3 g) is water. The ppt. of lead chromate (chrome yellow) is filtered, washed with water (by decantation) and boiled with sodium hydroxide solution (5 %, 7.8 mL) till the colour of the precipitate changes to bright red. The chrome red thus obtained is filtered (suction) and dried. Yield about 3 g.

7.2.3 Ferric Alum, [(NH₄)₂SO₄.Fe₂(SO₄)₃.24H₂O]

A double salt, ferric alum is obtained by mixing a solution of ammonium sulphate and ferric sulphate in presence of H_2SO_4. The ferric sulphate is obtained by the oxidation of ferrous sulphate with HNO_3.

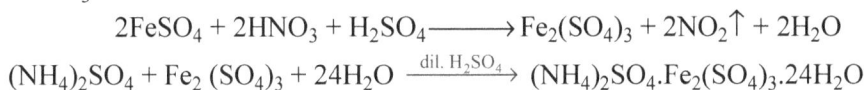

$$2FeSO_4 + 2HNO_3 + H_2SO_4 \longrightarrow Fe_2(SO_4)_3 + 2NO_2\uparrow + 2H_2O$$

$$(NH_4)_2SO_4 + Fe_2(SO_4)_3 + 24H_2O \xrightarrow{\text{dil. } H_2SO_4} (NH_4)_2SO_4.Fe_2(SO_4)_3.24H_2O$$

Chemicals:

Ferrous sulphate ($FeSO_4.7H_2O$)	3.75 g
Dil. H_2SO_4	18.8 mL
Conc. HNO_3	3.7 mL
Ammonium sulphate	2.1 g

Procedure

To a hot solution of $FeSO_4$ (3.75 g) in dil. H_2SO_4 (18.8 mL) (heated on a water bath) is added conc. HNO_3 (3.7 mL) in small lots till no more brown vapours of NO_2 are evolved (HNO_3 converts ferrous ions to ferric ions). The solution is evaporated to almost dryness and a solution of ammonium sulphate (2.1 g) in water (25 mL) added with stirring. A pale yellow solution is obtained. In case the solution is brown in colour few drops of conc. H_2SO_4 are added. The solution is allowed to stand for 2-3 days. The separated crystals of ferric alumn are filtered and dried in air. Yield about 3 g.

7.2.4 Ferrous Ammonium Sulphate [FeSO₄.(NH₄)₂SO₄.6H₂O]

Also known as Mohr's salt, ferrous ammonium sulphate, a double salt, is prepared by mixing aqueous solutions of ferrous sulphate and ammonium sulphate.

$$FeSO_4 + (NH_4)_2SO_4 + 6H_2O \longrightarrow FeSO_4(NH_4)_2SO_4 \cdot 6H_2O$$

Chemicals:

Ferrous sulphate $FeSO_4.7H_2O$	3 g
Ammonium sulphate	1.5 g
Dil. H_2SO_4	5.0 mL

Procedure

Ferrous sulphate (3 g) and ammonium sulphate (1.5 g) are dissolved in water (5 mL) containing dil H_2SO_4 (5 mL) by heating to 80-90°C. The solution is boiled for 2-3 min and then cooled. The separated ferrous ammonium sulphate is filtered and dried. Yield about 3.2 g.

Alternatively, ferrous ammonium sulphate can be prepare by dissolving a known quantity of iron filings in dil. H_2SO_4 and adding calculated amount of ammonium sulphate to the solution.

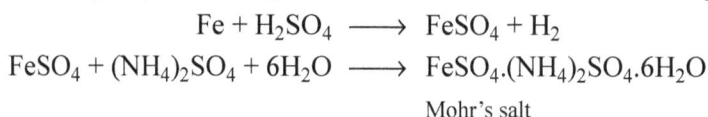

$$Fe + H_2SO_4 \longrightarrow FeSO_4 + H_2$$
$$FeSO_4 + (NH_4)_2SO_4 + 6H_2O \longrightarrow FeSO_4.(NH_4)_2SO_4.6H_2O$$
Mohr's salt

The method consists in adding iron filings (2.5 g) to dilute H_2SO_4 [obtained by careful addition of 1.4 mL conc. H_2SO_4 to 15 mL water (exothermic reaction)]. The mixture is heated slowly using a slow flame on a wire guage (Avoid contact of the evolved H_2 gas with flame in order to prevent explosion). The solution is cooled and filtered. To the filtrate is added a saturated solution of ammonium sulphate (2 g) in water. The solution is concentrated and cooled. The separated ferrous ammonium sulphate is filtered and dried. Yield about 2g.

7.2.5 Potash Alum, [K$_2$SO$_4$.Al$_2$(SO$_4$)$_3$.24H$_2$O]

Known as simply alum, potash alum is mostly used during the purification of water and is obtained by mixing aqueous solutions of potassium sulphate and aluminium sulphate.

$$K_2SO_4 + Al_2(SO_4)_3 + 24H_2O \longrightarrow K_2SO_4.Al_2(SO_4)_3.24H_2O$$

Chemicals:	Potassium sulphate	2 g
	Aluminium sulphate	8 g

Procedure

Aluminium sulphate (8 g) is dissolved in water (about 20 mL). If the solution is turbid add 2-3 drops of conc. H_2SO_4. If the solution is still turbid, it is filtered. To the clear filtrate is added potassium sulphate solution (2 g in 20 mL water). The solution is stirred and allowed to stand over night. The separated potash alumn is filtered and dried. Yield about 7 g.

7.3   COMPLEX SALTS

7.3.1 Aluminium Isopropoxide, Al[OCH(CH$_3$)$_2$]$_3$

Aluminium isopropoxide is mostly used in Meerwein-Ponndorf Varley reduction and is prepared by the reaction of aluminium with isoproyl alcohol in presence of mercuric chloride.

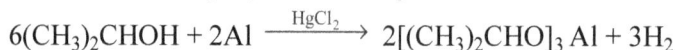

$$6(CH_3)_2CHOH + 2Al \xrightarrow{\text{HgCl}_2} 2[(CH_3)_2CHO]_3 Al + 3H_2$$

Chemicals:	Aluminium wire	20 g
	Isopropyl alochol (anhydrous)	240 mL
	Mercuric chloride	1 g

Procedure

A mixture of aluminium wire or turnings (20 g), anhydrous isoproyl alcohol (240 mL) and mercuric chloride (1 g) is gently refluxed on a water bath for 10-20 min until a vigorous exothermic reaction sets in. The reaction mixture is cooled at this stage. The mixture is gently refluxed for 6-7 hr. The reaction mixture is distilled in vacuum using an oil bath, aluminium isopropoxide passing over as a colourless viscous liquid at 140-150°/1-2 mm. The distillate solution solidifies to a colourless solid m.p. 118°. Yield 130 g.

7.3.2 Ammonium Tetrathiocyanato Diamine Chromate [NH$_4$. (NH$_2$)$_2$ Cr (CNS)$_4$]

Also known as Reineckes salt, ammonium tetrathionato diamine chromate, a complex salt is obtained by melting finely powdered ammonium thiocyanate and mixing the melt with ammonium dichromate.

Chemicals: Ammonium thiocyanate   6 g

Ammonium dichromate   2.25 g

Alcohol   40 mL

Procedure

Ammonium thiocyanate (6 g) is finely powdered and heated in a silica dish till it melts. To the molten mass is added finely powdered ammonium dichromate (2.25 g). Violent evolution of ammonia occurs. The dish is cooled, the mass obtained is powdered. It is suspened in hot alcohol (to dissolve any excess ammonium thiocyanate), filtered and washed with cold water to dissolve any chromate. The resultant red plates of Reinecke's salt are dried. Yield about 3 g.

7.3.3 Copper Phthalocyanine

Copper phthalocyanine, a blue pigment used in the manufacture of inks and paints and is prepared by heating phthalonitrile with copper bronze.

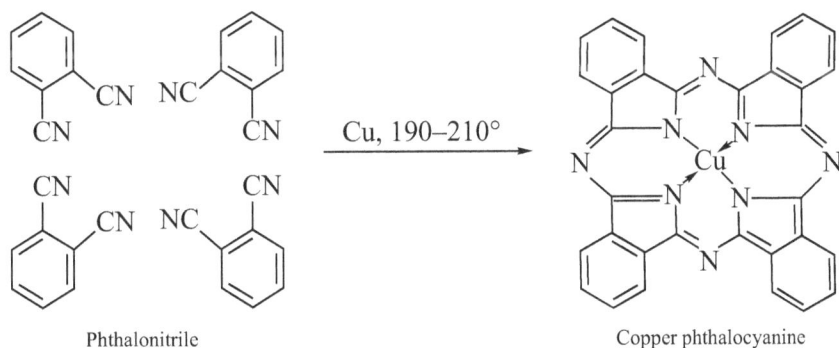

Phthalonitrile   Cu, 190–210°   Copper phthalocyanine

Chemicals: Phthalonitrite   12.8 g

Copper bronze   1.6 g

Procedure

A mixture of phthalonitrile (12.8 g) and copper bronze (1.6 g) is heated in an oil bath in 100 mL RB flask, the mixture being stirred with a thermometer. The product turns green at 190°C and becomes a paste at 220°C. The heating is continued at 220°C for about 15 min. The mass is allowed to cool and then grounded with alcohol (10 mL). It is refluxed with alcohol and filtered hot (this treatment removes excess of phthalonitrile).

The solid product obtained above is stirred with conc. H_2SO_4 (55 mL), allowed to stand for 1 hr. and filtered using a sintered glass funnel and the product washed with conc. H_2SO_4. The combined filtrate and washings are poured into crushed ice (100 g) with stirring. The flocculent blue precipitate is allowed to stand for 2-3 hour, filtered (suction) and washed with boiling water. The pigment is finally refluxed with alcohol filtered and dried at 110°C.

7.3.4 Cuprammonium Sulphate [$Cu(NH_3)_4SO_4.H_2O$]

A double salt, cuprammonium sulphate is used for agricultural purposes and is obtained by adding excess ammonia to copper sulphate solution.

$$CuSO_4 + 4NH_3 + 4H_2O \longrightarrow [Cu(NH_3)_4SO_4.H_2O]$$

The complex cuprammonium sulphate has a blue colour and is formed during the identification of copper in qualitative analysis. In fact a confirmatory test for the detection of Cu^{2+} involves formation of a deep blue colour by the addition of ammonia solution to $CuSO_4$ solution. This is due to the formation is complex ion $[Cu(NH_3)_4]^{2+}$, which has blue colour.

Chemicals:	Copper sulphate	1 g
	Ammonia solution (1 : 1)	2 mL
	Alcohol	4 mL

Procedure

Concentrated ammonia solution (1:1, 2 mL) is added to a solution of $CuSO_4$ (1 g) in water (about 5 mL) till a blue colour is obtained. To the clear solution is added dropwise alcohol (4 mL) with stirring. A dark blue precipitate of cuprammonium sulphate separates. The solution is heated at 50°C (water bath) to dissolve the ppt. (10 min). The solution on cooling gives crystals of cuprammonium sulphate. These are filtered, washed with alcohol and dried, yield 1 g.

7.3.5 Cuprous Mercuric Iodide, Cu_2HgI_4

Cuprous mercuric iodide is used for the detection of overheating of machine bearings etc.; the red colour of Cu_2HgI_4 changes to brownish-blue at 60-70°C, which again becomes red on cooling.

Cuprous mercuric iodide is prepared by the reaction of mercuric chloride with potassium iodide, followed by dissolving the formed scarlet red precipitate of K_2HgI_4 in excess potassium iodide and adding $CuSO_4$ solution to the formed K_2HgI_4.

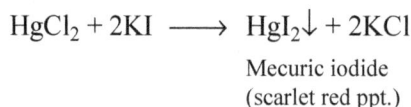

$$HgCl_2 + 2KI \longrightarrow HgI_2\downarrow + 2KCl$$

Mecuric iodide
(scarlet red ppt.)

$$HgI_2 + 2KI \longrightarrow K_2HgI_4$$

$$K_2HgI_4 + CuSO_4 \longrightarrow Cu_2HgI_4 + K_2SO_4$$
Cuprous mercuric
iodide

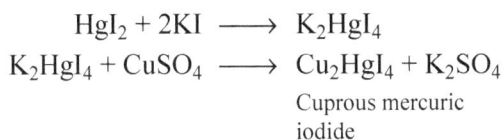

Chemicals:	Mercuric chloride	3 g
	Potassium iodide	6.6 g
	Copper sulphate	4 g
	SO$_2$ gas	

Procedure

To a solution of KI (3.3 g) in water (20 mL) is added a solution of mercuric chloride (3 g) in water (15 mL). The formed scarlet red precipitate of HgI$_2$ is filtered (Buchner funnel), washed with water and dried in an oven. The HgI$_2$ thus obtained is added to 3.3 g KI is 20 mL water. A soluble complex salt solution of K$_2$HgI$_4$ is obtained.

Finally, to the solution of K$_2$HgI$_4$ (obtained above) is added CuSO$_4$ solution (4 g in 30 mL water) and a stream of SO$_2$ is gently passed into the solution [SO$_2$ reduces Cu(ic) to Cu(ous) state]. A scarlet ppt. of Cu$_2$HgI$_4$ is obtained. The precipitate is filtered and dried at about 50°C in an oven.

Note: Sulphur dioxide is generated by the reaction of sodium sulphite in water with HCl.

7.3.6 Ferrocene

Ferrocene, dicyclopentadienyl iron (or biscyclopentadienyl iron), a sandwich compound has been used as antiknock additive for gasoline and also as a catalyst.

Ferrocene is synthesized from cyclopentadiene by treatment with diethyl amine in presence of ferrous chloride.

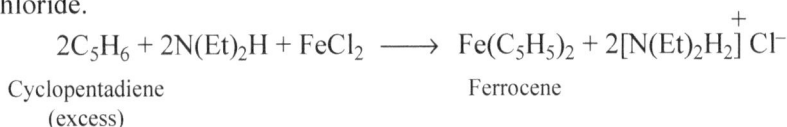

$$2C_5H_6 + 2N(Et)_2H + FeCl_2 \longrightarrow Fe(C_5H_5)_2 + 2[N(Et)_2H_2]^+ Cl^-$$
Cyclopentadiene (excess)   Ferrocene

Ferrocene is also prepared by the treatment of cyclopentadienyl magnesium bromide (Grignard reagent of cyclopentadiene) with ferrous chloride.

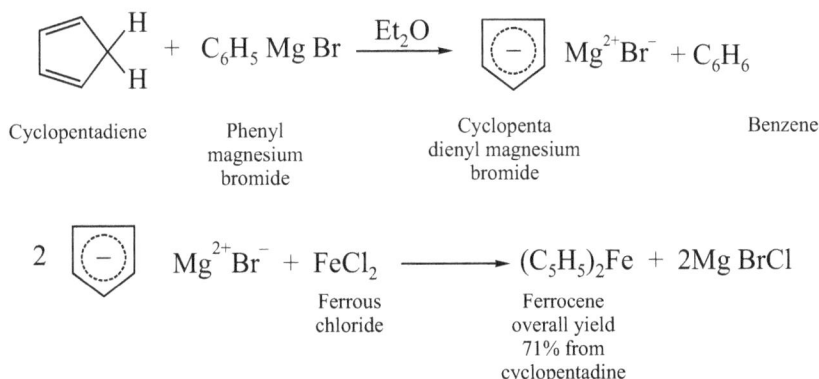

Cyclopentadiene	Phenyl magnesium bromide	Cyclopenta dienyl magnesium bromide	Benzene

$$2 \quad \boxed{\ominus} \quad Mg^{2+}Br^- + FeCl_2 \longrightarrow (C_5H_5)_2Fe + 2Mg\,BrCl$$

                          Ferrous chloride   Ferrocene overall yield 71% from cyclopentadine

The structure of ferrocene is as represented

Ferrocene

7.3.7 Potassium Trioxalatochromate, [K₃[Cr(C₂O₄)₃]]

The complex potassium trioxalatochromate is prepared by treating potassium dichromate with oxalic acid and potassium oxalate.

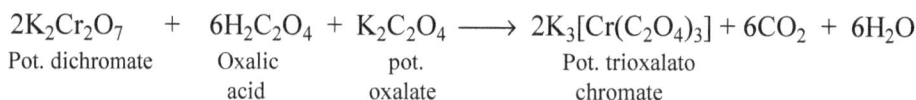

$$2K_2Cr_2O_7 \ + \ 6H_2C_2O_4 + K_2C_2O_4 \longrightarrow 2K_3[Cr(C_2O_4)_3] + 6CO_2 \ + \ 6H_2O$$

Pot. dichromate	Oxalic acid	pot. oxalate	Pot. trioxalato chromate

Chemicals:

Potassium oxalate (monohydrate)	1.5 g
Oxalic acid (dihydrate)	3.35 g
Potassium dichromate	1.35 g
acetone	7 mL

Procedure

Potassium dichromate (1.35 g) is added is small lots with stirring to a solution of potassium oxalate (1.5 g) and oxalic acid (3.35 g) in 45 mL water. The solution is evaporated to about one-third of the original volume and allowed to stand (5-6 hrs). The separated crystals are filtered (suction), washed with acetone (about 7 mL) and dried. Yield about 3.2 g.

7.3.8 Prussian Blue [Fe₄[Fe(CN)₆]₃]

A complex salt, Prussian blue is formed in qualitative analysis for the detection of ferric irons. It is prepared by mixing ferric chloride with potassium ferrocyanide solution.

$$4FeCl_3 + 3K_4[Fe(CN)_6] \longrightarrow Fe_4[Fe(CN)_6]_3 + 12 \, KCl$$

Prussian blue

Chemicals:

Ferric chloride (anhydrous)	2 g
Potassium ferrocyanide	3.3 g

Procedure

To a solution of ferric chloride (2 g) is water (2-3 mL) is added a solution of potassium ferrocyanide (3.3 g) in water. A blue coloured paste of prussian blue is obtained. It is filtered by decantation and dried in between the folds of filter paper. Yield about 3 g.

7.3.9 Sodium Ammonium Hydrogen Phosphate [Na(NH₄) HPO₄.4H₂O]

Also known as microcosmic salt, sodium ammonium hydrogen phophate is used as a fertiliser and is prepared by dissolving ammonium chloride and disodium hydrogen phosphate in water.

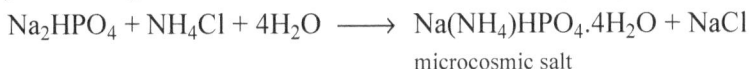

$$Na_2HPO_4 + NH_4Cl + 4H_2O \longrightarrow Na(NH_4)HPO_4.4H_2O + NaCl$$
microcosmic salt

Chemicals: Disodium hydrogen phosphate 3.75 g
(Na₂HPO₄)
Ammonium chloride   1.2 g

Procedure

Ammonium chloride (1.2 g) is added to a hot solution of disodium hydrogen phosphate (3.75 g) in minimum amount of water. The solution is allowed to stand (4-5 hr.). The separated product is filtered, washed with cold water (to remove NaCl) and dried. Yield about 2.4 g.

7.3.10 Sodium Ferric Oxalate, [Na₃[Fe(C₂O₄)₃].9H₂O

It is prepared by mixing aqueous solutions of ferric chloride and oxalic acid in presence of sodium hydroxide.

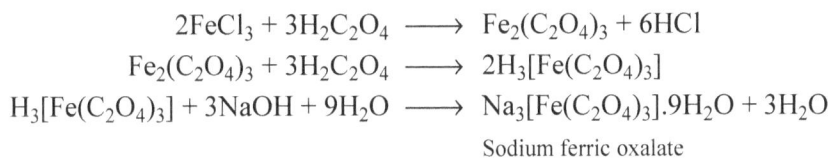

$$2FeCl_3 + 3H_2C_2O_4 \longrightarrow Fe_2(C_2O_4)_3 + 6HCl$$
$$Fe_2(C_2O_4)_3 + 3H_2C_2O_4 \longrightarrow 2H_3[Fe(C_2O_4)_3]$$
$$H_3[Fe(C_2O_4)_3] + 3NaOH + 9H_2O \longrightarrow Na_3[Fe(C_2O_4)_3].9H_2O + 3H_2O$$
Sodium ferric oxalate

Chemicals: Oxalic acid   3 g
Ferric chloride   2.4 g
Sodium hydroxide 4 g

Procedure

Sodium hydroxide pellets (3.1 g) is added slowly to a solution of FeCl₃ (2.4 g) in water (3 mL). To the above solution, a hot solution of oxalic acid (3 g) in water (12 mL) and sodium hydroxide pellets (0.9 g). The combined solution is allowed to stand (4-5 hrs). The separated product is filtered, washed with cold water and alcohol and dried. Yield about 3.6 g.

7.4 AMALGAMS

Amalgams are solutions of metal and mercury.

7.4.1 Aluminium-mercury Couple (Al-Hg) (Aluminium Amalgam)

Aluminium-mercury couple should be prepared just before use. The aluminium turnings or aluminium foil is washed first with alcohol and then with ether and dried in an air oven. It is placed in a saturated solution of mercuric chloride when the aluminium gets coated with a film of mercury. The liquid is pourd off, Al-Hg couple washed with cold water, alcohol and finally with benzene and used.

7.4.2 Sodium Amalgam, Na-Hg

Sodium amalgam is used for the reduction of double bonds in organic synthesis and replacement of acetylenic hydrogens with sodium. It is prepared from sodium and mercury (by mixing) as follows.

Mercury (10 g) is placed in a iron mortar (having a lid of asbestos sheet). To this is added clean dry sodium (5 g) cut into small pieces in one lot to the mortor covered with the asbestos sheet. There is instaneous reaction, the amalgam formed becomes liquid due to evolution heat. On cooling the amalgam solidifies. It is powdered using an iron pestle and kept in a stoppered bottle for use.

7.4.3 Zinc Amalgam, Zn-Hg

Zinc amalgam is used in the Clemmensen reduction of carbonyl groups. It is prepared from zinc and mercuric chloride.

Chemicals:	Granulated zinc	20 g
	Hydrochloric acid	(1 : 1)
	conc. HCl	
	Mercuric chloride	1.5 g

Procedure

To granulated zinc (20 g) placed in a 100 mL RB flask is added warm 1:1 HCl. The acid is decanted and the granulated zinc washed twice more. It is then wahsed with water. To the granulated zinc is added mercuric chloride (about 1.5 g) and conc. HCl (about 5-10 mL). The mixture is stirred for about 10 min. The aqueous solution is decanted off, the amalgamted zinc is washed once with distilled water and covered with 20 mL water and 20 mL conc. HCl.

7.5   ACTIVATED METALS

7.5.1 Copper Activated

Activated copper is useful for a number of reactions like the Ullmann coupling. It is prepared *in situ* by the sonication of copper powder in DMF at 60°C to give much better yield of the coupled product (*J.L. Luche, Ultrasonics, 1987,* **25**, *40*).

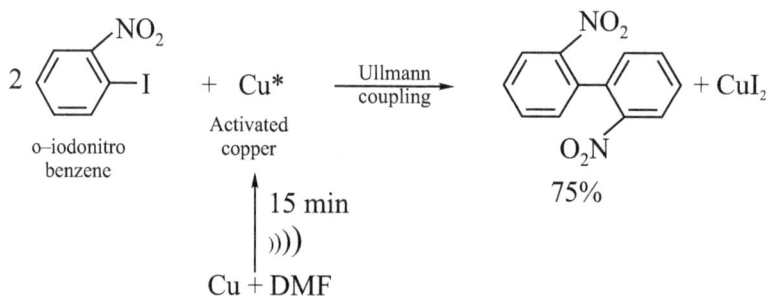

7.5.2 Potassium Very Fine Particles

Very fine potassium particles are used in Dieckmann and Thorpe Zieglar cyclisation. The ultrasonically dispersed potassium is obtained as silvery blue suspension by sonication in toluene. A typical application is given below (*J.L. Luche, C. Petrier and C. Duputy, Tetrahedron Lett., 1985, 26, 753*).

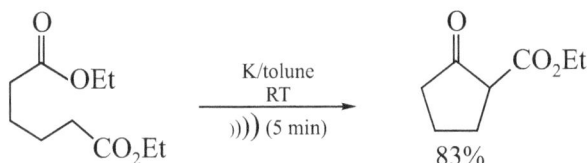

7.5.3 Zinc Activated

Zinc is used in Reformatsky reaction. Quantitative yields are obtained by using activated zinc which is obtained by sonication of zinc powder in presence of iodine. A typical reaction is given below (*B. Han and P.J. Boudjouk, J. Org. Chem., 1982, 47, 5030*).

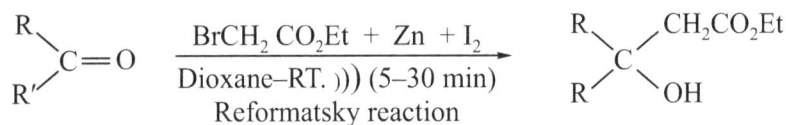

7.6   ORGANOMETALLICS

7.6.1 Alkyl Lithium

Alkyl lithium, an organometallic reagent used for organic synthesis is obtained by the sonication of aryl halides with lithium. A typical reaction is treatment of alkyl lithium with dimethyl formamide to give the corresponding aldehyde (*C. Petrier. A.L. Gemal and J.L. Luche, Tetrahedron Lett, 1982, 33, 3361*).

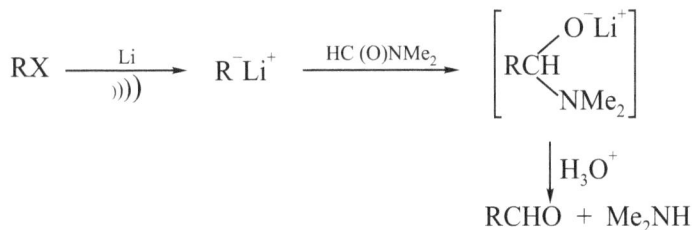

7.6.2 Methyl Magnesium Bromide, CH_3MgBr

Methyl magnesium bromide, a well known Grignard reagent which finds use in numerous synthesis is obtained by sonication of alkyl/aryl halide with magnesium (*J.D. Sprich and G.S. Lewondos, Inorg. Chim. Acta, 1982, 76, 1241*).

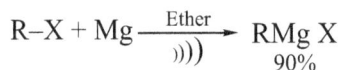

7.7   MISCELLANEOUS INORGANIC PREPARATIONS

7.7.1 Sodium Cobaltinitrite, [Na$_3$Co(NO$_2$)$_6$]

Sodium cobaltinitrite is used as a reagent for the detection of potassium. It is prepared from cobalt nitrate by reacting with sodium nitrite in present of acetic acid.

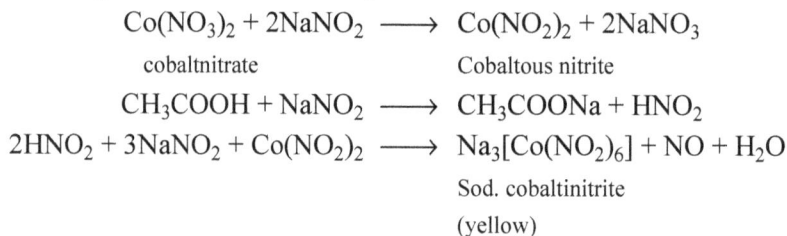

$$Co(NO_3)_2 + 2NaNO_2 \longrightarrow Co(NO_2)_2 + 2NaNO_3$$

cobaltnitrate   Cobaltous nitrite

$$CH_3COOH + NaNO_2 \longrightarrow CH_3COONa + HNO_2$$

$$2HNO_2 + 3NaNO_2 + Co(NO_2)_2 \longrightarrow Na_3[Co(NO_2)_6] + NO + H_2O$$

Sod. cobaltinitrite

(yellow)

Chemicals:	Cobalt nitrate	3 g
	Sodium nitrite	10 g
	Glacial acetic acid	2 mL

Procedure

Sodium nitrite (10 g) is dissolved in water (about 10 mL) in a conical flask fitted with a side tube (a filtration flask can be used). The solution is cooled to room temperature and cobalt nitrate (3 g) added with stirring. The apparatus is fitted up as shown in the Fig. 7.2 and a current of air drawn through the solution for about 30 min by using a pump. The ppt. formed is filtered. To the clear solution is added alcohol (about 10 mL) and solution allowed to stand (4-5 hrs). The separated crystals of sodium cobaltinitrite are filtered, washed with alcohol and dried. Yield about 4.5 g.

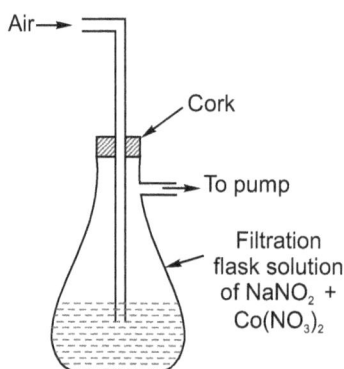

Fig. 7.2. Apparatus used for preparation of Na$_3$[Co (NO$_2$)$_6$].

7.7.2 Tetrabutyl Ammonium Hydrogen Sulphate, Bu$_4$N$^+$HSO$_4^-$

Tetrabutyl ammonium hydrogen sulphate, a phase transfer catalyst used for the transfer of anions into the organic phase and is a versatile reagent used extensively in organic synthesis. It is prepared (*C.M. starks, J. Am. Chem. Soc., 1971, 93, 195*) by the reaction of tetrabutyl ammonium bromide with dimethyl sulphate is chlorobenzene.

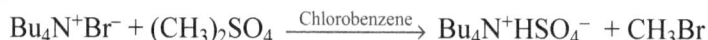

$$Bu_4N^+Br^- + (CH_3)_2SO_4 \xrightarrow{\text{Chlorobenzene}} Bu_4N^+HSO_4^- + CH_3Br$$

Chemicals:	Tetrabutylammonium bromide	9.8 g
	Chlorobenzene	15 mL
	Dimethyl sulphate	4.6 g

Procedure

To a mixture of tetrabutylammonium bromide (9.8 g) and chlorobenzene (15 mL) taken in a two necked RB flask fitted with a distillation column and a dropping funnel (heated to 80–85°C) is added dropwise dimethyl sulphate (4.6 g). The methyl bromide distils from the flask and is collected using a trap cooled with acetone-dry ice. When the rate of distillation of CH_3Br slows down, the heating is increased and a solution of conc. H_2SO_4 (0.08 mL) in water (30 mL) is cautiously added. The mixture is refluxed (40 hrs.) and then evaporated to dryness under reduced pressure. The residue is dissolved in CH_2Cl_2 (25 mL), solution washed with water (2 × 10 mL), dried (anhyd. Na_2SO_4) and distilled to give tetrabutylammonium hydrogen sulphate. Yield about 10 g. It can be recrystallised from isobutyl methyl ketone.

7.7.3 Tetrabutylammonium Chromate

A phase transfer reagent, tetrabutylammonium chromate in an excellent oxidising agent for all oxidations which involve the use of dichromate. It is prepared by the reaction of tetrabutyl-ammonium chloride with chromium trioxide.

$$Bu_4N^+Cl^- + Cr_2O_3 \longrightarrow Bu_4NCrO_4$$

| **Chemicals:** | Tetrabutylammonium chloride | 2.92 g |
| | Chromium trioxide | 1 g |

Procedure

To a stirred solution of chromium trioxide (1 g) in water (25 mL), an aqueous solution of tetrabutyl-ammonium chloride (2.92 g) in water (50 mL) is added at room temperature. The mixture is cooled (0°C), the separated solid filtered (using a sintered glass funnel), washed with cold water and dried in vacuum over P_2O_5. Yield 2.7 g.

7.7.4 Tetrabutylammonium Permanganate, Bu_4NMnO_4

A phase transfer reagent, tetrabutylammonium permanganate is an excellent reagent for all oxidations involving the use of $KMnO_4$. It is prepared by adding excess of a concentrated aqueous solution of tetrabutylammonium bromide to a stirred aqueous solution of $KMnO_4$. The purple reagent is filtered, washed will water and dried in vacuum.

7.7.5 Polystyrene-aluminium Chloride

Aluminium chloride is used as catalysis in various organic synthesis. Use of polystyrene-aluminium chloride is more advantageous as it reduces the work up procedure which in the normal course leads to environmental problems.

Polystrene-aluminium chloride is obtained by the reaction of polystrene-divinylbenzene copolymer beads with aluminium chloride (*D.C. Neckers, D.A. Kouistra and G.W. Green, J. Am. Chem. Soc., 1972, 9284*).

Chemicals:	Polystyrene-divinylbenzene	3.1 g (1.8 %)
	Copolymer beads	50-100 mesh
	Anhyd. aluminium chloride	3.7 g

Procedure

Carbon disulphide (5mL) is added to polystyrene-divinylbenzene copolymer beads (3.1 g) contained in a three necked 100 mL RB flask equipped with a stirrer, condenser and dropping funnel. This is followed by the addition of anhyd. $AlCl_3$(3.7g). The mixture is stirred and refluxed for about 30 min, cooled, cold water (40 mL) added and mixture stirred. The polymer beads are filtered, washed with water (100 mL), ether (15 mL), acetone, hot propyl alcohol and ether. The polystrene-aluminium chloride thus obtained is dried in vacuum for about 15 hr. in an oven.

7.7.6 Urea Hydrogen Peroxide Adduct

Urea-hydrogen peroxide adduct (UHP) is commercially available and can be prepared from urea and H_2O_2.

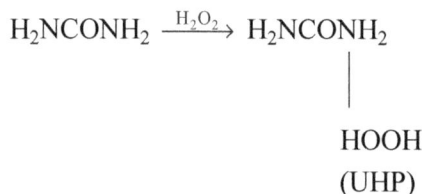

$$H_2NCONH_2 \xrightarrow{H_2O_2} H_2NCONH_2$$

$$|$$

$$HOOH$$

(UHP)

The reagent has been used in solid state for the oxidation of hydroxy aldehydes, and ketones to the corresponding hydroxy component (*R.S. Varma and K.P. Naik Ker, Org. Lett., 1999, 1, 189-191*).

EXERCISES

1. How is cupric acetate prepared. What are its uses.
2. Give preparation and uses of
   (*a*) Curprous oxide
   (*b*) Cuprous chloride
   (*c*) Lead chromate
   (*d*) Lead tetracetate
   (*e*) Active manganese dioxide
   (*f*) Silver oxide.
3. How is potash alum prepared? Give its uses.
4. Give preparation and uses of:
   (*a*) Aluminium isopropoxide

(b) Copper phthalocyanine

(c) Ferrocene

(d) Prussian blue

5. What are amalgams? How are they prepared? Give an example.

6. How are activated metals obtained, give two examples?

7. How are organometallic compounds obtained? Give preparation of methyl magnesium bromide.

8. What are phase transfer catalyst? What are their uses. Give preparation of tetrabutyl ammonium hydrogen sulphate.

9. How is polystrene aluminium chloride obtained. Give its uses?

10. How is urea-hydrogen peroxide adduct obtained. Give its uses?

OOO

Index

M

Macro-Scale Analysis 9

Magnesium 64, 81, 87, 90, 250

Magnesium Sulphate 259

Manganese 61, 81, 87, 90, 250

Manganese Dioxide 260

Mercuric Oxide 260

Mercury 46, 49, 81, 87, 90, 248

Methyl Magnesium Bromide 273

Methyl Orange 135

Methylene Blue 23

pH-Metric Titrations 192

Mohr's Method 157, 160

Mohr's Salt 146

Molar Concentration 185

Molarity 126

Molybdate Ion 109

N

Nickel 60, 82, 87, 90, 249

Nitrate 29

Nitrite 24

1-Nitroso-2-Naphthol Reagent 78

Normality 126

O

Organometallics 273

Oxalate 26

Oxine Reagent 64

P

Palladium 112

Paper Chromatography 92

Phenolphthalein 135

1, 10-Phenanthroline Reagent 79

Phosphate 32

Phosphomolydic Acid Reagent 75, 83

Platinum 111

Polystyrene-Aluminium Chloride 275

Potash Alum 266

Potassium 64, 65, 69, 82, 87, 90, 250

Potassium Dichromate 260

Potassium Ferrocyanide Reagent 74

Potassium Permanganate 261

Potassium Thiocarbonate 65

Potassium Trioxalatochromate 270

Potassium Very Fine Particles 273

Potentiometric Titrations 175

Precipitation Titrations 156, 182, 189

Preparation of Solution of the Mixture 42

Prussian Blue 33, 270

Pyrrole Reagent 114

Q

Qualitative Analysis 9

Quinalizarin Reagent 74

R

Redox Titrations 142, 181

Reference Electrodes 177

Removal of Borate and Fluoride 55

Removal of Interferring Anions 54

Removal of Oxalate and Tartarate 55

Removal of Phosphate 55

Rhodamine Reagent 112

Rhodizonate Solution 81

Ring Test 24, 29

Rubeanic Acid 111

Rubeanic Acid Reagent 78, 82

S

Selenium 114

Self-Indicator 143

Semi-Micro Scale Analysis 10

Silicate 35

Silver 46, 82, 87, 90, 248

Silver Oxide 262

Single Pan Electrical Balance 133

Sodium 87

Sodium Amalgam 272

Sodium Ammonium Hydrogen Phosphate 271

Sodium Bismutate 81

Sodium Chloride 262

Sodium Cobaltinitrite 274

Sodium Cobaltinitrite-Silver Nitrate Reagent 82

Sodium Ferric Oxalate 271

Sodium Nitropusside 23

Sodium Perborate 263

Sodium Rhodizonate Reagent 75

For Product Safety Concerns and Information please contact our EU
representative GPSR@taylorandfrancis.com
Taylor & Francis Verlag GmbH, Kaufingerstraße 24, 80331 München, Germany

www.ingramcontent.com/pod-product-compliance
Lightning Source LLC
Chambersburg PA
CBHW081056220326
41598CB00038B/7118

9 781032 789927